Chemical Ecology of Insects

Applications and Associations with Plants and Microbes

Chemical Ecology of Insects

Applications and Associations with Plants and Microbes

Editor

Jun Tabata

National Agriculture and Food Research Organization
Tsukuba
Japan

CRC Press
Taylor & Francis Group
Boca Raton London New York

CRC Press is an imprint of the
Taylor & Francis Group, an **informa** business
A SCIENCE PUBLISHERS BOOK

Cover illustration reproduced by kind courtesy of Scott Chimileski.

CRC Press
Taylor & Francis Group
6000 Broken Sound Parkway NW, Suite 300
Boca Raton, FL 33487-2742

Printed on acid-free paper
Version Date: 20170719

International Standard Book Number-13: 978-1-4987-6940-2 (Hardback)

Library of Congress Cataloging-in-Publication Data

Names: Tabata, Jun, editor.
Title: Chemical ecology of insects : applications and associations with plants and microbes / editor, Jun Tabata, National Agriculture and Food Research Organization, Tsukuba, Japan.
Description: Boca Raton, FL : CRC Press, Taylor & Francis Group, 2017. | "A science publishers book." | Includes bibliographical references and index.
Identifiers: LCCN 2017030459 | ISBN 9781498769402 (hardback : alk. paper)
Subjects: LCSH: Insects--Ecophysiology. | Animal chemical ecology. | Chemical ecology. | Plant chemical ecology.
Classification: LCC QL495 .C4725 2017 | DDC 595.7--dc23
LC record available at https://lccn.loc.gov/2017030459

Visit the Taylor & Francis Web site at
http://www.taylorandfrancis.com

and the CRC Press Web site at
http://www.crcpress.com

MIX
Paper from responsible sources
FSC™ C013985

Printed in the United Kingdom
by Henry Ling Limited

Preface

It has long been known that organisms including insects and plants communicate with each other via chemical stimuli, and such interactions often play an essential role in the ecosystem. However, the first stimulus compound was isolated and identified from the silk moth, by Butenandt and co-workers in only 1959, less than 60 years ago. Since then, the field of chemical ecology has been recognized and become a focus of attention for a broad range of biologists and chemists. Thus studies of chemical ecology have started on insects and their related chemical signals, although growing attention is now paid to many other materials, including terrestrial and aquatic animals and plants. I believe that insects are still one of the most exciting targets to illuminate the complex and profound forms of chemical signaling in living creatures, because insects show great diversity and adaptation to various environments in association with other organisms.

In this book, therefore, I would like to introduce various recent studies on chemical ecology in insects and their associated plants and microbes. Fortunately, this book contains contributions from authors who are at the foremost of this field as well as up-and-coming experts in the study of insects or insect-plant/microbe interactions. Chapters cover butterflies, moths, beetles, mosquitoes, flies, and many other insects as well as their associated plants and microbes including nematodes, yeasts, bacteria, and viruses. Thus, readers can follow their own interests to start or expand their own studies on chemical ecology. In addition, some chapters emphasize the application of chemical ecological knowledge for the conservation of nature or management of pests threatening agriculture and public health.

As is well known, insects and their life-histories are so divergent that it is impossible to cover this topic in its entirety in the limited page space available here. Nevertheless, this book covers impressive topics on insects and associated life forms that make use of chemical information in their various habitats. I am convinced that these topics will help readers to gain an overview of the chemical ecology of insects, plants, and microbes, and to further explore studies of natural products, biochemistry, ecology, evolution, and their applications.

I am very much indebted to a great number of colleagues who kindly helped to review the manuscript and discussed the content of this book.

Jun Tabata
Tsukuba, Japan
March 2017

Contents

Preface *v*

Part A: Chemical Ecology of Insects and Associated Plants and Microbes

1. Plant Secondary Metabolites in Host Selection of Butterfly 3
Hisashi Ômura

2. Function of the Lepidopteran Larval Midgut in Plant Defense Mechanisms 28
Naoko Yoshinaga and Naoki Mori

3. Chemically-mediated Interactions among Cucurbits, Insects and Microbes 55
Lori R. Shapiro & Kerry E. Mauck

4. Chemoecology and Behavior of Parasitic Nematode—Host Interactions: Implications for Management 91
Denis S. Willett, Xavier Martini, and Lukasz L. Stelinski

5. Microbial Endosymbionts and Chemical Ecology 114
Daisuke Kageyama

6. Chemical Ecology of Yeasts Associated with Insects 131
Jun Tabata and Hiroko Kitamoto

Part B: Applications of Insect Chemical Ecology to Agriculture, Environment Conservation, and Public Health

7. Application of Trail Pheromones to Management of Pest Ants 159
Eiriki Sunamura

8. Female Sex Pheromones and Mating Behavior in Diurnal Moths: Implications for Conservation Biology 170
Hideshi Naka

9. Mating Disruption: Concepts and Keys for Effective Application 197
Jun Tabata

10. Applied Chemical Ecology to Enhance Insect Parasitoid Efficacy in the Biological Control of Crop Pests 234
Ezio Peri, Rihem Moujahed, Eric Wajnberg, and Stefano Colazza

11. Challenges in Chemical Ecology for the Management of Vector-borne Diseases of Humans and Livestock 268
Jun Tabata

Index 294

PART–A

Chemical Ecology of Insects and Associated Plants and Microbes

CHAPTER

1

Plant Secondary Metabolites in Host Selection of Butterfly

Hisashi Ômura

Graduate School of Biosphere Science, Hiroshima University
1-4-4 Kagamiyama, Higashi-Hiroshima City, Hiroshima 739-8528, Japan
E-mail: homura@hiroshima-u.ac.jp

Abstract

Plant secondary metabolites have a role in defense against various herbivores, but they also often attract particular species and stimulate their oviposition and feeding behavior. Most butterflies are phytophagous in the larval stage and have a strong relationship with a narrow range of plants with a phylogenetic relationship and chemical similarity. Because larvae and females can respond to various plant chemicals using chemotactile (gustatory) receptors, behavioral and sensory adaptation to particular compounds are presumed to lead to the evolution of host specificity. From a limited range of butterfly species in the Pieridae, Papilionidae and Nymphalidae, a variety of plant components have been identified as semiochemicals for host selection, the majority of which are associated with female oviposition. The oviposition stimulants identified so far are (i) single compounds in relation to plant defensive chemicals or (ii) multiple components including relatively ubiquitous derivatives from primary metabolites. The finding that several closely related butterfly species use the same or corresponding compounds in different plants for host acceptance in part supports a putative evolutionary pathway in species diversification and host shift at the tribe or subfamily level.

1. Introduction

Plants biosynthesize a diverse and complex range of secondary metabolites as chemical barriers against herbivory (Rosenthal and Berenbaum 1991,

Mithöfer and Boland 2012). Plant defensive chemicals are not noxious to all herbivore species, and instead are often highly and selectively toxic to particular species or otherwise have some negative effects on their acceptability and performance. Therefore, most phytophagous insects are specialized to feed on a narrow range of plant species as a result of their physiological adaptations to particular defense chemicals (Städler 1992, Bernays and Chapman 1994). From chemical interactions between angiosperms and butterflies, Ehrlich and Raven (1964) proposed a well-known hypothesis of stepwise coevolution between plant chemical defense to herbivory and herbivore counter-adaptation to plant chemicals, which is regarded as a principal driving force of species diversification from each other. Since then, plant–butterfly interactions have received the most attention as an attractive model for ecological and evolutionary studies of plant chemicals implicated in host preference and specificity of phytophagous insects.

Most butterflies are phytophagous insects in the larval stage, in which they feed on particular plant species from only one or a few genera or a single family or subfamily (Ehrlich and Raven 1964, Janz and Nylin 1998). Of a wide variety of physiological and ecological factors, non-volatile plant chemicals play a significant role to limit the host range of butterflies because these substances directly mediate and regulate the behaviors of larval feeding and female oviposition (Thompson and Pellmyr 1991, Honda 1995). Therefore, suitable host plants are chemically characterized by possessing compounds that act as behavioral stimulants and lacking those act as behavioral deterrents. An array of host plants for a particular herbivore species, in general, exhibit chemical similarity, which is often strongly associated with historical patterns of host shifts within herbivore lineages (Becerra 1997). Comparative phylogenetic studies have proposed several possible pathways of diversification and switching of butterfly hosts at the within-family or subfamily level (Janz et al. 2001, Fordyce 2010, Ferrer-Paris et al. 2013). However, only a limited number of semiochemicals in relation to host selection have been identified in a few butterfly families. The purpose of this review is to compile available knowledge on plant chemicals that serve as chemotactile (gustatory) cues for host selection by butterflies and to discuss chemical trends in their host usage.

2. Contact Chemoreception of Plant Chemicals

Typical gustatory sensilla of butterflies appear as hair- or peg-like structures consisting of a thick cuticular wall surrounding an inner lumen with the dendrites of 2–4 receptor neurons (Mitchell et al. 1999, Chapman

2003, Kvello et al. 2006). The apical part of the cuticle is perforated, allowing non-volatile compounds to enter the lumen and stimulate the receptor neurons. Each receptor neuron responds to a broad range of chemicals or only to a specific type of molecule, and conveys information concerning these particular substances to the central nervous system (CNS) separately from other kinds of information. The information from single receptor neurons receiving particular molecules serves as a stimulatory or deterrent trigger for particular behaviors, such as female oviposition and larval feeding. This is called labeled line coding in the insect taste receptor system, in which particular plant secondary metabolites serve as host-indicating signals.

In butterflies, female adults encounter more plant species than larvae because of their high mobility. Therefore, host selection (oviposition) by females is more critical for offspring than that (feeding) by larvae. Gravid females depend on a wide variety of sensory cues to recognize available host plants (Renwick and Chew 1994, Carrasco et al. 2015). In searching and orientation at long distance, olfactory and visual cues from host plants play important roles. After alighting on a plant, the forelegs make contact with the leaf and other parts of a plant in order to receive both physical and chemical cues. This behavior is called drumming, in which they respond to various plant chemicals mainly with hair-like sensilla (sensilla trichodea) on the surface of foretarsi.

Like ovipositing females, larvae also use various sensory cues to assess plant food quality. Olfaction is used for initial orientation to and discrimination of possible hosts, whereas gustation is used to determine and initiate feeding (Hanson and Dethier 1973). Freshly hatched larvae feed first on its eggshell and then on plant tissues around the egg. Because plant tissues, even in their host plants, sometimes contain toxic or growth-inhibitive components, larvae need to avoid intake of these noxious substances. They mainly use peg-like sensilla (sensilla styloconica) on the maxilla for the assessment of plant chemistry (Frazier 1992).

3. Host Selection in Pieridae

The family Pieridae is a relatively small group in butterflies, consisting of about 1100 species in four subfamilies: Pseudopontiinae, Dismorphiinae, Coliadinae, and Pierinae (Ehrlich 1958, Braby 2005). The subfamilies of the Pieridae show a distinct host specialization, in which their host plants are mainly distributed in the Fabales, Brassicales, and Santalales (Ferrer-Paris et al. 2013). Of the three plant orders, Fabales is assumed to be the ancestral host of the Pieridae, and is mainly utilized by the subfamilies Dismorphiinae and Coliadinae. Within the subfamily Pierinae, a host shift from Fabales to Brassicales, followed by further shifts from Brassicales to

Santalales, had promoted diversification and adaptive radiation (Braby and Trueman 2006). The resting subfamily Pseudopontiinae, consisting of only one species, exploits the family Opiliaceae in the Santalales. The plant secondary metabolites involved in host selection have been identified to date in the genus *Pieris* in the Pierinae and the genera *Colias* and *Eurema* in the Coliadinae.

3.1 Pierinae and Brassicaceae

The plant family Brassicaceae is well known to produce glucosinolates, which have been studied intensively as mediators of various insect–plant interactions (Hopkins et al. 2009). When plant tissues are damaged, glucosinolates are hydrolyzed by myrosinases, and breakdown products such as isothiocyanates, nitriles, and oxazolidinethiones serve as a chemical barrier against herbivores (Winde and Wittstock 2011). The genus *Pieris* is a Brassicaceae-feeding specialist and closely associated with glucosinolates for their host selection. Gravid females, in the recognition of host plants, detect glucosinolates with tarsal contact chemoreceptors (Ma and Schoonhoven 1973, Du et al. 1995, Städler et al. 1995). *Pieris* butterflies evaluate the quality and quantity of plant glucosinolates and differ in oviposition preference and acceptance within Brassicaceae hosts (Huang and Renwick 1993, 1994). For example, *Pieris rapae* preferentially lays eggs on cultivated cruciferous plants such as cabbages, whereas *Pieris oleracea* (formerly *Pieris napi oleracea*) mainly exploits wild cruciferous plants such as *Arabis* spp. An aromatic glucosinolate, 3-indoylmethyl glucosinolate (glucobrassicin, **1**) contained in cabbage, is stimulatory for the oviposition of *P. rapae* and *Pieris brassicae* (Traynier and Truscott

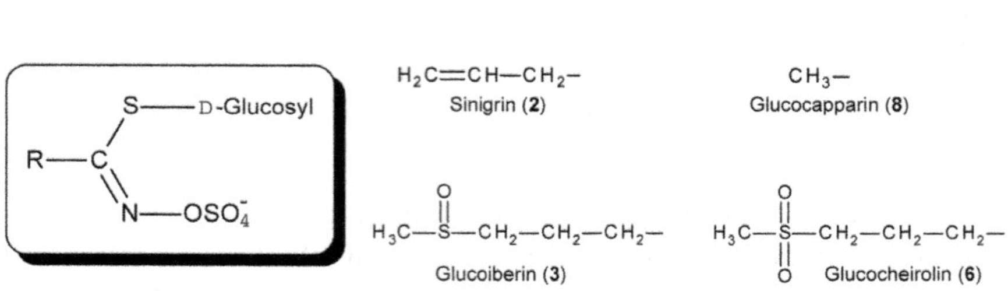

1991, Loon et al. 1992, Renwick et al. 1992). On the other hand, particular alkyl and thioalkyl glucosinolates, allyl glucosinolate (sinigrin, **2**) and (*R*)-3-methylsulfinylpropyl glucosinolate (glucoiberin, **3**), serve as strong oviposition stimulants for *P. oleracea* (Huang and Renwick 1994). In an oviposition choice between cabbage and wintercress (*Barbarea vulgaris*), *P. rapae* shows no preference, whereas *P. oleracea* prefers *B. vulgaris*. The oviposition preference of *P. oleracea* for *B. vulgaris* is attributed to a higher abundance of (2*R*)-hydroxy-2-phenylethyl glucosinolate (glucobarbarin, **4**) (Huang et al. 1994).

Several cruciferous plants possess oviposition deterrents for *Pieris* butterflies. Candytuft *Iberis amara* is omnipresent in grain fields in Europe and is sometimes observed to have eggs of *P. oleracea*. In addition to sinigrin, this plant is found to biosynthesize cucurbitacins (**5**), which are strong oviposition deterrents for *P. rapae* but only weak deterrents for *P. oleracea* (Huang et al. 1993a). Wormseed mustard *Erysimum cheiranthoides* is a preferential host for *P. oleracea* due to the presence of thioalkyl glucosinolates, e.g., 3-methylsulfonylpropyl glucosinolate (glucocheirolin, **6**) and glucoiberin (Huang et al. 1993b). However, *P. rapae* never oviposits on *E. cheiranthoides* because its cardenolide components including erysimoside (**7**) serve as strong oviposition deterrents (Renwick et al. 1989, Sachedev-Gupta et al. 1990).

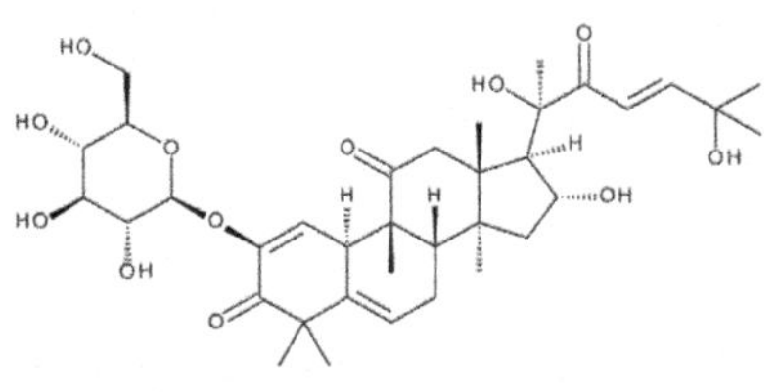

2-O-β-D-Glucopyranosyl cucurbitacin I (**5**)

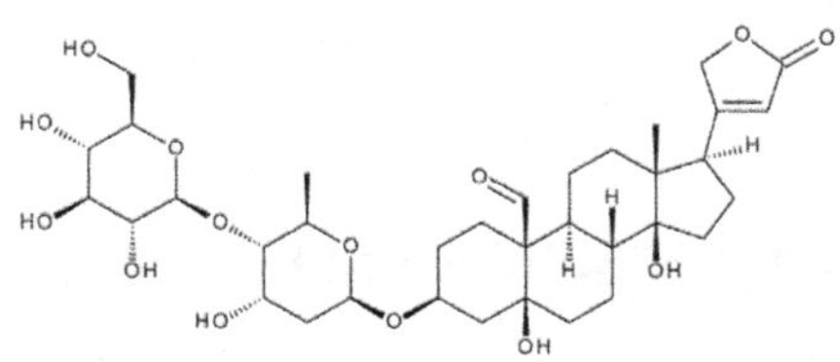

Erysimoside (**7**)

Pieris larvae show feeding responses to filter papers treated with cruciferous plant sap (Verschaffelt 1910). Larval feeding is stimulated by plant glucosinolates, which are received by chemoreceptors on the maxillary sensilla (Schoonhoven 1969). Particular alkyl and thioalkyl glucosinolates, e.g., methyl glucosinolate (glucocapparin, **8**) and glucoiberin, elicit larval feeding by *P. brassicae* (David and Gardiner 1966), whereas an aromatic glucosinolate, 2-phenethylglucosinolate (gluconasturtiin, **9**), stimulates feeding behavior of *P. rapae* (Miles et al. 2005). In several *Pieris* butterflies, the preference for glucosinolates is similar in both feeding and oviposition, suggesting that a common chemosensory system is involved in host selection at both larval and adult stages (Chew and Renwick 1995, Renwick 2002).

Particular cruciferous plants have feeding deterrents and sometime toxic substances for *Pieris* larvae. Glucosinolates, when present in plant tissues at high concentrations, are toxic to neonate *Pieris* larvae (Rotem et al. 2003). Several fractions of plant extract of *E. cheiranthoides* contain particular cardenolides, which strongly deter larval feeding of *P. rapae* (Dimock et al. 1991, Sachdev-Gupta et al. 1993a). Garlic mustard (*Alliaria petiolate*) is resistant to *P. oleracea*, in which a cyanoallyl glucoside (alliarinoside, **10**) and an apigenin flavonoid (**11**) are responsible for blocking larval feeding (Haribal and Renwick 2001, Haribal et al. 2001, Renwick et al. 2001). This plant has invaded the habitat of *Pieris virginiensis* in North America. Because sinigrin and alliarinoside are present together at high concentrations, *A. petiolate* is a preferential host for *P. virginiensis* females but a toxic plant for its larvae (Davis and Cipollini 2014, Davis et al. 2015). *Pieris* larvae receive feeding deterrents with different chemical structures, particular glucosinolates and flavonoids, with the same chemoreceptor (Zhou et al. 2009).

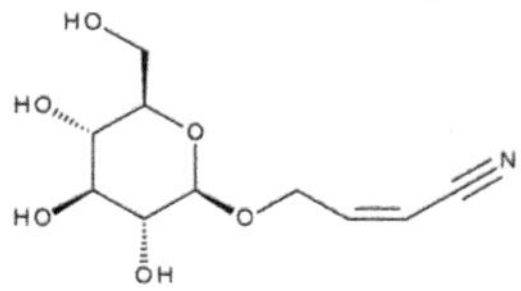

Alliarinoside (**10**)

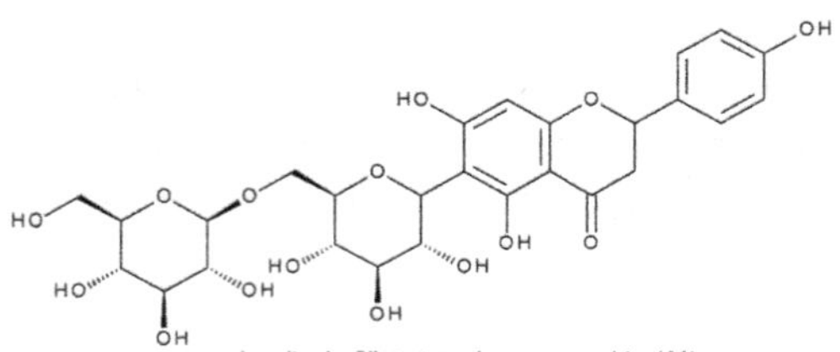

Isovitexin-6"-O-β-D-glucopyranoside (**11**)

3.2 Coliadinae and Fabaceae

Most species belonging to the subfamily Coliadinae are Fabaceae-feeders. *Colias erate poliographus* mainly exploits herbal fabaceous plants, especially white clover *Trifolium repens*. In the oviposition of *C. erate poliographus* on this plant, D-(+)-pinitol (**12**) is a principal stimulant, whereas cyanogenic glucosides (**13**, **14**), methyl-D-glucoside, and glycerin synergistically enhanced the stimulatory effect of D-(+)-pinitol (Honda et al. 1997a, 2012). Of note, D-(+)-pinitol also stimulates the oviposition behavior of *Eurema mandarina* on leaves of *Albizia julibrissin* and *Lespedeza cuneata* (Mukae et al. 2016). Because D-pinitol is widely distributed in the Fabaceae, it is likely that other plant components play key roles in different host usage between *C. erate* and *E. mandarina*.

With respect to larval feeding stimulants for *E. mandarina*, D-(+)-pinitol, *myo*-inositol, D-fructose, and cellulose have been identified from *L. cuneata* (Numata et al. 1985). Soyasaponin was found to stimulate larval feeding of *C. erate poliographus* (Matsuda et al. 1998).

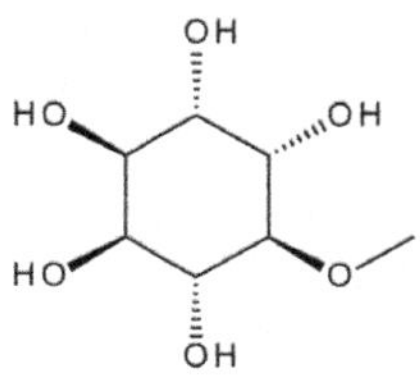

D-(+)-Pinitol (**12**)

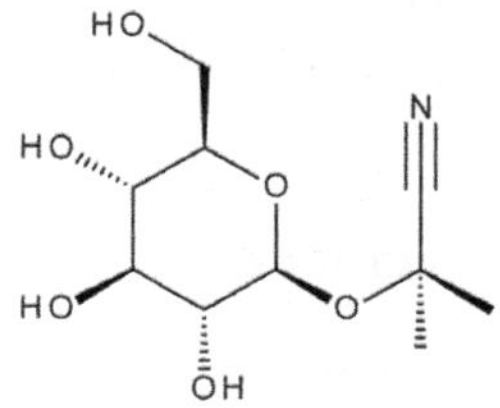

Linamarin (**13**)

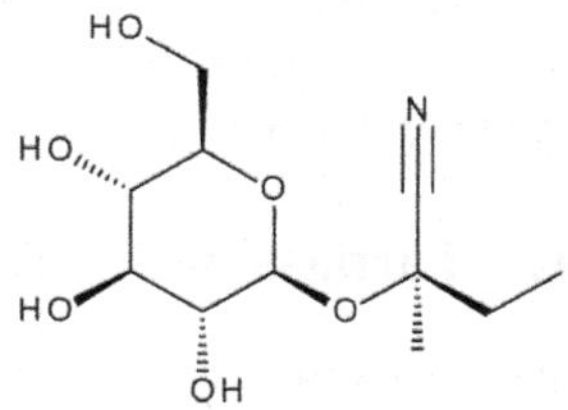

Lotaustralin (**14**)

4. Host Selection in Papilionidae

The family Papilionidae comprises 500–700 species belonging to three subfamilies: the Baroniinae with only one species *Baronia brevicornis*, the Parnassinae with ca 50 species in two tribes (Parnassinii and Zerynthiini), and the Papilioninae with over 500 species in three tribes (Graphiini, Troidini, and Papilinonini) (Caterino et al. 2001, Zakharov et al. 2004). The Baroniinae has been suggested to be a sister lineage to all other Papilionidae, and to have originated from an ancestral species between the Papilionidae and Pieridae. This species feeds on *Acacia cochliacantha* (Fabaceae). On the other hand, the Parnassinae and Papilinoninae are highly variable in their larval host plant utilization, ranging from a restricted selection of one or two plant families to polyphagy of up to eight different plant families. However, most species in these subfamilies are specialists using a narrow range of host plants in one plant family (Scriber 1984, Brown et al. 1995). For approximately 80% of swallowtail species, the main hosts occur within five plant families, namely, the Aristolochiaceae, Annonaceae, Lauraceae, Rutaceae, and Apiaceae (Berenbaum 1995, Aubert et al. 1999). Of these plant families, the Aristolochiaceae is assigned to be the putative ancestral host for the genus *Parnassius* in the Parnassinae and the genus *Papilio* in the Papilioninae (Fordyce 2010).

The Papilionidae is one of most extensively studied butterfly groups with regards to semiochemical-mediated host selection by gravid females, in which several different classes of compounds such as flavonoides, hydroxycinnamates, and polyols stimulate female oviposition only through combinations of these compounds (Nishida 1995, 2014). In this multiple-component oviposition system of swallowtail butterflies, particular plant components are common stimulants for different papilionid species, whereas several stimulants identified from different host plants have structural similarity (e.g., Haribal and Feeny 2003, Nakayama et al. 2003). These findings support the possibility that plant chemistry in host acceptance (oviposition) by gravid females may facilitate host shifts and diversification in the Papilionidae (Berenbaum 1995, Feeny 1995). In contrast, limited information is available for plant chemicals involved in the larval feeding.

4.1 Parnassinae and Aristolochiaceae

Luehdorfia japonica in the tribe Zerynthiini is an Aristolochiaceae-feeder. *Asarum asperum* (*Heterotropa aspera*) in the Aristolochiaceae is one of the main hosts for this species, and was found to contain a flavonol triglucoside (**15**) as an oviposition stimulant (Nishida 1994). Of note, *Luehdorfia puziloi*, a sibling species of *L. japonicum*, never exploits *A. asperum*, from which a neolignoid (**16**) was identified as a larval feeding-deterrent (Honda et al. 1995a). *Parnassius* spp. in the tribe Parnassini has an inferred host shift from the Aristolochiaceae to the Papaveraceae and Fumiriaceae (Fordyce 2010), in which some flavonoids might be involved.

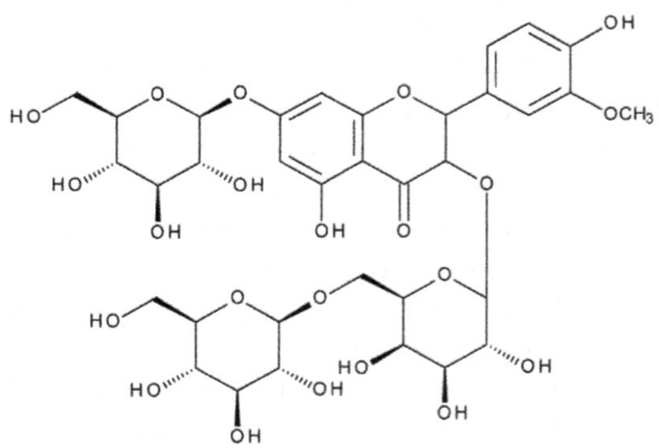

Isorhamnetin 3-*O*-glucopyranosyl-(1→6)-galactopyranoside-7-*O*-glucopyranoside (**15**)

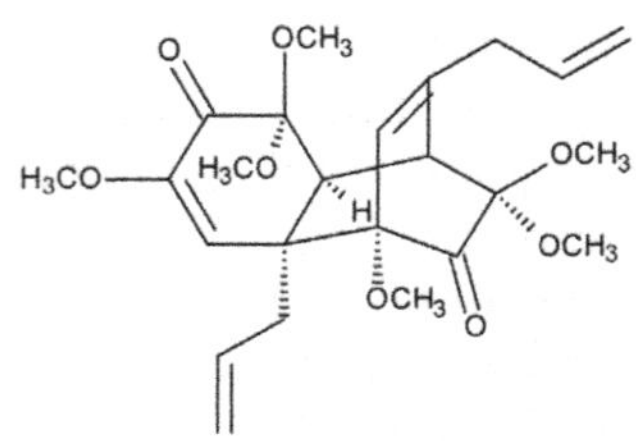

Asatone (**16**)

4.2 Papilioninae and Aristolochiaceae

Of the three tribes of the subfamily Papilioninae, the tribe Troidini feeds mainly on Aristolochiaceae (Aubert et al. 1999). In the genus *Parides*, no major host shifts from Aristolochiaceae have been assumed to occur throughout its history (Fordyce 2010), suggesting that troidine species have strong relationships with aristolochiaceous plants. Aristolochiaceous plants are known to have unique toxic compounds, aristolochic acids (AAs, **17**). Several troidine butterflies have been found to be associated with AAs in host selection. Pipevine swallowtail *Battus philenor* in North America and Chinese windmill *Byasa* (*Atrophaneura*) *alcinous* in East Asia depend on AAs and cyclitols, D-(+)-pinitol (**12**) and sequoyitol (**18**), in host selection for aristolochiaceous plants (Nishida and Fukami 1989, Papaj et al. 1992, Sachdev-Gupta et al. 1993b). AAs also serve as larval feeding stimulants for *B. alcinous* (Nishida and Fukami 1989).

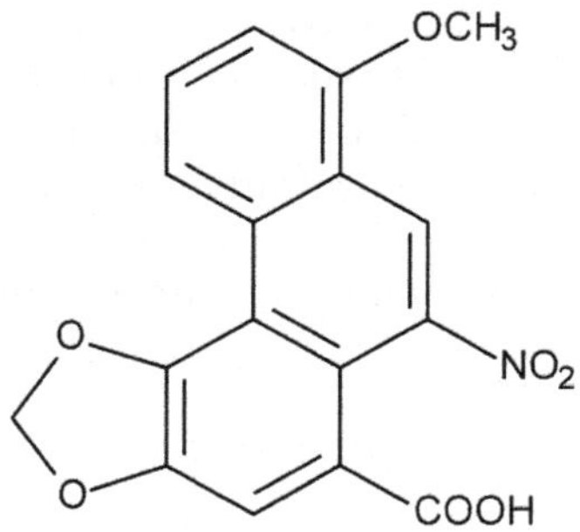

Aristolochic acid I (**17**)

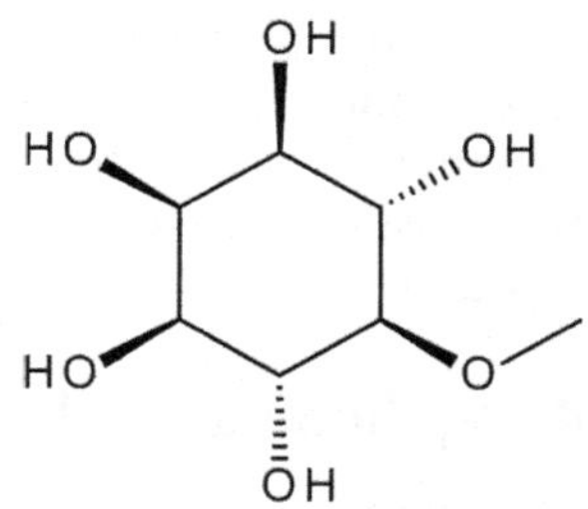

Sequoyitol (**18**)

4.3 Papilioninae and Annonaceae

The tribe Graphiini in the Papilioninae contains a wide range of Annonaceae-feeding species (Aubert et al. 1999). In this tribe, the genera *Graphium* and *Eurytides* are presumed not to have changed their usage of Annonaceae as a major host (Fordyce 2010). Although pawpaw, *Asimina triloba* (Annonaceae), is known to have toxic polyketides acetogenins

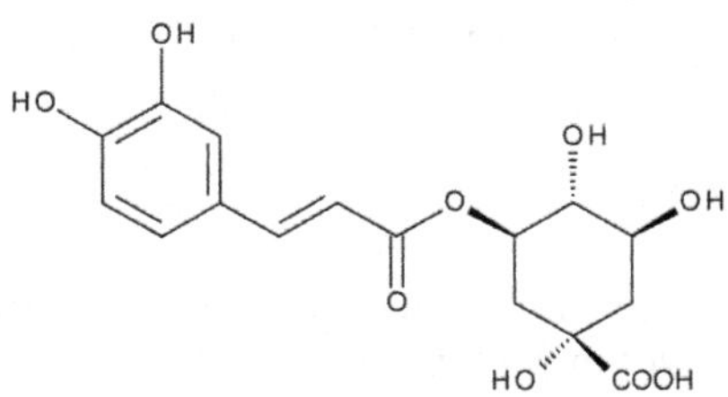

3-Caffeoyl-*muco*-quinic acid (**19**)

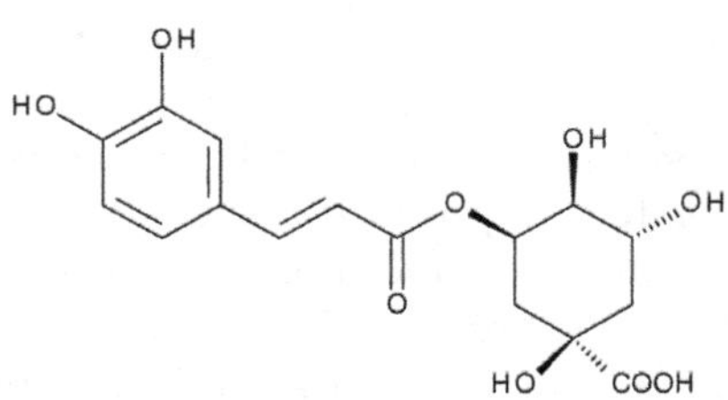

5-Caffeoylquinic acid (**20**)

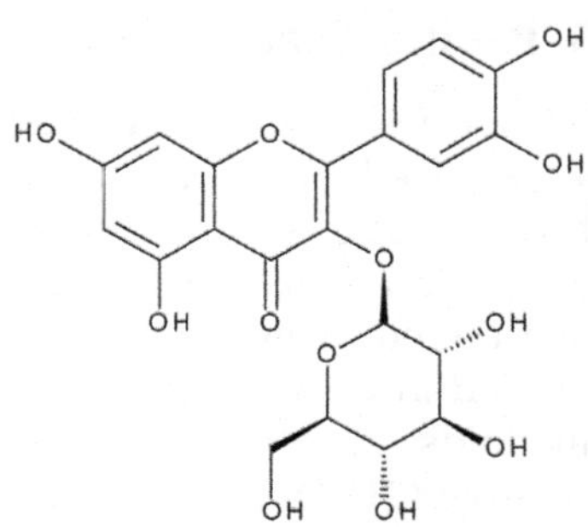

Quercetin 3-*O*-β-glucopyranoside (**21**)

(Martin et al. 1999), zebra swallowtail *Protographium* (*Eurytides*) *marcellus* prefers to oviposit on the terminal (young) leaves of pawpaw, in which 3-caffeoy-*muco*-quinic acid (**19**) is present at high concentrations and strongly stimulates female oviposition (Haribal and Feeny 1998). As this plant grows, the concentration of this hydroxycinnamic acid derivative decreases gradually. Alternatively, several flavonoids such as rutin (**24**) and nicotiflorine have been found to occur at significant concentrations. The fact that these flavonoids suppress the oviposition-stimulatory effect of 3-caffeoy-*muco*-quinic acid at low doses suggests that *P. marcellus* females assess host quality on the basis of plant chemistry (Haribal and Feeny 2003).

4.4 Papilioninae and Lauraceae

Lauraceae-feeders in the Papilioninae are placed mostly in the tribe Papilionini and partly in the tribe Graphiini (Aubert et al. 1999). A North American papilionine species, spicebush swallowtail *Papilio troilus*, lays eggs on the foliage of *Sassafras albidum*, from which 3-caffeoyl-*muco*-quinic acid (**19**) was identified as an oviposition stimulant that synergistically acts together with other plant components (Carter et al. 1999). Notably, *P. troilus* also oviposits on lauraceaous plants lacking this hydroxycinnamic acid derivative, e.g., *Lindera benzoin*, *Persea borbonia*, and *Cinnamomum camphora*, suggesting that unidentified plant components play pivotal roles in host recognition of this species. An East Asian graphinine species, common bluebottle *Graphium sarpedon*, feeds on camphor tree *C. camphora* as a host, in which 5-caffeoylquinic acid (**20**) and quercetin 3-*O*-β-glucopyranoside (**21**) have been identified as larval feeding stimulants (Zhang et al. 2015).

4.5 Papilioninae and Rutaceae

Most species in the tribe Papilionini, especially the genus *Papilio*, are Rutaceae-feeders (Aubert et al. 1999). However, *Papilio* species differ greatly in host acceptance and preference within the Rutaceae because rutaceous plants have a wide variety of plant chemistries. *Papilio xuthus* mainly feeds on various cultivated *Citrus* plants. The oviposition stimulants of *P. xuthus* identified in *Citrus unshiu* included 10 compounds: 4 flavonoids (**22-25**); 3 alkaloids (**26-28**); 1 cyclitol (**29**); 1 nucleoside (**30**); and 1 amino acid derivative (**31**) (Nishida et al. 1987, Ohsugi et al. 1985, 1991). However, none of the individual components elicited oviposition responses. The specific activity was provoked only when these compounds were applied as a mixture, indicating the significance of a synergistic effect of multiple components in host recognition. *Papilio protenor* is also a Rutaceae-feeding swallowtail and exerts oviposition responses to a mixture of six compounds (**22, 28, 31-34**) from *C. unshiu* leaves (Honda 1986, 1990, Honda and Hayashi 1995a). Of the six compounds, hesperidin (**22**), (−)-synephrine (**28**), and (−)-stachydrine (**31**) were common oviposition stimulants between *P. protenor* and *P. xuthus*. In addition, chlorogenic acid (**35**) present in *Zanthoxylum (Fagara) ailanthoides* synergistically stimulated oviposition by *P. protenor* (Honda, 1990). The gustatory receptor gene involved in the recognition of synephrine, PxutGr1, has been identified from the tarsal sensilla of *P. xuthus* females (Ozaki et al. 2011). On the other hand, *P. xuthus* larvae also depend on a mixture of 11 plant components in feeding on *C. unshiu* (Murata et al. 2011). Except for (−)-stachydrine, 10 of the larval feeding stimulants were not coincident with the female oviposition stimulants of *P. xuthus*.

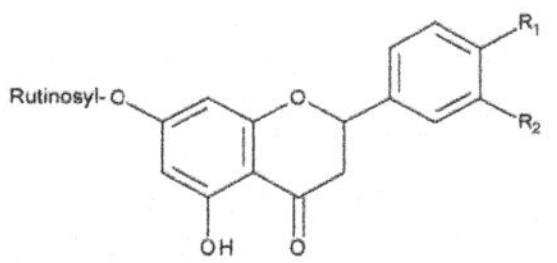

Hesperidin (**22**) R1: OCH3 R2: OH
Narirutin (**23**) R1: OH R2: H

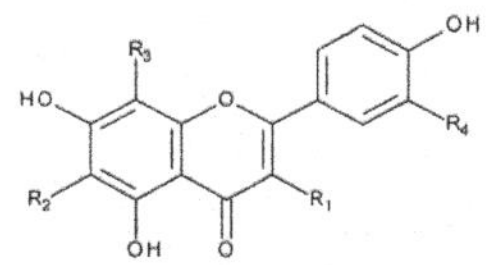

Rutin (**24**) R1: O-Rutinosyl R2,R3: H R4: OH
Vicenin-2 (**25**) R1,R4: H R2,R3: C-Glucosyl

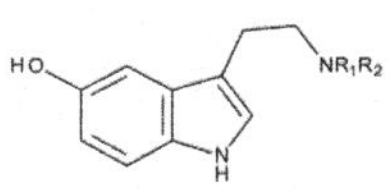

5-Hydroxy-*N*ω-methyltryptamine (**26**) R1: H R2: CH3
Bufotenin (**27**) R1,R2: CH3

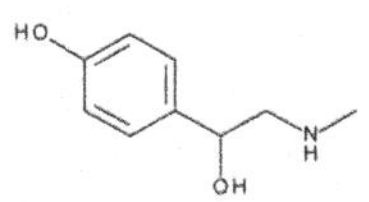

(-)-Synephrine (**28**)

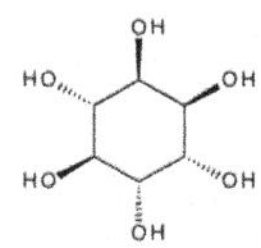

D-*chiro*-Inositol (**29**)

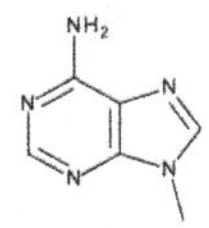

Adenosine (**30**)

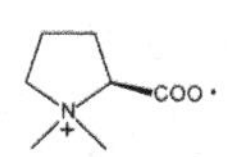

L-(-)-Stachydrine (**31**)

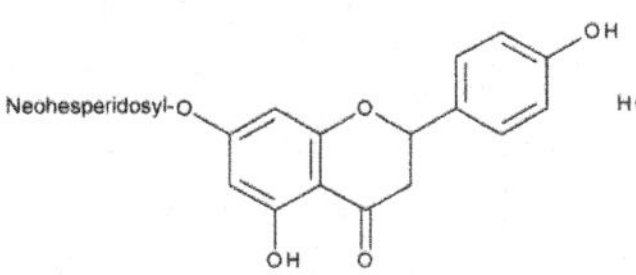

Naringin (**32**)

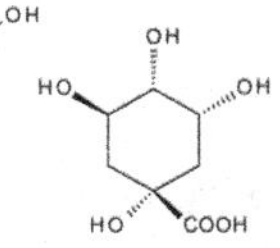

(-)-Quinic acid (**33**)

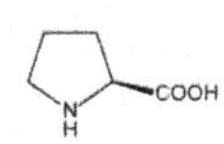

L-(-)-Proline (**34**)

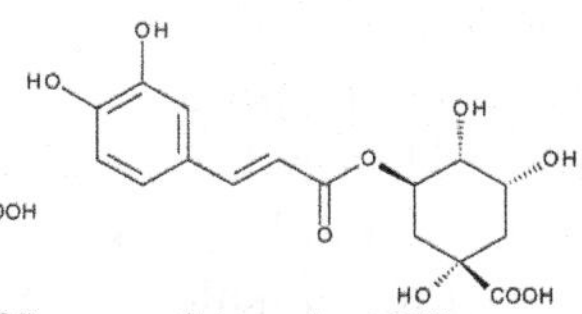

Chlorogenic acid (**35**)

Orixa japonica is a preferential host for *Papilio bianor* but a non-host for *P. xuthus*. Of the plant components, a mixture of hydroxylated γ-lactone (**36**) and hydroxycinnamoyl ester (**37**) strongly elicits oviposition responses from *P. bianor* (Ono et al. 2000a, b), whereas hydroxybenzoic acid derivatives (**38, 39**) deter both female oviposition and larval feeding in *P. xuthus* (Ono et al. 2004).

Phellodendron amurense includes a characteristic flavonoid glycoside, phellamurin (**40**), at high levels. This compound singly stimulates oviposition by *Papilio maackii*, but has little influence on that by *P. xuthus*, and even strongly deters that by *P. protenor* (Honda and Hayashi 1995b, Honda et al. 2011). Therefore, of the three *Papilio* species, only *P. protenor* never lays eggs on the foliage of *P. amurense*.

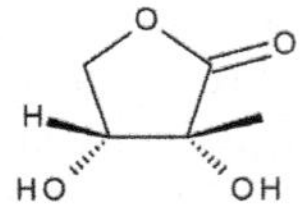

(-)-2-*C*-Methyl-D-*erythrono*-1,4-lactone (**36**)

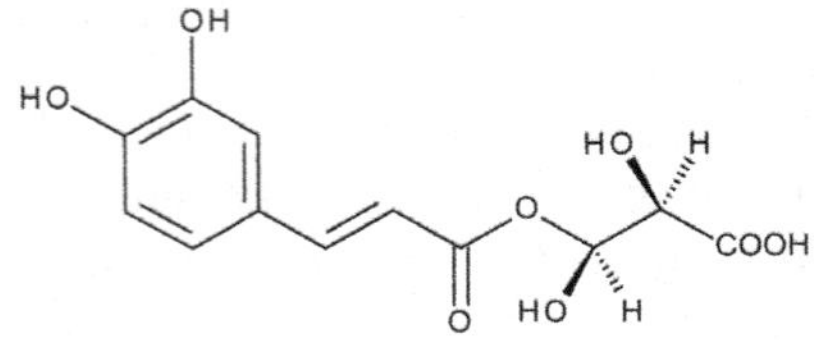

(-)-4-(*E*)-Caffeoyl-L-threonic acid (**37**)

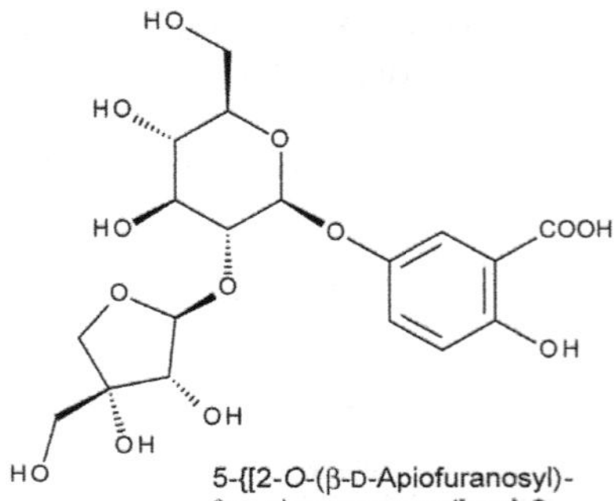

5-{[2-*O*-(β-D-Apiofuranosyl)-β-D-glucopyranosyl]oxy}-2-hydroxybenzoic acid (**38**)

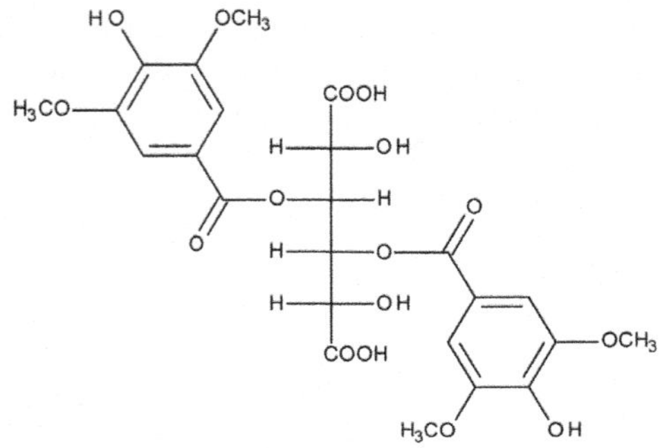

3,4-*O*-Disyringoyl-aldaric acid (**39**)

Toddalia asiatica and *Murraya paniculata* are sympatric rutaceous plants in the Okinawa Islands, Japan. *Papilio polytes* uses *T. asiatica* as a major host but never oviposits on the leaves of *M. paniculata*. In *T. asiatica*, the oviposition stimulant for *P. polytes* is a mixture of hydroxymethylproline (**41**) and trihydroxyaliphatic acid (**42**), which are also present in *Glycosmis citrifolia*, another host of *P. polytes* (Nakayama et al. 2003). In *M. paniculata*, a pyridine alkaloid (trigonelline, **43**) was identified as an oviposition deterrent for *P. polytes* (Nakayama and Honda 2004).

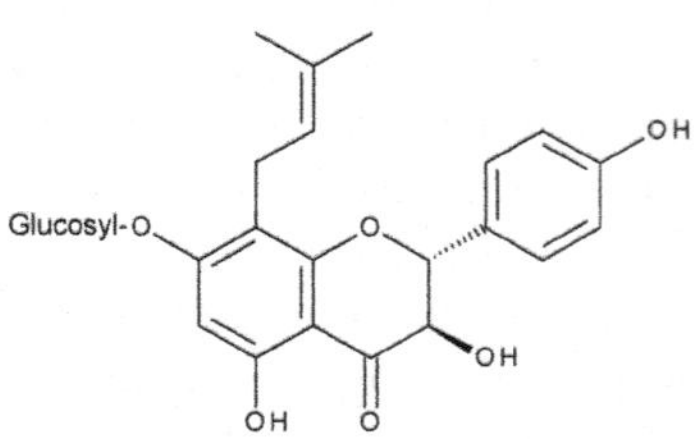

Phellamurin (**40**)

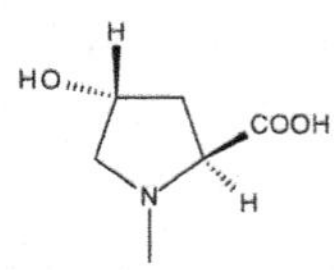

trans-4-Hydroxy-*N*-methyl-L-proline (**41**)

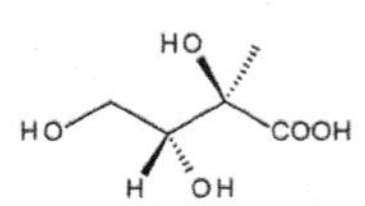

2-*C*-Methyl-D-erythronic acid (**42**)

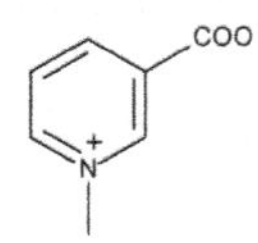

Trigonelline (**43**)

4.6 Papilioninae and Apiaceae

Females of Apiaceae-feeding swallowtails, e.g., several subspecies of *Papilio machaon*, often show mistaken oviposition on rutaceous plants, suggesting that common oviposition stimulants are included in both Apiacea and Rutaceae (Nishida 2014). Black swallowtail *Papilio polyxenes* was found to oviposit in response to a blend of chlorogenic acid (**35**), luteorin 7-*O*-(6″-*O*-malonyl)-β-D-glucopyranoside (**44**), and tyramine (**45**) in wild carrot *Daucus carota* and wild parsnip *Pastinaca sativa* (Feeny et al. 1988, Carter et al. 1998). Chlorogenic acid and luteorin glycoside elicit sensory responses from female fore-tarsal sensilla (Roessingh et al. 1991). Tyramine is an analog to synephrine, the oviposition stimulant for Rutaceae-feeders, whereas luteorin glucoside contains a naringin-like aglycone. Chlorogenic acid is also an oviposition stimulant for Rutaceae-feeders. These structural relationships in oviposition stimulants support a possible host shift of swallowtail butterflies from the Rutaceae to the Apiaceae.

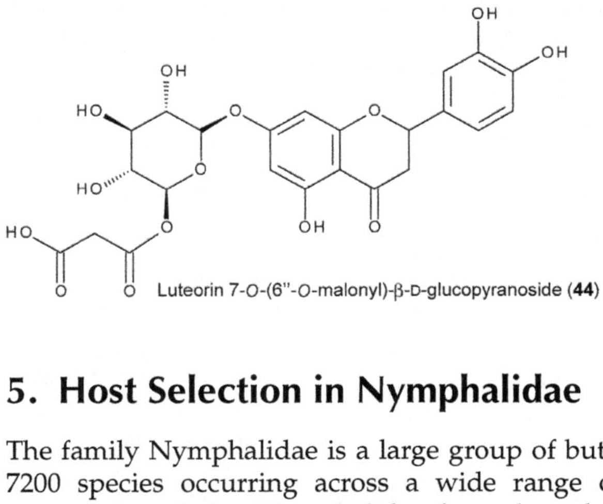

Luteorin 7-*O*-(6"-*O*-malonyl)-β-D-glucopyranoside (**44**)

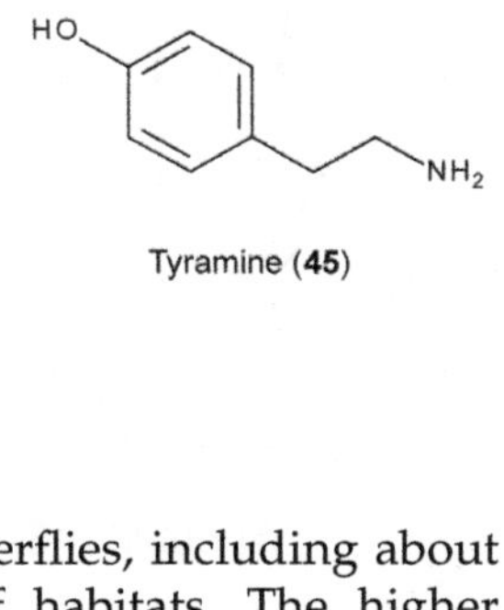

Tyramine (**45**)

5. Host Selection in Nymphalidae

The family Nymphalidae is a large group of butterflies, including about 7200 species occurring across a wide range of habitats. The higher systematics of the Nymphalidae have long been discussed, and the phylogenetic relationships of over 10 subfamilies are still unclear. Recent studies using molecular data and phylogenetic methods classified them into several subgroups, i.e., one basal branch (Libytheinae) and either three clades (danaoid, nymphaloid, and satyroid *sensu* Freitas and Brown 2004) or four clades (danaine, satyrine, heliconiine, and nymphaline *sensu* Wahlberg et al. 2003). Within the Nymphalidae, larval host usage is also highly complicated, but distinct associations with particular plant orders were shown at the subfamily level (Ferrer-Paris et al. 2013). In the subfamily Nymphalinae, the ancestral host clade is regarded as the 'urticalean rosids' including the Urticaceae, Ulmaceae, and Cannabaceae, and species diversity is higher in part of the Nymphalinae, which have exploited novel host plants in the orders Lamiales and Asterales (Janz et al. 2001, Weingartner et al. 2006, Nylin and Wahlberg 2008). However, within this family, there is little information about the plant chemicals involved in female oviposition and larval feeding.

5.1 Nymphalinae and Plantaginaceae

Iridoid glycosides are widespread in over 50 different plant families of the subclass Asteridae, in which aucubin (**46**) and catalpol (**47**) are especially prominent in the Plantaginaceae (Dobler et al. 2011). Although iridoid glycosides are highly toxic for generalist insect herbivores, several specialist herbivores belonging to the subfamily Nymphalinae are attracted to iridoid-containing plants. Within the tribe Junonini, the genus *Junonia* shares much of the same host plant repertoire as closely related plant groups (Nylin and Wahlberg 2008), from which a strong association with plant iridoid glycosides has been discovered. Catalpol contained in *Plantago lanceolate* (Plantaginaceae) was found to stimulate female oviposition and larval feeding from common buckeye *Junonia coenia* (Pereyra and Bowers 1988, Klockars et al. 1993).

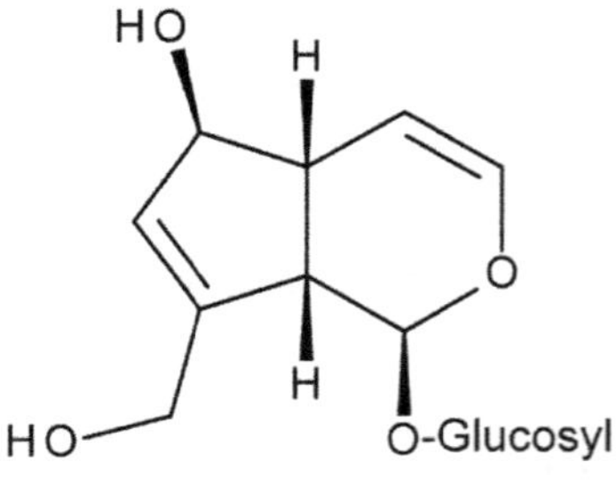

Aucubin (**46**)

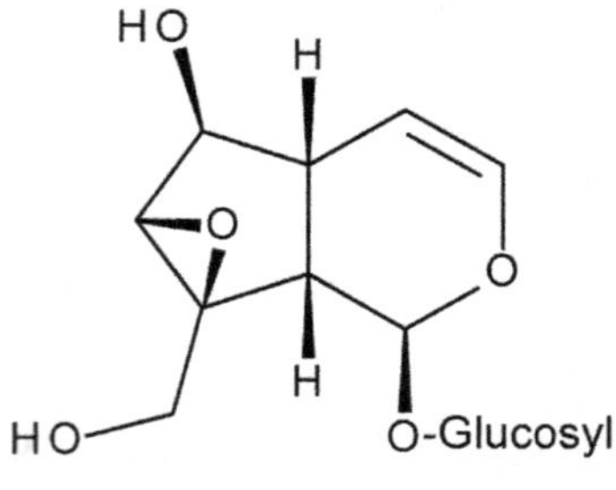

Catalpol (**47**)

The tribe Melitaeini feeds primarily on Asteridae plants, most of which contain iridoids. Catalpol serves as a feeding attractant and stimulant for larvae of the checkerspot butterfly *Euphydryas chalcedona* (Browers 1983). Larvae of Glanville fritillary *Melitaea cinxia* exploit leaves of *P. lanceolate* and *Veronica spicata* (Plantaginaceae) in Finland. Although both host plants contain aucubin and catalpol, females of *M. cinxia* show preferential oviposition on *P. lanceolate* because this host has higher concentrations of iridoid glycosides than *V. spicata* (Nieminen et al. 2003). Of note, there have been host shifts within the Melitaeini to non-iridoid containing plants in the families Asteraceae and Acanthaceae in the Asteridae, indicating that species diversification within the Melitaeini followed this host shift (Fordyce 2010).

5.2 Danainae and Apocynaceae

The subfamily Danainae includes many species feeding on particular host plants in the order Gentianales, especially the family Apocynaceae containing the subfamily Asclepiadoideae (formerly family Asclepiadacea in APG I). Pyrrolizidine alkaloids (PAs) are typical secondary plant metabolites that are widely distributed in this family as a defense against herbivores (Hartmann 1999). Living danaine butterflies have a specific relationship with PAs in their life history (e.g., pharmacophagy of PAs), which is considered to originate from their host usage of the genus *Parsonsia*, an ancestral group containing PAs in the family Apocynaceae (Edger 1984, Honda et al. 1997b). However, it has so far been revealed that four Apocynaceae-feeding danaine butterflies depend on structurally different plant compounds for oviposition. Monarch butterfly *Danaus plexippus* lays eggs on the leaves of milkweed *Asclepias* spp. in which the oviposition stimulants are flavonol (quercetin) glycosides (**48**) (Haribal and Renwick 1996, 1998). *Danaus plexippus* show low oviposition preference

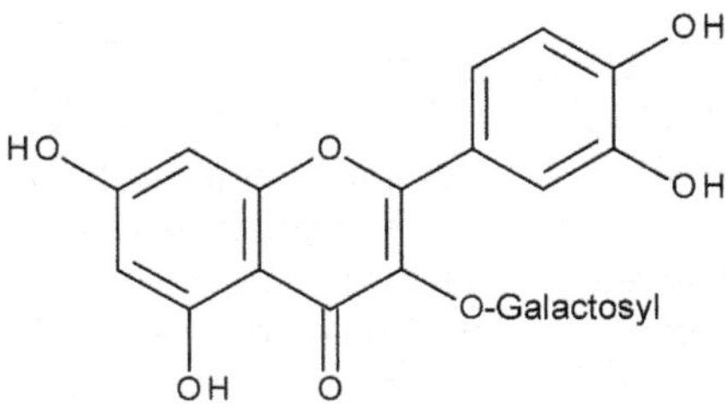

Quercetin-3-*O*-β-D-galactoside (**48**)

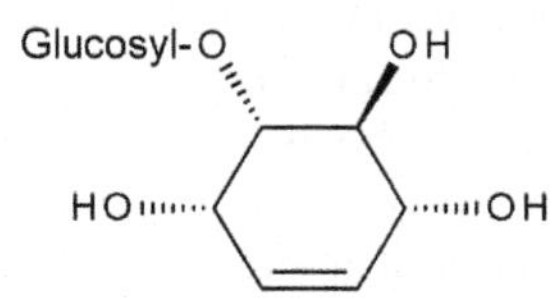

Conduritol F 2-β-glucoside (**49**)

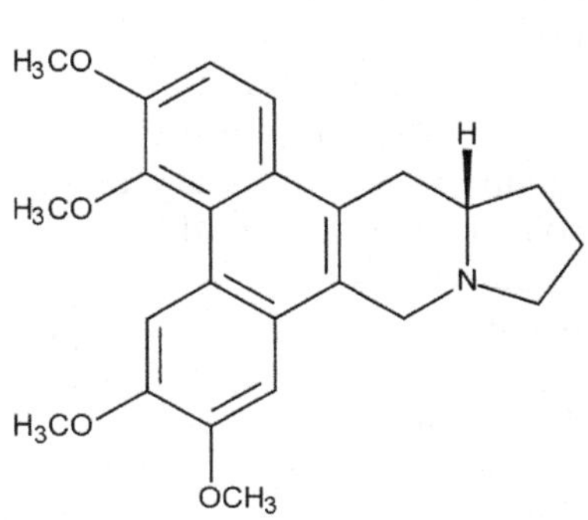

(+)-Isotylocrebrine (**50**)

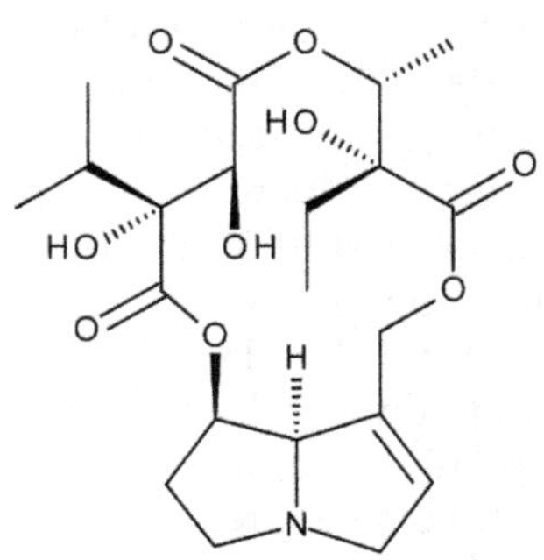

Parsonsianine (**51**)

for milkweed hosts containing cardiac glycosides (cardenolides) at higher concentrations, which have a negative impact on the survival rate of neonate larvae (Oyeyele et al. 1990, Zalucki et al. 1990). Host-plant extracts and fractions containing active flavonoids elicited contact chemosensory responses from not only fore-tarsi but also mid-tarsi and antennal tips in females (Baur et al. 1998). *Parantica sita* showed oviposition responses to condulitols (**49**) in the foliage of *Marsdenia tomentosa* (Honda et al. 2004), whereas *Ideopsis similis* depended on phenanthroindolizidine alkaloids (**50**) for oviposition on *Tylophora tanakae* (Honda et al. 1995b, 2001). The oviposition stimulants for giant danaid butterfly *Idea leuconoe* in *Parsonsia alboflavescens* (*Parsonsia laevigata*) are macrocyclic pyrrolizidine alkaloids (**51**) (Honda et al. 1997b).

6. Conclusion and Perspectives

Host selection depends on the balance of stimulatory and deterrent inputs with, in oligophagous and monophagous herbivores, a dominating role of host-specific plant chemicals (Chapman 2003). Since Verschaffelt's (1910) ground-breaking discovery of chemical associations between *Pieris* larvae and its cruciferous hosts, early studies have focused on the behavioral and sensory responses of butterflies to plant defense chemicals present in their hosts. Indeed, several defense substances categorized into the glucosinolates, iridoids, flavonoids, or AAs have been found to stimulate solely female oviposition and/or larval feeding of particular butterfly species. The usage of defense chemicals for host selection could presumably be achieved by expressing an appropriate receptor on a stimulatory neuron instead of a deterrent neuron (Chapman 2003). Emphasis of the neural pathway from a stimulatory cell to the CNS could lead to the pivotal role of glucosinolates in host selection for *Pieris* species. On the other hand, a series of intensive studies have revealed that host selection by butterfly is regulated in most cases by multiple components, with a key role of a particular compound (its partial structure) involved in a plant primary metabolite. This finding suggests the co-existence of redundant and synergistic mechanisms for host selection, probably in order to adjust for chemical fluctuation in host plants and maintain evolutionary potential for host shift in the future. From this point of view, it would be worth investigating which derivatives of the primary metabolites are involved in butterfly–plant relationships between *Pieris* species and Brassicaceae and between *J. coenia* and Plantaginaceae, in which host-specific defense chemicals serve as major oviposition and feeding stimulants.

Female preference in oviposition is not necessarily in accordance with larval feeding preference and developmental performance; this discrepancy is often prominent in the relationship with alien invasive plants (e.g., DiTommaso and Losey 2003, Davis and Cipollini 2014). In general, the potential host range of butterfly larvae is often wider than the range of plants actually utilized by females (Wiklund 1975, Smiley 1978). In several papilionid species, female oviposition preference is controlled primarily by one or more loci on the X chromosome, which are not linked to larval feeding preference and developmental performance (Thompson et al. 1990, Scriber et al. 1991). Therefore, female oviposition, especially host recognition and acceptance, is critically important in determining host plant range (Janz and Nylin 1997). In the chemical (olfactory and taste) sense of *Drosophila*, although the CNS is quite similar between adults and larvae, differences in the peripheral chemosensory system are relatively prominent, e.g., the number of gustatory receptor neurons (GRNs) on the head is estimated to be 300 for adults and 80 for larvae (Vosshall et al. 2007). It is highly likely that female butterflies have a larger number of GRNs than larvae in order to conduct precise host selection for oviposition. Comparative studies between adults and larvae are needed for certain butterfly species to reveal the variety of peripheral receptor neurons and the structure of the CNS involved in host selection.

Recent progress in molecular phylogeny has indicated the possible evolutionary history of host plant associations for several butterfly species (Janz et al. 2001, Fordyce 2010, Ferrer-Paris et al. 2013). However, host shift depending on plant chemistry has so far been discussed only within a narrow range of butterfly species in the same subfamily or tribe. In spite of accumulating ecological data on butterfly host usage, there has been slow progress in the structural elucidation of the semiochemicals because bioactivity-guided fractionation to identify active compounds is time-consuming and labor intensive. With respect to semiochemicals for host selection in other butterfly families, oxalic acid, a larval feeding stimulant of *Zizeeria* (*Pseudozizeeria*) *maha* (Yamaguchi et al. 2016), has been identified only in the Lycaenidae, and there has been no structural elucidation in the Hesperiidae. To further understand host specialization, diversification, and switching within and across butterfly families, further studies should be performed on behavioral and sensory responses to host plant chemicals using various butterfly species.

REFERENCES

Aubert, J., L. Legal, H. Descimon and F. Michel. 1999. Molecular phylogeny of swallowtail butterflies of the tribe Papilionini (Papilionidae, Lepidoptera). Mol. Phylogenet. Evol. 12: 156–167.

Baur, R., M. Haribal, J.A.A. Renwick and E. Städler. 1998. Contact chemoreception related to host selection and oviposition behaviour in the monarch butterfly, *Danaus plexippus*. Physiol. Entomol. 23: 7–19.

Becerra, J.X. 1997. Insects on plants: macroevolutionary chemical trends in host use. Science 276: 253–256.

Berenbaum, M.R. 1995. Chemistry and oligophagy in the Papilionidae. pp. 27–38. *In*: J.M. Scriber, Y. Tsubaki and R.C. Lederhouse [eds.]. Swallowtail Butterflies: Their Ecology and Evolutionary Biology. Scientific Publishers, Gainesville, FL, USA.

Bernays, E.A. and R.F. Chapman. 1994. Host-Plant Selection by Phytophagous Insects. Chapman & Hall, New York, NY, USA.

Braby, M.F. 2005. Provisional checklist of genera of the Pieridae (Lepidoptera: Papilionoidea). Zootaxa 832: 1–16.

Braby, M.F. and J.W.H. Trueman. 2006. Evolution of larval host plant associations and adaptive radiation in pierid butterflies. J. Evol. Biol. 19: 1677–1690.

Browers, M.D. 1983. The role of iridoid glycosides in host-plant specificity of checkerspot butterflies. J. Chem. Ecol. 9: 475–493.

Brown, K.S. Jr., C.F. Klitzke, C. Berlingeri and P. Euzébio Rubbo dos Santos. 1995. Neotropical swallowtails: chemistry of food plant relationships, population ecology, and biosystematics. pp. 405–445. *In*: J.M. Scriber, Y. Tsubaki and R.C. Lederhouse [eds.]. Swallowtail Butterflies: Their Ecology and Evolutionary Biology. Scientific Publishers, Gainesville, FL, USA.

Carrasco, D., M.C. Larsson and P. Anderson. 2015. Insect host plant selection in complex environments. Curr. Opin. Insect Sci. 8: 1–7.

Carter, M., K. Sachdev-Gupta and P. Feeny. 1998. Tyramine from the leaves of wild parsnip: a stimulant and synergist for oviposition by the black swallowtail butterfly. Physiol. Entomol. 23: 303–312.

Carter, M., P. Feeny and M. Haribal. 1999. An oviposition stimulant for spicebush swallowtail butterfly, *Papilio troilus*, from leaves of *Sassafras albidum*. J. Chem. Ecol. 25: 1233–1245.

Caterino, M.S., R.D. Reed, M.M. Kuo and F.A.H. Sperling. 2001. A partitioned likelihood analysis of swallowtail butterfly phylogeny (Lepidoptera: Papilionidae). Syst. Biol. 50: 106–127.

Chapman, R.F. 2003. Contact chemoreception in feeding by phytophagous insects. Annu. Rev. Entomol. 48: 455–484.

Chew, F.S. and J.A.A. Renwick. 1995. Host plant choice in *Pieris* butterflies. pp. 214–238. *In*: R.T. Cardé and W.J. Bell [eds.]. Chemical Ecology of Insects 2. Chapman & Hall, New York, NY, USA.

David, W.A.L. and B.O.C. Gardiner. 1966. Mustard oil glucosides as feeding stimulants for *Pieris brassicae* larvae in a semi-synthetic diet. Entomol. Exp. Appl. 9: 247–255.

Davis, S.L. and D. Cipollini. 2014. Do mothers always know best? Oviposition mistakes and resulting larval failure of *Pieris virginiensis* on *Alliaria petiolate*, a novel, toxic host. Biol. Invasions 16: 1941–1950.

Davis, S.L., T. Frisch, N. Bjarnholt and D. Cipollini. 2015. How does garlic mustard kill the West Virginia white butterfly? J. Chem. Ecol. 41: 948–955.

Dimock, M.B., J.A.A. Renwick, C.D. Radke and K. Sachdev-Gupta. 1991. Chemical constituents of an unacceptable crucifer, *Erysimum cheiranthoides*, deter feeding by *Pieris rapae*. J. Chem. Ecol. 17: 525–533.

DiTommaso, A. and J.E. Losey. 2003. Oviposition preference and larval performance of monarch butterflies (*Danaus plexippus*) on two invasive swallow-wort species. Entomol. Exp. Appl. 108: 205–209.

Dobler, S., G. Petschenka and H. Pankoke. 2011. Coping with toxic plant compounds—the insect's perspective on iridoid glycosides and cardenolides. Phytochemistry 72: 1593–1604.

Du, Y.-J., J.A.A. van Loon and J.A.A. Renwick. 1995. Contact chemoreception of oviposition-stimulating glucosinolates and an oviposition-deterrent cardenolide in two subspecies of *Pieris napi*. Physiol. Entomol. 20: 167–174.

Edgar, J.A. 1984. Parsonsieae: ancestral larval foodplants of the Danainae and Ithomiinae. pp. 91–93. *In*: R.I. Vane-Wright and P.R. Ackery [eds.]. The Biology of Butterflies. Academic Press, London, UK.

Ehrlich, P.R. 1958. The comparative morphology, phylogeny and higher classification of the butterflies (Lepidoptera: Papilionoidea). Univ. Kansas Sci. Null. 39: 305–370.

Ehrlich, P.R. and P.H. Raven. 1964. Butterflies and plants: a study in coevolution. Evolution 18: 586–608.

Feeny, P. 1995. Ecological opportunism and chemical constraints on the host associations of swallowtail butterflies. pp. 9–15. *In*: J.M. Scriber, Y. Tsubaki and R.C. Lederhouse [eds.]. Swallowtail Butterflies: Their Ecology and Evolutionary Biology. Scientific Publishers, Gainesville, FL, USA.

Feeny, P., K. Sachdev, L. Rosenberry and M. Carter. 1988. Luteolin 7-*O*-(6″-*O*-malonyl)-β-D-glucoside and *trans*-chlorogenic acid: oviposition stimulants for the black swallowtail butterfly. Phytochemistry 27: 3439–3448.

Ferrer-Paris, J.R., A. Sánchez-Mercado, Á.L. Viloria and J. Donaldson. 2013. Congruence and diversity of butterfly-host plant associations at higher taxonomic levels. PLoS ONE 8: e63570.

Fordyce, J.A. 2010. Host shifts and evolutionary radiations of butterflies. Proc. R. Soc. B 277: 3735–3743.

Frazier, J.L. 1992. How animals perceive secondary plant compounds. pp. 89–134. *In*: G.A. Rosenthal and M.R. Berenbaum [eds.]. Herbivores—Their Interactions with Secondary Plant Metabolites. Second Edition. Volume II: Ecological and Evolutionary Processes. Academic Press, San Diego, CA, USA.

Freitas, A.V.L. and K.S. Brown Jr. 2004. Phylogeny of the Nymphalidae (Lepidoptera). Syst. Biol. 53: 363–383.

Hanson, F.E. and V.G. Dethier. 1973. Role of gustation and olfaction in food plant discrimination in the tobacco hornworm, *Manduca sexta*. J. Insect Physiol. 19: 1019–1034.

Haribal, M. and P. Feeny. 1998. Oviposition stimulant for the zebra swallowtail butterfly, *Eurytides marcellus*, from the foliage of pawpaw, *Asimina triloba*. Chemoecology 8: 99–110.

Haribal, M. and P. Feeny. 2003. Combined roles of contact stimulant and deterrents in assessment of host-plant quality by ovipositing zebra swallowtail butterflies. J. Chem. Ecol. 29: 653–670.

Haribal, M. and J.A.A. Renwick. 1996. Oviposition stimulants for the monarch butterfly: flavanol glycosides from *Asclepias curassavica*. Phytochemistry 41: 139–144.

Haribal, M. and J.A.A. Renwick. 1998. Identification and distribution of oviposition stimulants for monarch butterflies in hosts and non-hosts. J. Chem. Ecol. 24: 891–904.

Haribal, M. and J.A.A. Renwick. 2001. Seasonal and population variation in flavonoid and alliarinoside content of *Alliaria petiolate*. J. Chem. Ecol. 27: 1585–1594.

Haribal, M., Z. Yang, A.B. Attygalle, J.A.A. Renwick and J. Meinwald. 2001. A cyanoallyl glucoside from *Alliaria petiolate*, as a feeding deterrent for larvae of *Pieris napi oleracea*. J. Nat. Prod. 64: 440–443.

Hartmann, T. 1999. Chemical ecology of pyrrolizidine alkaloids. Planta 207: 483–495.

Honda, K. 1986. Flavanone glycosides as oviposition stimulants in a papilionid butterfly, *Papilio protenor*. J. Chem. Ecol. 12: 1999–2010.

Honda, K. 1990. Identification of host-plant chemicals stimulating oviposition by swallowtail butterfly, *Papilio protenor*. J. Chem. Ecol. 16: 325–337.

Honda, K. 1995. Chemical basis of differential oviposition by lepidopterous insects. Arch. Insect Biochem. Physiol. 30: 1–23.

Honda, K. and N. Hayashi. 1995a. Chemical factors in rutaceous plants regulating host selection by two swallowtail butterflies, *Papilio protenor* and *P. xuthus* (Lepidoptera: Papilionidae). Appl. Entomol. Zool. 30: 327–334.

Honda, K. and N. Hayashi. 1995b. A flavonoid glucoside, phellamurin, regulates differential oviposition on a rutaceous plant, *Phellodendron amurense*, by two sympatric swallowtail butterflies, *Papilio protenor* and *P. xuthus*: the front line of a coevolutionary arms race? J. Chem. Ecol. 21: 1531–1548.

Honda, K., T. Saitoh, S. Hara and N. Hayashi. 1995a. A neolignoid feeding deterrent against *Luehdorfia puziloi* larvae (Lepidoptera: Papilionidae) from *Heterotropa aspera*, a host plant of sibling species, *L. japonica*. J. Chem. Ecol. 21: 1541–1548.

Honda, K., A. Tada, N. Hayashi, F. Abe and T. Yamauchi. 1995b. Alkaloidal oviposition stimulants for a danaid butterfly, *Ideopsis similis* L., from a host plant, *Tylophora tanakae* (Asclepiadaceae). Experientia 51: 753–756.

Honda, K., W. Nishii and N. Hayashi. 1997a. Oviposition stimulants for sulfur butterfly, *Colias erate poliographys*: cyanoglucosides as synergists involved in host preference. J. Chem. Ecol. 23: 323–331.

Honda, K., N. Hayashi, F. Abe and T. Yamauchi. 1997b. Pyrrolizidine alkaloids mediate host-plant recognition by ovipositing females of an old world danaid butterfly, *Idea leuconoe*. J. Chem. Ecol. 23: 1703–1713.

Honda, K., H. Ômura, N. Hayashi, F. Abe and T. Yamauchi. 2001. Oviposition-stimulatory activity of phenanthroindolizidine alkaloids of host-plant origin to a danaid butterfly, *Ideopsis similis*. Physiol. Entomol. 26: 6–10.

Honda, K., H. Ômura, N. Hayashi, F. Abe and T. Yamauchi. 2004. Conduritols as oviposition stimulants for the danaid butterfly, *Parantica sita*, identified from a host plant, *Marsdenia tomentosa*. J. Chem. Ecol. 30: 2285–2296.

Honda, K., H. Ômura, M. Chachin, S. Kawano and T.A. Inoue. 2011. Synergistic or antagonistic modulation of oviposition response of two swallowtail butterflies, *Papilio maackii* and *P. protenor*, to *Phellodendron amurense* by its constitutive prenylated flavonoid, phellamurin. J. Chem. Ecol. 35: 575–581.

Honda, K., H. Minematsu, K. Muta, H. Ômura and W. Nishii. 2012. D-Pinitol as a key oviposition stimulant for sulfur butterfly, *Colias erate*: chemical basis for female acceptance of host- and non-host plants. Chemoecology 22: 55–63.

Hopkins, R.J., N.M. van Dam and J.J.A. van Loon. 2009. Role of glucosinolates in insect-plant relationships and multitrophic interactions. Annu. Rev. Entomol. 54: 57–83.

Huang, X. and J.A.A. Renwick. 1993. Differential selection of host plants by two *Pieris* species: the role of oviposition stimulants and deterrents. Entomol Exp. Appl. 68: 59–69.

Huang, X. and J.A.A. Renwick. 1994. Relative activities of glucosinolates as oviposition stimulants for *Pieris rapae* and *P. napi oleracea*. J. Chem. Ecol. 20: 1025–1037.

Huang, X., J.A.A. Renwick and K. Sachdev-Gupta. 1993a. Oviposition stimulants and deterrents regulating differential acceptance of *Iberis amara* by *Pieris rapae* and *P. napi oleracea*. J. Chem. Ecol. 19: 1645–1663.

Huang, X., J.A.A. Renwick and K. Sachdev-Gupta. 1993b. A chemical basis for differential acceptance of *Erysimum cheiranthoides* by two *Pieris* species. J. Chem. Ecol. 19: 195–210.

Huang, X., J.A.A. Renwick and K. Sachdev-Gupta. 1994. Oviposition stimulants in *Barbarea vulgaris* for *Pieris rapae* and *P. napi oleracea*: isolation, identification and differential activity. J. Chem. Ecol. 20: 423–438.

Janz, N. and S. Nylin. 1997. The role of female search behavior in determining host plant range in plant feeding insects: a test of the information processing hypothesis. Proc. R. Soc. Lond. B 264: 701–707.

Janz, N. and S. Nylin. 1998. Butterflies and plants: a phylogenetic study. Evolution 52: 486–502.

Janz, N., K. Nyblom and S. Nylin. 2001. Evolutionary dynamics of host-plant specialization: a case study of the tribe Nymphalini. Evolution 55: 783–796.

Klockars, G.K., M.D. Bowers and B. Cooney. 1993. Leaf variation in iridoid glycoside content of *Plantago lanceolate* (Plantaginaceae) and oviposition of the buckeye, *Junonia coenia* (Nymphalidae). Chemoecology 4: 72–78.

Kvello, P., T.J. Almaas and H. Mustaparta. 2006. A confined taste area in a lepidopteran brain. Arthro. Struct. Develop. 35: 35–45.

Loon, J.J.A. van, A. Blaakmeer, F.C. Griepink, T.A. van Beek, L.M. Schoonhoven and A. de Groot. 1992. Leaf surface compound from *Brassica oleracea* (Cruciferae) induces oviposition by *Pieris brassicae* (Lepidoptera: Pieridae). Chemoecology 3: 39–44.

Ma, W.C. and L.M. Schoonhoven. 1973. Tarsal contact chemosensory hairs of the large white butterfly *Pieris brassicae* and their possible role in oviposition behaviour. Entomol. Exp. Appl. 16: 343–357.

Martin, J.M., S.R. Madigosky, Z.-M. Gu, D. Zhou, J. Wu and J.L. McLaughlin. 1999. Chemical defense in the zebra swallowtail butterfly, *Eurytides marcellus*, involving annonaceous acetogenins. J. Nat. Prod. 62: 2–4.

Matsuda, K., M. Kaneko, K. Kusaka, T. Shishido and Y. Tamaki. 1998. Soyasaponins as feeding stimulants to the oriental clouded yellow larva, *Colias erate poliographus* (Lepidoptera: Pieridae). Appl. Entomol. Zool. 33: 255–258.

Miles, C.I., M.L. del Campo and J.A.A. Renwick. 2005. Behavioral and chemosensory responses to a host recognition cue by larvae of *Pieris rapae*. J. Comp. Physiol. A 191: 147–155.

Mitchell, B.K., H. Itagaki and M.-P. Rivet. 1999. Peripheral and central structures involved in insect gustation. Microsc. Res. Tech. 47: 401–415.

Mithöfer, A. and W. Boland. 2012. Plant defense against herbivores: chemical aspects. Annu. Rev. Plant Biol. 63: 431–450.

Mukae, S., T. Ohashi, Y. Matsumoto, S. Ohta and H. Ômura. 2016. D-Pinitol in Fabaceae: an oviposition stimulant for common grass yellow, *Eurema mandarina*. J. Chem. Ecol. 42: 1122–1129.

Murata, T., N. Mori and R. Nishida. 2011. Larval feeding stimulants for a Rutaceae-feeding swallowtail butterfly, *Papilio xuthus* L. in *Citrus unshiu* leaves. J. Chem. Ecol. 37: 1099–1109.

Nakayama, T. and K. Honda. 2004. Chemical basis for differential acceptance of two sympatric rutaceous plants by ovipositing females of a swallowtail butterfly, *Papilio polytes* (Lepidopteram Papilionidae). Chemoecology 14: 199–205.

Nakayama, T., K. Honda, H. Ômura and N. Hayashi. 2003. Oviposition stimulants for the tropical swallowtail butterfly, *Papilio polytes*, feeding on a rutaceous plant, *Toddalia asiatica*. J. Chem. Ecol. 29: 1621–1634.

Nieminen, M., J. Suomi, S. van Nouhuys, P. Sauri and M.-L. Riekkola. 2003. Effect of iridoid glycoside content on oviposition host plant choice and parasitism in a specialist herbivore. J. Chem. Ecol. 29: 823–844.

Nishida, R. 1994. Oviposition stimulant of a zerynthiine swallowtail butterfly, *Luehdorfia japonica*. Phytochemistry 36: 873–877.

Nishida, R. 1995. Oviposition stimulants of swallowtail butterflies. pp. 17–26. *In*: J.M. Scriber, Y. Tsubaki and R.C. Lederhouse [eds.]. Swallowtail Butterflies: Their Ecology and Evolutionary Biology. Scientific Publishers, Gainesville, FL, USA.

Nishida, R. 2014. Chemical ecology of insect-plant interactions: ecological significance of plant secondary metabolites. Biosci. Biotech. Biochem. 78: 1–13.

Nishida, R. and H. Fukami. 1989. Oviposition stimulants of an Aristolochiaceae-feeding swallowtail butterfly, *Atrophaneura alcinous*. J. Chem. Ecol. 15: 2565–2575.

Nishida, R., T. Ohsugi, S. Kokubo and H. Fukami. 1987. Oviposition stimulants of a *Citrus*-feeding swallowtail butterfly, *Papilio xuthus* L. Experientia 43: 342–344.

Numata, A., H. Yamaguchi, K. Hokimoto, M. Ohtani and K. Takaishi. 1985. Host-plant selection by the yellow butterfly larvae, *Eurema hecabe mandarina* (Lepidoptera: Pieridae): attractants and arrestants. Appl. Entomol. Zool. 20: 314–321.

Nylin, S. and N. Wahlberg. 2008. Does plasticity drive speciation? Host-plant shifts and diversification in nymphaline butterflies (Lepidoptera: Nymphalidae) during the tertiary. Biol. J. Linn. Soc. 94: 115–130.

Ohsugi, T., R. Nishida and H. Fukami. 1985. Oviposition stimulant of *Papilio xuthus*, a *Citrus*-feeding swallowtail butterfly. Agric. Biol. Chem. 49: 1897–1900.

Ohsugi, T., R. Nishida and H. Fukami. 1991. Multi-component system of oviposition stimulants for a Rutaceae-feeding swallowtail butterfly, *Papilio xuthus* (Lepidoptera: Papilionidae). Appl. Entomol. Zool. 26: 29–40.

Ono, H., R. Nishida and Y. Kuwahara. 2000a. Oviposition stimulant for a Rutaceae-feeding swallowtail butterfly, *Papilio bianor* (Lepidoptera: Papilionidae): hydroxycinnamic acid derivative from *Orixa japonica*. Appl. Entomol. Zool 35: 119–123.

Ono, H., R. Nishida and Y. Kuwahara. 2000b. A dihydroxy-γ-lactone as an oviposition stimulant for the swallowtail butterfly, *Papilio bianor*, from the rutaceous plant, *Orixa japonica*. Biosci. Biotechnol. Biochem. 64: 1970–1973.

Ono, H., Y. Kuwahara and R. Nishida. 2004. Hydroxybenzoic acid derivatives in a nonhost rutaceous plant, *Orixa japonica*, deter both oviposition and larval feeding in a Rutaceae-feeding swallowtail butterfly, *Papilio xuthus* L. J. Chem. Ecol. 30: 287–301.

Oyeyele, S.O. and M.P. Zalucki. 1990. Cardiac glycosides and oviposition by *Danaus plexippus* on *Asclepias fruticose* in south-east Queensland (Australia), with notes on the effect of plant nitrogen content. Ecol. Entomol. 15: 177–185.

Ozaki, K., M. Ryuda, A. Yamada, A. Utoguchi, H. Ishimoto, D. Calas et al. A gustatory receptor involved in host plant recognition for oviposition of a swallowtail butterfly. Nat. Comm. 2: 542.

Papaj, D.R., P. Feeny, K. Sachdev-Gupta and L. Rosenberry. 1992. D-(+)-Pinitol, an oviposition stimulant for the pipevine swallowtail butterfly, *Battus philenor*. J. Chem. Ecol. 18: 799–815.

Pereyra, P.C. and M.D. Browers. 1988. Iridoid glycosides as oviposition stimulants for the buckeye butterfly, *Junonia coenia* (Nymphalidae). J. Chem. Ecol. 14: 917–928.

Renwick, J.A.A. 2002. The chemical world of crucivores: lures, treats and traps. Entomol. Exp. Appl. 104: 35–42.

Renwick, J.A.A. and F.S. Chew. 1994. Oviposition behavior in Lepidoptera. Annu. Rev. Entomol. 39: 377–400.

Renwick, J.A.A., C.D. Radke and K. Sachdev-Gupta. 1989. Chemical constituents of *Erysimum cheiranthoides* deterring oviposition by the cabbage butterfly, *Pieris rapae*. J. Chem. Ecol. 15: 2161–2169.

Renwick, J.A.A., C.D. Radke, K. Sachdev-Gupta and E. Städler. 1992. Leaf surface chemicals stimulating oviposition by *Pieris rapae* (Lepidoptera: Pieridae) on cabbage. Chemoecology 3: 33–38.

Renwick, J.A.A., W. Zhang, M. Haribal, A.B. Attygalle and K.D. Lopez. 2001. Dual chemical barriers protect a plant against different larval stages of an insect. J. Chem. Ecol. 27: 1575–1583.

Roessingh, P., E. Städler, R. Schöni and P. Feeny. 1991. Tarsal contact chemoreceptors of the black swallowtail butterfly *Papilio polyxenes*: responses to phytochemicals from host- and non-host plants. Physiol. Entomol. 16: 485–495.

Rosenthal, G.A. and M.R. Berenbaum. 1991. Herbivores—Their Interactions with Secondary Plant Metabolites. Second edition. Volume I: The Chemical Participants. Academic Press, San Diego, CA, USA.

Rotem, K., A.A. Agrawal and L. Kott. 2003. Parental effects in *Pieris rapae* in response to variation in food quality: adaptive plasticity across generations? Ecol. Entomol. 28: 211–218.

Sachdev-Gupta, K., J.A.A. Renwick and C.D. Radke. 1990. Isolation and identification of oviposition deterrents to cabbage butterfly, *Pieris rapae*, from *Erysimum cheiranthoides*. J. Chem. Ecol. 16: 1059–1067.

Sachdev-Gupta, K., C.D. Radke, J.A.A. Renwick and M.B. Dimock. 1993a. Cardenolides from *Erysimum cheiranthoides*: feeding deterrents to *Pieris rapae* larvae. J. Chem. Ecol. 19: 1355–1369.

Sachdev-Gupta, K., P.P. Feeny and M. Carter. 1993b. Oviposition stimulants for the pipevive swallowtail butterfly, *Battus philenor* (Papilionidae), from an *Aristolochia* host plant: synergism between inositols, aristolochic acids and a monogalactosyl diglyceride. Chemoecology 4: 19–28.

Schoonhoven, L.M. 1969. Sensitivity changes in some insect chemoreceptors and their effect on food selection behavior. Proc. K. Ned. Akad. Wetensch. Amsterdam 72: 491–498.

Scriber, J.M. 1984. Larval food plant utilization by the world Papilionidae (Lep.): latitudinal gradients reappraised. Tokurana (Acta Rhopalocera) 6/7: 1–50.

Scriber, J.M., B.L. Giebink and D. Snider. 1991. Reciprocal latitudinal clines in oviposition behaviour of *Papilio glaucus* and *P. canadensis* across the Great Lakes hybrid zone: possible sex-linkage of oviposition preferences. Oecologia 87: 360–368.

Smiley, J. 1978. Plant chemistry and the evolution of host specificity: new evidence from *Heliconius* and *Passiflora*. Science 201: 745–747.

Städler, E. 1992. Behavioral responses of insects to plant secondary compounds. pp. 45–88. *In*: G.A. Rosenthal and M.R. Berenbaum [eds.]. Herbivores—Their Interactions with Secondary Plant Metabolites. Second Edition. Volume II: Ecological and Evolutionary Processes. Academic Press, San Diego, CA, USA.

Städler, E., J.A.A. Renwick, C.D. Radke and K. Sachdev-Gupta. 1995. Tarsal contact chemoreceptor response to glucosinolates and cardenolides mediating oviposition in *Pieris rapae*. Physiol. Entomol. 20: 175–187.

Thompson, J.N. and O. Pellmyr. 1991. Evolution of oviposition behavior and host preference in Lepidoptera. Annu. Rev. Entomol. 36: 65–89.

Thompson, J.N., W. Wehling and R. Podolsky. 1990. Evolutionary genetics of host use in swallowtail butterflies. Nature 344: 148–150.

Traynier, R.M.M. and R.J.W. Truscott. 1991. Potent natural egg-laying stimulant for cabbage butterfly *Pieris rapae*. J. Chem. Ecol. 17: 1371–1380.

Verschaffelt, E. 1910. The causes determining the selection of food in some herbivorous insects. Proc. Acad. Sci. Amst. 13: 536–542.

Vosshall, L.B. and R.F. Stocker. 2007. Molecular architecture of smell and taste in *Drosophila*. Annu. Rev. Neurosci. 30: 505–533.

Wahlberg, N., E. Weingartner and S. Nylin. 2003. Towards a better understanding of the higher systematics of Nymphalidae (Lepidoptera: Papilionoidea). Mol. Phylogenet. Evol. 28: 473–484.

Weingartner, E., N. Wahlberg and S. Nylin. 2006. Dynamics of host plant use and species diversity in *Polygonia* butterflies (Nymphalidae). J. Evol. Biol. 19: 483–491.

Wiklund, G. 1975. The evolutionary relationship between adult oviposition preferences and larval host plant range in *Papilio machaon* L. Oecologia 18: 185–197.

Winde, I. and U. Wittstock. 2011. Insect herbivore counteradaptations to the plant glucosinolate-myrosinase system. Phytochemistry 72: 1566–1575.

Yamaguchi, M., S. Matsuyama and K. Yamaji. 2016. Oxalic acid as a larval feeding stimulant for the pale grass blue butterfly *Zizeeria maha* (Lepidoptera: Lycaenidae). Appl. Entomol. Zool. 51: 91–98.

Zakharov, E.V., M.S. Caterino and F.A.H. Sperling. 2004. Molecular phylogeny, historical biography, and divergence time estimates for swallowtail butterflies of the genus *Papilio* (Lepidoptera: Papilionidae). Syst. Biol. 53: 193–215.

Zalucki, M.P., L.P. Brower and S.B. Malcolm. 1990. Oviposition by *Danaus plexippus* in relation to cardenolide content of three *Asclepias* species in the southeastern U.S.A. Ecol. Entomol. 15: 231–240.

Zhang, Y., Z.-H. Zhan, S. Tebayashi, C.-S. Kim and J. Li. 2015. Feeding stimulants for larvae of *Graphium sarpedon nipponum* (Lepidoptera: Papilionidae) from *Cinnamomum camphora*. Z. Naturforschung C 70: 145–150.

Zhou, D.-S., C.-Z. Wang and J.A.A. van Loon. 2009. Chemosensory basis of behavioural plasticity in response to deterrent plant chemicals in the larva of the small cabbage white butterfly *Pieris rapae*. J. Insect Physiol. 55: 788–792.

CHAPTER

2

Function of the Lepidopteran Larval Midgut in Plant Defense Mechanisms

Naoko Yoshinaga* **and Naoki Mori**

Division of Applied Life Sciences, Graduate School of Agriculture,
Kyoto University, Kitashirakawa-Oiwake, Sakyo, Kyoto 606-8502, Japan
E-mail: yoshinaga.naoko.5v@kyoto-u.ac.jp (NY)
E-mail: mori.naoki.8a@kyoto-u.ac.jp (NM)

Abstract

Most lepidopteran families appeared in the Jurassic period and evolved alongside angiosperms (flowering plants). During the long history of the arms race between these adversaries, the larval midgut has been the 'front line' of plant chemical defense and larval metabolic counter adaptations, and various unique strategies have developed in both insects and plants. Metabolic reactions such as the detonation of a "mustard bomb" or successful detoxification are often associated with larval digestion and absorption during food intake. In this chapter, recent progress in the understanding of plant–insect interactions, especially direct chemical defense (isothiocyanates, benzoxazinoids, and iridoids) and indirect defense mediated by larval elicitors, are reviewed in the context of the total performance of the larval midgut during mastication/digestion. In addition, the evolutionary background of each system will be discussed.

1. Introduction

Polypod type lepidopteran larvae, so-called caterpillars, can be characterized by a soft and flexible cuticle-covered body with short legs,

*Corresponding author

which are more suitable for crawling rather than running or jumping (Chapman 1998). Within the simple internal structure, the alimentary canal occupies a large part of the body cavity. Thus, caterpillars can probably best be described as a "walking stomach". As suggested by their morphology, most species inhabit a limited area or certain plants where their food source is very close and focus on mastication/digestion, and they migrate only when conditions are unacceptable. Such a lifestyle may be relevant to their adaptability to host-plant chemical defenses. Plants tissues contain various secondary metabolites, many of which function as deterrent/anti-feedants, toxins, or precursors of physical defense systems against herbivores. For individual polyphagy, insects such as herbivorous grasshoppers that actively locomote to switch host plants, it seems relatively easy to cope with chemicals that exhibit cumulative toxicity (Bernays and Minkenberg 1997). Most herbivorous lepidopteran larvae, however, can only choose between deleterious tissues or less-deleterious tissues, and they have to cope with a cocktail of allelochemicals. This is why larval adaptation to host plant secondary metabolites, as well as host recognition by oviposition stimulants (Nishida 2005), are equally crucial for survival. Studies have revealed how host plant allelochemicals are degraded, detoxified, absorbed, or detonated in the larval midgut. In a sense, their midgut tract is the front line of the long-standing evolutionary battle between plants and insects. First of all, the insect midgut simultaneously plays a central role in digestion and absorption. Some allelochemicals affect midgut function by interfering directly with nutrition (dietary protein) or indirectly with the digestion of nutrients in the gut. Furthermore, digestive enzymes involved in detoxifying allelochemicals (e.g. β-glucosidase) are in other cases hijacked to activate the toxicity (as discussed later). In this sense, a digestive enzyme can be a double-edged sword. An adaptation to a new allelochemical requires not only gaining a mutated enzyme to neutralize it but also with an enzyme that can uptake nutrients efficiently enough to pay for cost of coping with the toxicity (Slansky 1992). Secondly, recent studies found that the insect midgut releases chemical signals that allow plants to detect insect attacks (so-called elicitors). Fatty acid–amino acid conjugates (FACs), one of the best studied herbivorous elicitors, are actively synthesized and may possibly enhance larval nitrogen assimilation (Yoshinaga et al. 2008). If the physiological function of FACs is sufficient to gain nutrients and enable the insect to grow faster, then the negative effect of triggering plant indirect responses – and consequently attract natural enemies – will be countered. Larvae in general benefit from maximizing their growth rate, because shortening of the total larval stage will diminish the risk of parasitism (Loader and Damman 1991). Whether having FACs act for better or worse depends on the balance of nutrition and offending allelochemicals in plant tissues, as it also crucially affects the larval growth rate. Thus again, plant defense has a highly complex interaction with insect nutritional absorption.

How did such a complicated interaction develop? Angiosperms have produced a series of secondary metabolites through occasional mutations and recombination, and some of these compounds might have incidentally served to reduce or destroy the palatability of the plant. Such a plant could have escaped from herbivore attack, and the evolutionary radiation of plants subsequently generated an entire family or group of related families. The evolution of plant chemical defenses was closely followed by biochemical adaptation in insect herbivores, and the newly evolved detoxification mechanism, "key innovation", resulted in the adaptive radiation of herbivore lineages. This concept of stepwise co-evolution was introduced using diversified patterns of interactions between butterflies and their host plants (Ehrlich and Raven 1964). According to this theory, butterflies should have radiated soon after the major angiosperm radiation (140 to 100 million years ago), but the timing is currently not wholly supported because of some contradictions with regard to the ages found in fossil records (Labandeira et al. 1994, Vane-Wright 2004). Recently, several studies elucidated the molecular basis of insect adaptation to plant allelochemicals, and some have successfully described co-evolutionary interactions in host plant and herbivore systems (Li et al. 2004, Berenbaum and Zangerl 2008, Berenbaum 2011). Because the adaptation occurred over a time scale of millions of years, direct comparisons with non-adapted conspecific races are almost impossible in most cases, so too is the identification of the initial genetic changes that enabled a host shift. The third section in this chapter will be used to introduce recent progress on how plant defenses and insect adaptations have been shaped as we see them today.

2. Multiple Counter Adaptation Strategies against Plant Defense Substances

Lepidopteran larval midguts are equipped with various systems that function as the first line of defense against ingested plant allelochemicals. The gut organs in different lepidopteran species have almost identical structures, but the adaptation strategies vary between different species or even within species depending on the type of allelochemicals. One major type of defense substances is the glycosides, which are activated by plant glucosidase and release toxic aglycones upon plant tissue damage. So-called "two-component plant defenses" are widespread in the plant kingdom, and well-known classes of compounds include alkaloid, benzoxazinoid, glucosinolates, cyanogenic, and iridoid glucosides as well as salicinoids. To overcome the glycosides, insects have developed various different physiological adaptations, including deactivating plant β-glucosidases or metabolic enzymatic strategies in the digestive system of insects. Of note, some species share the same system for detoxifying a group of allelochemicals, whereas others have developed a completely different strategy. Such differences cannot always be attributed to a simple classification of generalists versus specialists, although in general, adaptations are often inducible in generalists whereas in specialists they are often constitutive (Pentzold et al. 2014). In the following section three classes of allelochemicals have been selected to introduce recent studies on how lepidopteran insects have developed novel methods to overcome or circumvent them.

2.1 Isothiocyanates

Glucosinolates, which are common in Brassicae plants, release highly reactive isothiocyanates as defense compounds upon tissue damage (Rask et al. 2000). These are the best-studied example of a two-component system that requires both a substrate (glucosinolate) and a detonator (myrosinase) to invoke the defensive bioactive function. To avoid toxicity to the plant itself, these two factors are stored separately in intact plant tissues (Koroleva et al. 2000, Andréasson and Jørgensen 2003). As shown in Fig. 1, the core structure of glucosinolates is composed of a S-glucosylated thiohydroximate sulfate ester with a variable amino acid-derived side-chain. Cell disruption enables the myrosinase to reach the glucosinolate and cleave the glucose moiety, detonating the "mustard oil bomb" by formation of an unstable aglycone (Fig. 1-1). The aglycone spontaneously rearranges its structure depending on conditions such as pH or other biotic factors, and among the products the most toxic hydrolysis product, isothiocyanate (Fig. 1-2), causes insect growth inhibition or in some cases lethality, as summarized in the review by Winde and Wittstock (2011).

The lepidopteran pest *Pieris rapae* and some other Pierid species specialize in feeding on glucosinolate-containing plants and can redirect the hydrolysis reaction toward the formation of simple nitriles (Fig. 1-3) instead of isothiocyanates, and the nitriles are excreted in the feces. This specific metabolism is caused by a larval gut protein, designated nitrile-specifier protein (NSP), which by itself has no hydrolytic activity on glucosinolates and is unrelated to any functionally characterized protein (Wittstock et al. 2004). NSP was not found in other Pierid species that do not feed on glucosinolate-containing plants nor in generalist Noctuid species (Wheat et al. 2007). Another specialist *Plutella xylostella* (Plutellidae) has a unique adaptation mechanism that competes with myrosinase activity. The larval gut is equipped with a large excess of sulfatase activity compared with the amount of glucosinolates, and the sulfatase converts glucosinolates to desulfoglucosinolates (Fig. 1-4), which cannot be hydrolyzed by myrosinases (Ratzka et al. 2002). Although the mechanisms are unknown, generalist lepidoptera occasionally but successfully feed on plants of the Brassicales. Recently, glutathione-conjugate derivatives (cysteinylglycine- and cysteinyl-isothiocyanate conjugates) of the major glucosinolate hydrolysis product were identified from the feces of *Spodoptera littoralis* (Noctuidae) larvae feeding on *Arabidopsis thaliana* plants (Schramm et al. 2012). Conjugation of isothiocyanates to the nucleophilic thiol (-SH) group of glutathione, a cysteine-containing tripeptide (γ-Glu-Cys-Gly) (Fig. 1-5), is a common detoxification strategy found in animals, plant, and bacterial cells (Jeschke et al. 2016). Besides *S. littoralis*, other noctuid generalist species such as *S. exigua*, *Mamestra brassicae*, *Trichoplusia ni*, and *Helicoverpa armigera* employ this ubiquitous strategy (Schramm et al. 2012). However, this method seems adaptive only if the diet has a low content of glucosinolates. The conjugation depletes glutathione in midgut tissues and hemolymph, and reduces the concentration of cysteine, which is the precursor of glutathione (Jeschke et al. 2016). Feeding experiments

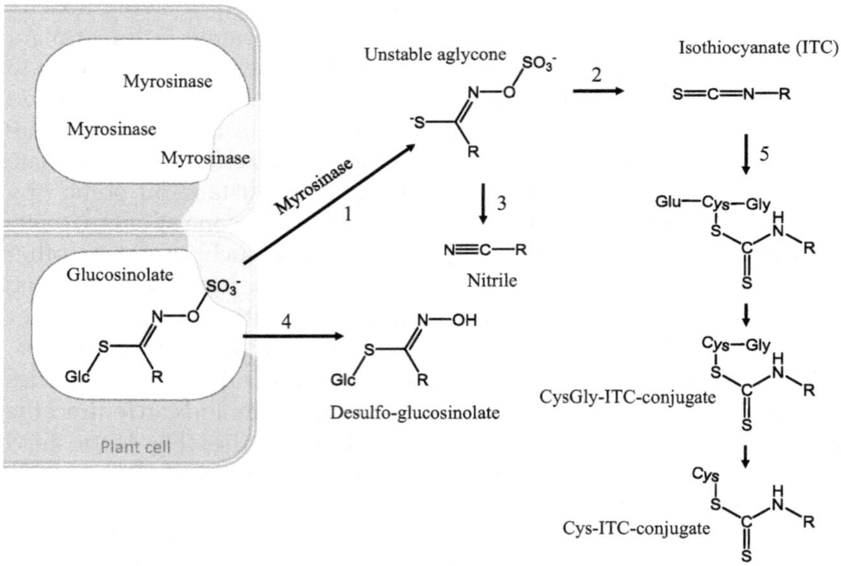

Fig. 1. Glucosinolate–myrosinase system and detoxifying metabolic pathways in different lepidopteran species. (1) Upon tissue damage, myrosinase from the plant cells converts glucosinolate to aglycone. (2) Unstable aglycone automatically forms toxic isothiocyanate. (3) Nitrile-specifier protein (NSP) in the Pierid larval midgut redirects the reaction to form non-toxic nitrile. (4) Sulfatase in *P. xylostella* desulfates glucosinolates. (5) In the midgut of generalists *S. frugiperda*, *S. exigua*, *M. brassicae*, *T. ni*, and *H. armigera*, glutathione is used to catch the isothiocyanate. CysGly-, and cys-ITC-conjugate are excreted into the frass.

involving ingesting isothiocyanates invoked active proteolysis, as well as deamination of amino acids with an increase in uric acid and elevated lipid content in *S. littoralis*, all of which resulted from an insufficient supply of cysteine for glutathione biosynthesis. The negative growth and protein hydrolysis effects were relieved by a dietary supply of cysteine (Jeschke et al. 2016). This is a good example of a plant–insect interaction that is closely related to larval nutrition and digestion.

2.2 Benzoxazinoids

Benzoxazinoids are indole-derived protective and allelophatic plant secondary metabolites found in species of the family Poaceae of monocot plants, including major crops such as maize (*Zea mays*), wheat (*Triticum aestivum*), and rye (*Secale cereale*). As summarized by Frey et al. (2009), the biosynthetic pathway and the various analogs have been elucidated in maize. The synthetic enzymes have also been identified in wheat (Nomura et al. 2002, 2005, 2008) and diploid Triticales (Nomura et al. 2007). The most abundant benzoxazinoid in maize and wheat is DIMBOA (Fig. 2), usually stored in the vacuole of plants as a stable glucoside. Upon herbivory damage, the hydrolyzed toxic aglycone forms cyclic hemiacetals that then form α-oxo-aldehydes (Fig. 2-2), which react with nucleophilic groups on various biomolecules (Dixon et al. 2012, Pérez et al. 1989). Non-adapted silkworm *Bombyx mori* (Bombycidae) larvae, fed on a toxic aglycone DIMBOA-containing diet, died within 3 days (Sasai et al. 2009). DIMBOA inhibited the activities of trypsin and chymotrypsin in the larval midgut of *Ostrinia nubilalis* (Crambidae) and digestive proteases and detoxification enzymes in the larval midgut of *Sesamia nonagrioides* (Noctuidae) (Houseman et al. 1992, Ortego et al. 1998). DIMBOA also seems to deplete cellular glutathione levels and leads to irreversible inactivation of enzymes with cysteine residues in their active site, as demonstrated *in vitro* (Dixon et al. 2012).

We found that the noctuid moth species *Mythimna separata*, known as pest of a range of crops such as maize, rice, and sorghum, can tolerate high levels of DIMBOA in artificial diets with no obvious negative effect (Sasai et al. 2009). The larval frass contained three glucosides, DIMBOA-2-*O*-Glc, HMBOA-2-*O*-Glc, and methoxy glucoside carbamate (Fig. 2-3). The incubation of DIMBOA with a midgut tissue suspension of *M. separata* in the presence of UDP-D-glucose generated DIMBOA-2-*O*-Glc, suggesting that the larvae can convert toxic DIMBOA to DIMBOA-Glc by glucosylation using an UDP-glucosyltransferase(s). The result was confirmed in more detailed analysis using a time course and sampling in *S. frugiperda* and *S. littoralis* larval tissues (Glauser et al. 2011). After a slow release of DIMBOA by plant-derived β-glucosidases, transformation of DIMBOA to DIMBOA-Glc was observed in both species. Of note, the glucoside in the frass of *Spodoptera* species was identified as (2*S*)-DIMBOA-Glc (Fig. 2-4), whereas the glucoside in plants is (2*R*)-DIMBOA-Glc, suggesting the reglucosylation is not the reverse reaction of plant-derived β-glucosidases but a stereoselective transformation that plant β-glucosidases cannot cleave off again (Wouters et al. 2014). Reglucosylation seems to be a common detoxification system in *S. frugiperda*, *S. littoralis*, and *S. exigua*, although it was not apparent in *Helicoverpa armigera*. Additionally, among the genus *Spodoptera*, the levels of counter-adaptation to DIMBOA seem to differ by species. DIMBOA had an anti-feedant effect on *S. exigua* but stimulated feeding in *S. frugiperda* in dual-choice experiments (Rostás 2007). In a no-choice setup, larvae of *S. exigua* gained less biomass and had prolonged development when fed on an artificial diet containing DIMBOA. However, pupal weight was not significantly different between treatments (Rostás 2007). The specialist *S. frugiperda* counteracts the

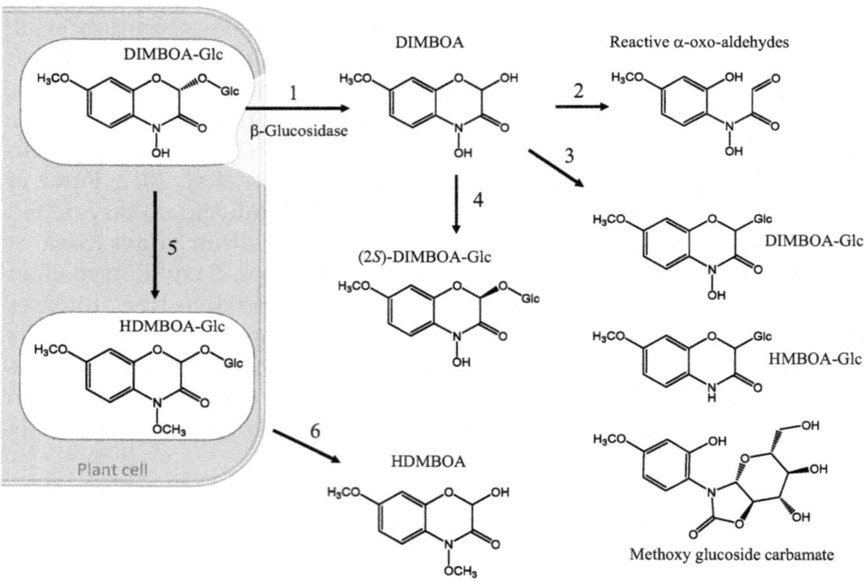

Fig. 2. Benzoxazinoid system in monocot plants and detoxifying pathways in lepidopteran insects. (1) DIMBOA-Glc mixed with β-glucosidase in insect gut release aglycone, DIMBOA. (2) DIMBOA forms reactive α-oxo-aldehydes that binds to nucleophilic groups of various biomolecules. (3) Three reglucosylated metabolites are found in frass of the specialist *M. separata*. (4) In the glucosylation process in *Spodoptera* larvae, stereoisomer (2*S*)-DIMBOA-Glc is formed, which is not used as substrate for the β-glucosidase. (5) In response to herbivory, plants synthesize HDMBOA-Glc using the synthase, DIMBOA-Glc 4-O-methyltransferase. (6) The released aglycone HDMBOA shows deterrent activity against the specialist *S. frugiperda*.

release of DIMBOA more successfully than the generalist *S. littoralis*, and accumulates relatively more DIMBOA-Glc and less DIMBOA in the gut, as well as much less DIMBOA in the frass (Glauser et al. 2011). According to this study, *S. frugiperda* has a higher capacity than *S. littoralis* to transform DIMBOA *in vitro*, which may explain the differences observed *in vivo*. The study also demonstrated that a new type of benzoxazinoid, HDMBOA-Glc, was actively induced in maize plants upon insect attack (Fig. 2-5). This glucoside released a highly unstable aglycone, HDMBOA (Fig. 2-6), which showed a pronounced deterrent effect even for the specialist *S. frugiperda*, suggesting that this single modification made it less-readily reglucosylated in the same conditions. The biosynthesis of DIMBOA-Glc to HDMBOA-Glc was found in wheat and Job's tears (*Coix lacryma-jobi*), but not in rye (Oikawa et al. 2002). Oikawa et al. identified DIMBOA-Glc 4-O-methyltransferase, a synthase of HDMBOA-Glc, from wheat leaves treated with jasmonic acid (JA). HDMBOA-Glc might be at the front line of plant defense where insects are currently struggling for an adaptation.

2.3 Iridoids

Iridoid glucosides are members of large group of terpene derivatives found in more than 50 plant families of the Asteridae (Dobler et al. 2011). All iridoid glucosides are eight-, nine-, or ten-carbon cyclopentanoids connected to an oxygenated heterocyclohexane, where normally a β-D-glucopyranose is attached at the C1 atom (Boros and Stermitz 1990) (Fig. 3). Glucosides are activated in the insect gut, either by co-ingested plant β-glucosidase or by endogenous larval β-glucosidase (Pankoke et al. 2012). The resulting iridoid aglycone is unstable, and the pyran ring opens to form a glutaraldehyde-like dial structure (Fig. 3-2) that covalently binds to nucleophilic side-chains via imine formation (Konno et al. 1999). Thus, the aglycones denature enzymes by causing non-specific crosslinks in proteins or reduce nutritive value of dietary proteins (Konno et al. 1997), consequently leading to a lower efficiency of conversion of ingested food, a lower relative growth rate, reduced larval weight, and higher larval mortality (Bowers 1991, Dobler et al. 2011). These toxic physiological effects are often dose-dependent and differ among individual iridoid glycosides depending on their chemical structure and the presence or absence of functional groups (Bowers and Puttick 1988, Puttick and Bowers 1988).

The concentrations of iridoid glycosides in plant tissues also are influenced by abiotic and biotic factors such as the availability of nutrients, genotype, plant age, competition, and herbivory (Pankoke et al. 2013). Therefore, the acquired chemical adaptations of insects that feed on these plants may vary as well, and this implicates possible selection pressure on herbivore populations and tritrophic interactions (Lampert and Bowers 2010). For example, *Lymantria dispar* (Lymantriidae) suffered a stronger reduction in larval growth on an artificial diet containing asperuloside than those with other iridoids, aucubin or catalpol, where aucubin was also more detrimental than catalpol (Fig. 3) (Bowers and Puttick 1988). However, in testing *S. eridania* larvae in similar experiments, asperuloside had no significant effect on larval growth, whereas aucubin showed a strong growth inhibition and catalpol was even stronger (Puttick and Bowers 1988). How such differences can be caused by structural changes and/or modifications to functional groups has not yet been explained clearly.

Privet leaves contain oleuropein as a defense substance, and privet-specialist lepidopteran larvae such as *Dolbina tancrei* (Sphingidae) and *Brahmaea wallichii* (Brahmaeidae) have been reported to retain a high concentration (>50 mM) of free glycine in their midgut contents (Konno et al. 1997). The amino group in glycine is hypothesized to inhibit the protein-denaturing activity of oleuropein by competing with the amino groups in the side chain of unspecific proteins in the gut (Konno et al. 1999, 2000) (Fig. 3-3). A survey of 43 lepidopteran and hymenopteran species revealed

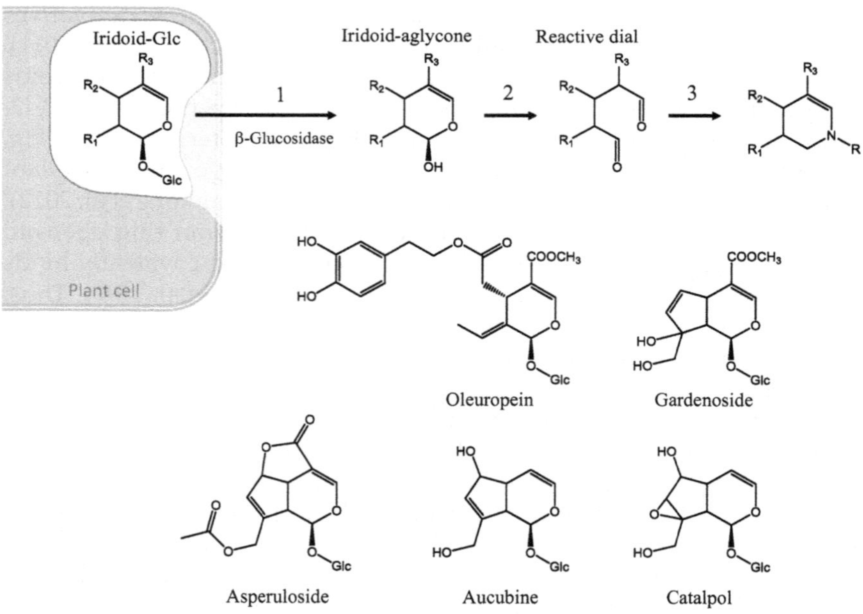

Fig. 3. Toxicity activation of iridoids upon herbivory. (1) Either plant β-glucosidase or insect endogenous β-glucosidase releases unstable aglycones. (2) The aglycone forms a glutaraldehyde-like dial compound, and (3) it easily binds to nucleophilic side-chains. Some insects accumulate an excess amount of amino acids in the midgut contents to provide amide groups as the reactive site for the aglycone.

that 11 privet-feeding species had higher glycine concentrations (40 mM, on average) compared to 32 non-privet-feeding species (2 mM, on average) (Konno et al. 2009). Additionally, among the privet-feeders surveyed, three species with lower glycine concentrations instead secreted GABA or β-alanine. Such free amino acid secretion into the midgut lumen could be a common adaptive trait against the oleuropein-based chemical defense of privet (Konno et al. 2009). In their survey, all the midgut contents were collected after the insects fed on their natural host plants, and the non-privet-feeding species very likely fed on plant tissues containing no iridoid glycosides. As a result, the glycine concentration in non-privet-feeding generalist *S. litura* seems to be at the level of other free amino acids in the gut contents. Our recent study feeding *S. litura* larvae with gardenoside, an iridoid glycoside from gardenia leaves, revealed notably higher concentrations of β-alanine in the gut, suggesting *S. litura* larvae can intensively induce amino acid accumulation as necessary (unpublished data). However, this strategy does not work in this case and the larvae suffered from delayed growth rate and irreversible digestive disorders leading to higher mortality. The mechanism of inducing β-alanine in response to ingested iridoid glycosides is under investigation. These data suggest that once an iridoid aglycone was released and activated, the toxicity might be beyond control and in the worst-case scenario it may cause death in a few days. Other groups reported a different adaptation strategy in the generalist species *Spilosoma virginica* (Arctiidae). In order to avoid releasing aglycones, the larvae decreased their endogenous β-glucosidase activity in response to the host plant (Pankoke et al. 2010). Similar data were obtained using another arctiid generalist *Grammia incorrupta*, a true generalist that feeds on a dozen or more different species during its lifetime. The adaptive decrease in β-glucosidase activity together with frequent host switching behavior might enable the caterpillars to tolerate toxic plant iridoid glycosides for short periods of time without suffering from detrimental postingestive effects (Pankoke et al. 2012). Of note, there are specialist herbivores that are attracted to iridoid-glycoside-containing plants. Larvae of several nymphalid species are reported to use iridoid glycosides as feeding stimulants (Bowers 1983), and females of the butterfly *Junonia coenia* use iridoid compounds as oviposition stimulants (Pereyra and Bowers 1988). These species are known to take up and sequester iridoid glycosides in hemolymph (Bowers and Stamp 1997). Such insects that sequester iridoid glycosides are well-studied and reported not only from lepidopteran insects but also from beetles, aphids, and sawflies (Dobler et al. 2011). As summarized in their review article, Dobler et al. (2011) discussed that although it is difficult to prove, a possible reason for this sequestration may be defense against predators such as birds and arthropods.

3. Plant Defense Elicitors Synthesized in the Midgut

As introduced in the previous section, the insect midgut is a protective organ where many plant defense substances are detoxified or absorbed to avoid toxicity. Although it may appear contradictory, midgut tissues also produce elicitors that trigger plant defense responses. Despite the wide range of inducible plant responses against herbivorous insects, the number of herbivore elicitors characterized chemically is still very limited. These comprise glucose oxidase (Eichenseer et al. 1999), alkaline phosphatase (Funk 2001), β-glucosidase (Mattiacci et al. 1995), FACs (Alborn et al. 1997, Paré et al. 1998, Pohnert et al. 1999, Mori et al. 2001, 2003, De Moraes and Mescher 2004, Spiteller et al. 2004), caeliferins (Alborn et al. 2007), inceptins (Schmelz et al. 2006, 2007), and bruchins (Doss et al. 2000) (Fig. 4). Among these, glucose oxidase and alkaline phosphatase were identified in the salivary gland. Bruchins, isolated from pea seed beetle *Bruchus pisorum* and cowpea seed beetle *Callosobruchus maculatus* (Coleoptera: Chrysomelidae), are elicitors that initiate neoplastic growth on pods at the site of egg oviposition; callus cells lift the eggs and hinder larval entry into the pod tissue (Doss et al. 2000). Other elicitors are found to be active in regurgitants, which have basically the same constituents as the midgut contents. These elicitors, except for the digestive enzymes, have drawn attention because they have no clear reason for them to be there.

Our recent study of FACs, a class of elicitor for plant volatile emission, is trying to elucidate the background for why insects synthesize these at a risk of a fatal attack, as a consequence of indirect plant defenses. The current findings on FACs in insects, mostly in Lepidoptera, are described in the following section.

3.1 Fatty Acid–Amino Acid Conjugates

Many plants respond to herbivory by inducing the release of volatile organic compounds (VOCs), which are important chemical cues for natural enemies of herbivores. Several elicitors in larval regurgitants have been identified to trigger this innovative plant defense system, and some of the best known elicitors are the FACs. The first FAC type elicitor, *N*-(17-hydroxylinolenoyl)-L-glutamine, identified from beet armyworm *S. exigua* was named volicitin and showed its strongest activity on maize seedlings (Alborn et al. 1997). Since then, the mechanism of how plants react to elicitors has been well documented, but studies of the biosynthesis and the physiological role of FACs in insect are still relatively limited. Collatz and Mommsen (1974) first reported FAC-like compounds from a cricket, spiders, and a crayfish, and suggested their role as biosurfactants because of their amphiphilic properties. However, their role may not be so simple. In *Spodoptera* larvae, the fatty acid moieties of FACs are derived from food

Fatty acid-amino acid conjugates (FACs)

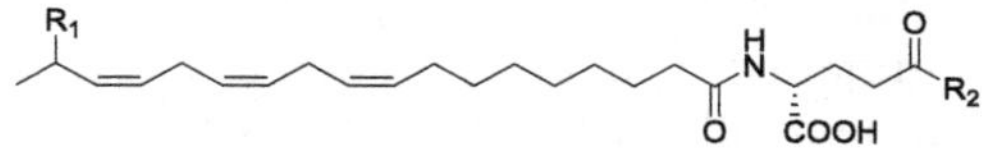

Caeliferin A 16:1

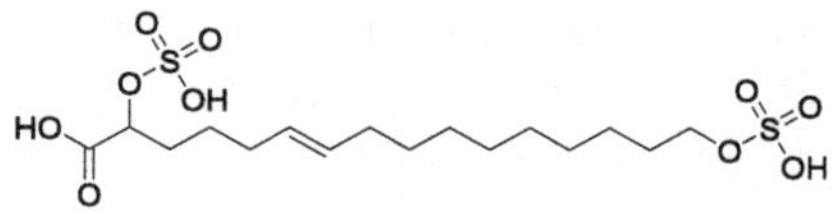

B 16:1

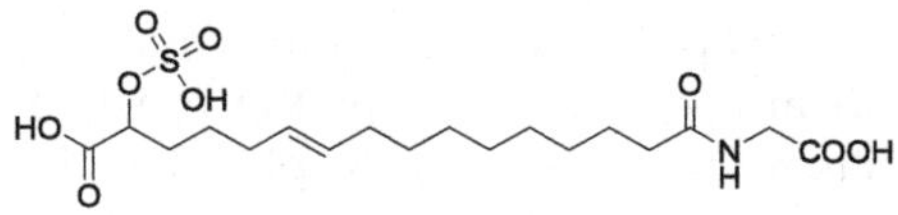

Inceptin

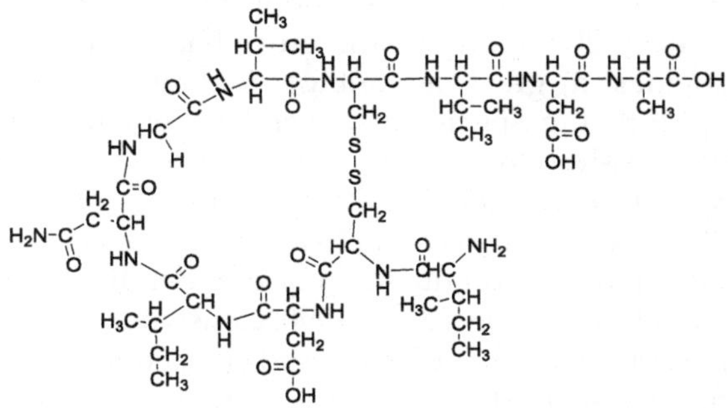

Bruchin

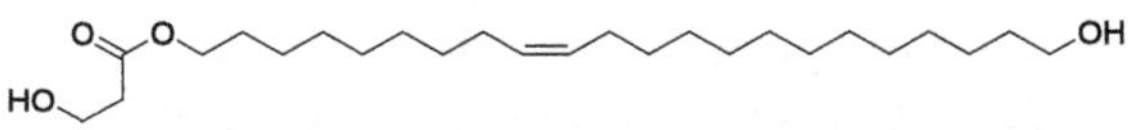

Fig. 4. Plant defense elicitors identified from insects. FACs have many analogs, R_1: H or OH, R_2: OH or NH_2, and the number of unsaturated bonds can differ reflecting the fatty acid composition in food plant tissues. Caeliferin also has other analogs, A16:0 and B16:0. Inceptin [+ICDINGVCVDA–], is a peptide derived from chloroplastic ATP synthase γ-subunit (cATPC) proteins. Bruchin also has several analogs varied with the carbon chain length and degree of unsaturation.

plants and approximately mimic the composition of unsaturated fatty acids in the plants, but glutamine in the diet was not used immediately for FAC synthesis (Paré et al. 1998, Aboshi et al. 2007). Furthermore, the amino acid components of FACs appear to be restricted to glutamine and glutamic acid in all insects reported so far (Yoshinaga et al. 2007). FACs accumulate in the gut lumen and can reach high concentrations, but degradative enzyme(s) in the midgut hydrolyzes FACs to yield free glutamine and fatty acids (Mori et al. 2001). The quantitative impact of FAC synthesis and degradation on glutamine metabolism raised the question of whether FACs might also function as a way to store glutamine in the gut lumen. We found a major source for the biosynthesis of FACs to be glutamine in midgut cells, a part of which was newly synthesized from glutamic acid and ammonia through the enzymatic reaction of glutamine synthetase (GS). The hypothesis that FACs in insects might improve glutamine synthetase productivity is based on the fact that nitrogen assimilation efficiency of *S. litura* larvae was improved by 40–>60%, when the diet was enriched with linolenic acid. Thus, glutamine-containing FACs in the gut lumen may function as a form of glutamine storage, a key compound of nitrogen metabolism (Yoshinaga et al. 2008). It is widely accepted that GS, which catalyzes the condensation of glutamic acid and ammonia, is one of the most important enzymes in nitrogen metabolism for most heterotrophs. This is especially true for herbivores who constantly suffer from a deficiency of nitrogen nutrients. Disorder in GS function and glutamine shortage can easily cause death (Kutlesa and Cavebey 2001), but on the other hand, successful nitrogen assimilation provides a great advantage for insects. As introduced previously, Lepidopteran species in general benefit from maximizing growth rate, because shortening of the total larval stage might diminish the risk of parasitism (Loader and Damman 1991), and minimize exposure to plant defensive substances such as phenol oxidase or proteinase inhibitor, which reduce food nutrition or prevent digestion in caterpillars (Felton et al. 1989, Farmer and Ryan 1990, Constabel et al. 1995). Although elicitation of induced plant defenses and attraction of natural enemies by FACs are obvious tradeoffs, the positive effect on nitrogen assimilation might be a strong enough incentive. Whether this holds true across a broad spectrum of species can be assessed by the fact that FACs appear to be significantly prevalent among lepidopteran species. Of 29 lepidopteran species screened, regardless of known tritrophic interactions, FACs were found in 19 of these species (Yoshinaga et al. 2010). All FAC-containing species had *N*-acyl-L-glutamine, which might be evolutionarily older than FACs. Although the sampling data were biased toward noctuid and sphingid species, a comparison of the diversity of FACs with lepidopteran phylogeny indicates that glutamic acid conjugates can be synthesized by relatively primitive species, whereas hydroxylation of fatty acids is mostly limited to larger and more developed macrolepidopteran species (Fig. 5). We have only shown that glutamine-FACs are involved in the nitrogen assimilation process, although the function of glutamic acid conjugates remain completely unknown. Despite the probable physiological benefit of FACs, we also found 10 lepidopteran species where the gut content did not contain detectable amounts of any FACs. What determined whether one species should have FACs or not will be an interesting question for future studies.

FACs are not limited to Lepidoptera only. Glutamic acid and glutamine conjugates have been identified in Teleogryllus crickets, larval fruit fly *Drosophila melanogaster* (Yoshinaga et al. 2007) and in katydids (Alborn et

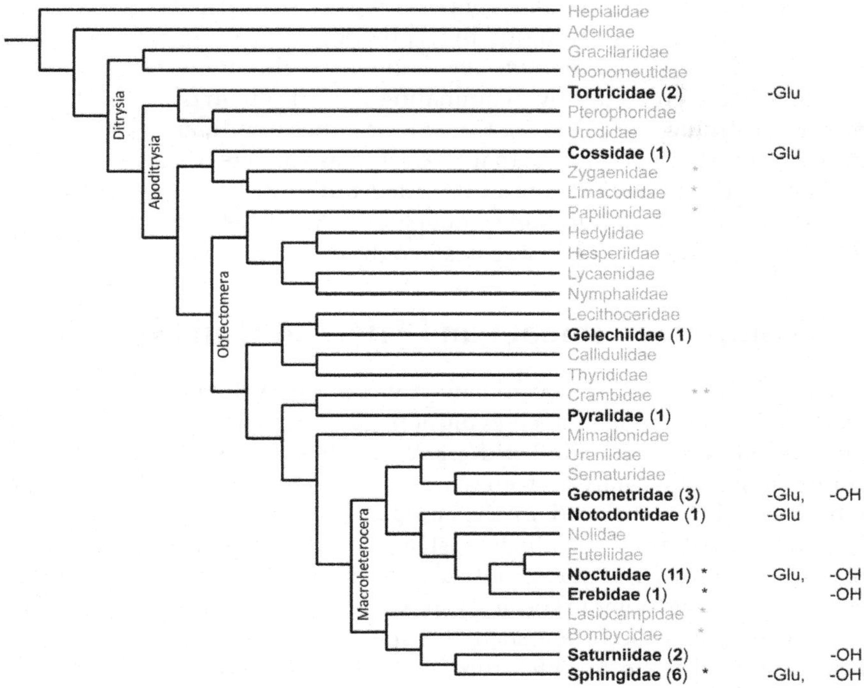

Fig. 5. Lepidopteran family tree and identified FAC analogs. Family names in bold represents families of FACs-containing species, and number after the family name is the number of FAC-containing species. "-Glu" indicates one or a few species with glutamic acid conjugates, and "-OH" indicates that some species have hydroxylated fatty acid conjugates. Asterisks indicate species that have no detectable FACs. The family tree is based on previous reports (Kawahara and Breinholt 2014, Zahari et al. 2012).

al. 2007). The profile of FAC analogs in these crickets and fruit flies was similar to that of tobacco hornworm *Manduca sexta* (Sphingidae), showing glutamic acid conjugates predominantly over glutamine conjugates, although it was the other way around in the case of katydids. Whether these insects share a common synthetic mechanism for FACs is unclear. Fresh midgut tissues of crickets, fruit fly larvae and several lepidopteran larvae all showed the ability to synthesize glutamine conjugates *in vitro* when incubated with substrates (glutamine and sodium linolenate). Such direct conjugation was also observed for glutamic acid conjugates in all the insects, but the amount of product was very small and did not reflect the *in vivo* FAC patterns in each species. In fruit fly larvae, the predominance of glutamic acid conjugates could be explained by a shortage of the substrate glutamine in midgut tissues, and in *M. sexta*, a rapid hydrolysis of glutamine conjugates has been reported. In crickets, we found an additional unique biosynthetic pathway for glutamic acid conjugates. *Teleogryllus taiwanemma* converted glutamine conjugates to glutamic acid conjugates by deaminating the side chain of the glutamine moiety (Yoshinaga et al. 2014). Considering these findings together with previous results, a possibility that FACs in these insects has been obtained by convergent evolution cannot be ruled out, but it is more likely that ancestral insects had glutamine conjugates, and crickets and other insects developed glutamic acid conjugates in a different way.

4. Evolutionary Impact and Effect of Domestication

The evolution of the glucosinolate–myrosinase system has been well-studied in the context of co-evolution. As introduced previously, the detoxification procedure in lepidopteran species differs in strategy, target compounds, and catalyzing enzymes. Pieris caterpillars adopted NSP, which is solely expressed in the midgut and promotes the formation of nitrile breakdown products instead of toxic isothiocyanates upon myrosinase catalyzed glucosinolate hydrolysis. Assaying larval midgut activity in 18 lepidopteran species, Wheat et al. (2007) showed glucosinolate feeding species exactly matched the species that have NSP activity in larvae. According to the literature, a single evolutionary origin of NSP protein in ancestral Pierid species is clearly suggested, and a subsequent loss of the protein seems to have happened after secondary host plant shifts to other plant in the order Santalales. Multiple independent temporal reconstructions provided a robust estimate that NSP has evolved recently (within the past 10 million years) after the appearance of the Brassicales plant order, which is estimated to be roughly 90 million years ago. Subsequently, the net diversification rates of Pierinae has increased (burst of speciation) compared with that of their sister clade the Coliadinae, whose members did not colonize Brassicales (Wheat et al. 2007). These findings provided a convincing example of co-evolution dynamics, and NSP might be a key innovation enabling the Pieridae species to detoxify glucosinolate and to inhabit Brassicae plants.

As already discussed, at least 120 different glucosinolates has been reported from sixteen families of the Brassicales (Fahey et al. 2001). They are classified into several groups based on their chemical structures, and continued innovation in glucosinolate biosynthesis was plotted on the phylogeny of the Brassicales; aromatic (Phe-based) and branched-chain amino acid (Leu, Ile, Val)-based glucosinolates are found universally in all Brassicales, whereas indole-based glucosinolates (i) are found basically in core Brassicales, and methionine-based glucosinolates (ii) are predominant in the Brassicaceae and some New World members of the Capparaceae

(Mithen et al, 2010). There are also novel structural elaborations of glucosinolates (iii) unique to the core Brassicaceae lineage, and the escalation of glucosinolate diversity using new amino acid substrates occurred via both single gene and whole genome duplications, involving the neofunctionalization of both regulatory and core biosynthesis genes (Edger et al. 2015). Steps (i-iii) are a key innovation in the Brassicales, because each step was synchronized by shifts in diversification rates. Of note, the timing of this evolutionary step only shortly preceded the host shift and diversification timing of Pieris butterfly species (Edger et al. 2015).

The other aspect of the glucosinolate–myrosinase system in Brassicale plants is defense against pathogens (Rask et al. 2000). Numerous papers have been published recently on this topic, and some show a remarkable similarity in both plant–pathogen and plant–herbivore interactions. The levels of some glucosinolates change upon challenge to the plant by pathogens (Doughty et al. 1991, Doughty et al. 1996), and negative effects have been shown of glucosinolate degradation products on fungi causing disease in Brassicaceae (Greenhalgh et al. 1976, Mithen et al. 1986, Sotelo et al. 2015). The Brassicaceae-specialist fungi *Alternaria brassicicola* has adapted to the presence of indolic glucosinolates and can cope with their hydrolysis products, whereas the generalist fungi *Botrytis cinerea*, which infects a variety of hosts, is more sensitive to these phytochemicals (Buxdorf et al. 2013). Although the glucosinolate–myrosinase system is likely to be the most important in plant interactions with pests and diseases, it also has physiological function for the plant itself, especially in sulfur nutrition storage (Rask et al. 2000) and stress resistance (Martínez-Ballesta et al. 2013). Studying the genetic basis of these side-effects is essential to understand the whole mechanism how such diversified interactions have evolved.

Compared with evolution over a time scale of millions of years, crop domestication has a very short history, yet crop plants have undergone considerable genotypic and phenotypic changes during the process of domestication (Doebley et al. 2006, Burger et al. 2008, Meyer et al. 2012, Olsen and Wendel 2013). Maize arose from a single domestication event of Balsas teosinte (*Zea mays* ssp. *parviglumis*), a grass from the Balsas river watershed in southern Mexico, about 9000 years ago (Matsuoka et al. 2002, Piperno et al. 2009). The number of genes affected by artificial selection during domestication is estimated to be only about 1200 (corresponding to 2–4%) throughout the maize genome (Wright et al. 2005). In general, domestication has consistently reduced chemical resistance against herbivorous insects (Chen et al. 2015). Maize plants are infested with fall armyworm *S. frugiperda* larvae at a significantly higher frequency than Balsas teosinte plants in the field, where both host plants coexist within a cultivated field (Takahashi et al. 2012). This defensive trait might be explained by the constitutive contents of benzoxazinoids, the predominant class of defensive secondary metabolites. Among the benzoxazinoids, the most toxic HDMBOA-Glc was the predominant compound in teosinte, whereas DIMBOA-Glc was the most abundant in maize hybrid B73 (Maag et al. 2015). Based on evidence of a metabolic tradeoff between high yield and better defense, the susceptibility of maize plants was thought to be the outcome of domestication from its ancestral teosinte (Rosenthal and Dirzo 1997). However, a quantitative difference was not clearly shown in the above analysis. There was no significant difference in the total amount of benzoxazinoids between cultivated maize and ancestral teosinte, because drastic changes in benzoxazinoid contents in both plants were observed through ontogenic stages, as well as large variability among the different genotypes of teosinte (Maag et al. 2015). Rather, a qualitative difference, as shown by the ratio of HDMBOA/DIMBOA, affected the tolerability/susceptibility to herbivores, and the ratio might have been influenced during domestication. Tropical maize germplasm had a higher constitutive HDMBOA-Glc/DIMBOA-Glc ratio and it was hypothesized to be caused by breeding to select plants with constitutively elevated HDMBOA-Glc accumulation in tropical regions where pressure from chewing arthropods and pathogens tends to be higher (Meihls et al. 2013). On the other hand, higher DIMBOA-Glc concentrations in maize cultivars in temperate zones have been associated with a latitudinal shift in insect pest pressure towards piercing-sucking herbivores such as aphids, because DIMBOA mediates resistance against aphids via the induction of callose deposition (Meihls et al. 2013).

There is another case where crop domestication affected a plant–insect interaction. (*E*)-β-caryophyllene is one of the main volatile components emitted from maize upon herbivory and is considered as an important chemical cue for natural enemies. In particular, when the roots are attacked by beetle larvae of *D. v. virgifera*, the plants also emit (*E*)-β-caryophyllene below-ground (Köllner et al. 2008). Terpene synthase 23 (TPS23), which produces (*E*)-β-caryophyllene from farnesyl diphosphate, is similar to (*E*)-β-caryophyllene synthases from dicotyledons, but they are the result of repeated evolution. The sequence of TPS23 is maintained by positive selection in maize and its closest wild relatives, teosinte species (*Zea* spp.). The gene encoding TPS23 is active in teosinte species and European maize lines, but decreased transcription in most North American lines has resulted in the loss of (*E*)-β-caryophyllene production. This was probably not because of a direct mutation in the tps23 gene itself but due to alteration of the regulatory network that results in its transcription (Köllner et al. 2008). The authors of the previous report pointed out that the failure of past efforts to control *D. v. virgifera* with nematodes in North America may be caused by the lack of (*E*)-β-caryophyllene release. As to the above-ground tritrophic interaction, however, there is no report of an obvious negative effect caused by the loss of (*E*)-β-caryophyllene. The other two major volatile components, (*E*)-β-farnesene and (*E*)-α-bergamotene, have already been identified as major constituents of a blend used by parasitic wasps to find their lepidopteran hosts (Schnee et al. 2006).

On the other hand, little has been elucidated on the evolutionary background of indirect defense systems. A recent study revealed the plant signal transduction triggered by FAC treatment or herbivory feeding towards VOC synthesis and other defense reaction are surprisingly similar to the cascades for plant defense against microbes (Mithöfer and Boland 2008, Bonaventure et al. 2011, Zipfel 2014, Yoshinaga 2015). In general, plant VOC functions are not limited to the urgent recruitment of natural enemies of herbivores, but as intermediates interacting with a broader range of organisms. The surfaces of flowers, leaves, and roots are known habitats of numerous microorganisms that affect plant fitness, either positively or negatively, and interfere with plant–insect interactions. VOCs such as terpenoids, benzenoid compounds, aliphatics, and sulfur-containing compounds emitted by plants strongly affect these colonizing bacteria (Junker and Tholl 2013). In particular, terpenoids have been shown to possess defensive roles in numerous plant species, and their antimicrobial properties have been reported in *Z. mays* and Arabidopsis (Gershenzon and Dudareva 2007, Huffaker et al. 2011, Huang et al. 2012). They prevent the establishment of detrimental microbes and also promote the growth of mutualists by serving as carbon sources (Junker and Tholl 2013). Considering that such interactions originated from ancient immune responses in primitive plants and bacteria, and that arthropod herbivory joined the interaction later, it would not be surprising if plants use the same or a similar pathways for elicitor perception, signaling, activation, or suppression in their defense against insects. Then, how did FACs in insect regurgitants become an elicitor of such plant responses? Previous reports indicated that *N*-acyl amino acids are quite commonly distributed in microorganisms and are similar in structure to FACs (Brady and Clardy 2000, Peypoux et al. 2004). Although some of them have been reported to function as biosurfactants, their functions remain mostly unknown. As expected, several microbes that can synthesize FACs were identified in the gut lumen of *S. exigua*, and recently, a hydrolase of FACs was identified to be a novel Dps protein (DNA binding protein from starved cell) (Spiteller et al. 2000, Ping et al. 2007). Although FACs in insects seem to be synthesized actively by the insect itself for their importance in nitrogen metabolism, the finding related to microbe-based synthesis is interesting considering the involvement of ancestral microbes in FAC-mediated plant–insect interactions. One possibility is that an ancestral plant receptor for microbial *N*-acyl amino acids had an affinity for insect-produced FACs, accidentally leading to plants releasing VOCs that serve as exploitable information for foraging parasitoids. The other possibility could be that the FAC conjugase in insects has its origin in bacteria that inhabited ancestral plants. Identification of an FAC conjugase in lepidopteran insects will provide an answer for this point and also offer a new avenue for phylogenetic studies of insect classes, including crickets, katydids, fruit flies, and so on.

5. Conclusion

The nonneural regulatory mechanisms in the human intestine are so sophisticated that some scientists have nicknamed it our "second brain". Insect midguts may also deserve the same nickname. Numerous hormonal peptides released from endocrine cells in the midgut regulate digestion, metabolic processes, diuresis, immunodefense, and feeding/foraging behavior in insects (Sehnal and Žitňan 1996). A novel peptide that modulates feeding behavior was recently identified from *B. mori* hemolymph and named HemaP (Nagata et al. 2011). Unlike most other neuropeptides, which are common throughout insects and even vertebrates, HemaP-like peptides are conserved across lepidopteran species but are not observed in other insect orders. Besides HemaP, there seem to be several other neuropeptides such as allatostatins, myosupressins, allatotropins, and short neuropeptide F that are altogether concerned and concerted in their regulation of feeding behavior (Audsley et al. 2008, Nagata et al. 2012a, b). These endogenous signals, in addition to plant allelochemicals, should have great effect on a decision whether to eat or not to eat by individual insects. Additionally, ultrastructural and immunocytological studies have shown some midgut endocrine cells, scattered amongst the principal midgut cells, have one side reaching and opening onto the lumen, where they have microvilli and can probably respond directly to substances in the lumen (Endo and Nishitsutsuji-Uwo 1981). Such an endocrine system in lepidopteran larvae might be involved in metabolic changes caused by allelochemicals; for example, in the detection of iridoid compounds and the consequent induction of β-alanine into gut lumen as shown in *S. litura*. Recent endocrinal approaches and chemical ecological studies have both made remarkable progresses in each field, but there are few interactions with each other. Future collaboration may shed new light on insect–plant interactions and then we may instead call caterpillars "walking and thinking stomachs".

REFERENCES

Aboshi, T., N. Yoshinaga, K. Noge, R. Nishida and N. Mori. 2007. Efficient incorporation of unsaturated fatty acids into volicitin-related compounds in *Spodoptera litura* (Lepidoptera: Noctuidae). Biosci. Biotechnol. Biochem. 71: 607–610.

Alborn, H.T., T.V. Hansen, T.H. Jones, D.C. Bennett, J.H. Tumlinson, E.A. Schmelz et al. 2007. Disulfooxy fatty acids from the American bird grasshopper *Schistocerca americana*, elicitors of plant volatiles. Proc. Natl. Acad. Sci. USA 104: 12976–12981.

Alborn, H.T., T.C.J. Turlings, T.H. Jones, G. Stenhagen, J.H. Loughrin and J.H. Tumlinson. 1997. An elicitor of plant volatiles from beet armyworm oral secretion. Science 276: 945–949.

Andréasson, E. and L.B. Jørgensen. 2003. Localization of plant myrosinases and glucosinolates. Recent Adv. Phytochem. 37: 79–99.

Audsley, N., H.J. Matthews, N.R. Price and R.J. Weaver. 2008. Allatoregulatory peptides in Lepidoptera, structures, distribution and functions. J. Insect Physiol. 54: 969–980.

Berenbaum, M.R. 2011. Chemical mediation of coevolution: phylogenetic evidence for Apiaceae and associates. Ann. Missouri Bot. Gard. 88: 45–59.

Berenbaum, M.R. and A.R. Zangerl. 2008. Facing the future of plant-insect interaction research: Le retour à la "raison d'être". Plant Physiol. 146: 804–811.

Bernays, E.A. and O.P.J.M. Minkenberg. 1997. Insect herbivores: different reasons for being a generalist. Ecology 78: 1157–1169.

Bonaventure, G., A. VanDoorn and I.T. Baldwin. 2011. Herbivore-associated elicitors: FAC signaling and metabolism. Trends Plant Sci. 16: 294–299.

Boros, C.A. and F.R. Stermitz. 1990. Iridoids. An updated review. Part I. J. Nat. Prod. 53: 1055–1147.

Bowers, M.D. 1983. The role of iridoid glycosides in host-plant specificity of checkerspot butterflies. J. Chem. Ecol. 9: 475–493.

Bowers, M.D. 1991. Iridoid glycosides. pp. 297–325. *In*: G.A. Rosenthal and M.R. Berenbaum [eds.]. Herbivores: Their Interactions with Secondary Plant Metabolites. Vol. I: The Chemical Participants. Academic Press, San Diego, CA, USA.

Bowers, M.D. and N.E. Stamp. 1997. Fate of host-plant iridoid glycosides in lepidopteran larvae of Nymphalidae and Arctiidae. J. Chem. Ecol. 23: 2955–2965.

Bowers, M.D. and G.M. Puttick. 1988. Response of generalist and specialist insects to qualitative allelochemical variation. J. Chem. Ecol. 14: 319–334.

Brady, S.F. and J. Clardy. 2000. Long-chain *N*-acyl amino acid antibiotics isolated from heterologously expressed environmental DNA. J. Am. Chem. Soc. 122: 12903–12904.

Burger, J.C., M.A. Chapman and J.M. Burke. 2008. Molecular insights into the evolution of crop plants. Am. J. Bot. 95: 113–122.

Buxdorf, K., H. Yaffe, O. Barda and M. Levy. 2013. The effects of glucosinolates and their breakdown products on necrotrophic fungi. PLoS ONE 8: e70771.

Chapman, R.F. 1998. The Insects: Structure and Function. Fourth Edition. Cambridge University Press, Cambridge, UK.

Chen, Y.H., R. Gols and B. Benrey. 2015. Crop domestication and its impact on naturally selected trophic interactions. Annu. Rev. Entomol. 60: 35–58.

Constabel, C.P., D.R. Bergey and C.A. Ryan. 1995. Systemin activates synthesis of wound-inducible tomato leaf polyphenoloxidase via the octadecanoid defense signaling pathway. Proc. Natl. Acad. Sci. USA 92: 407–411.

Collatz, K.G. and T. Mommsen. 1974. The structure of the emulsifying substances in several invertebrates. J. Comp. Physiol. 94: 339–352.

De Moraes, C.M. and M.C. Mescher. 2004. Biochemical crypsis in the avoidance of natural enemies by an insect herbivore. Proc. Natl. Acad. Sci. USA 101: 8993–8997.

Dixon, D.P., J.D. Sellars, A.M. Kenwright and P.G. Steel. 2012. The maize benzoxazinone DIMBOA reacts with glutathione and other thiols to form spirocyclic adducts. Phytochemistry 77: 171–178.

Dobler, S., G. Petschenka and H. Pankoke. 2011. Coping with toxic plant compounds—the insect's perspective on iridoid glycosides and cardenolides. Phytochemistry 72: 1593–1604.

Doebley, J.F., B.S. Gaut and B.D. Smith. 2006. The molecular genetics of crop domestication. Cell 127: 1309–1321.

Doss, R.P., J.E. Oliver, W.M. Proebsting, S.W. Potter, S.R. Kuy, S.L. Clement et al. 2000. Bruchins: insect-derived plant regulators that stimulate neoplasm formation. Proc. Natl. Acad. Sci. USA 97: 6218–6223.

Doughty, K.J., M.M. Blight, C.H. Bock, J.K. Fieldsend and J.A. Pickett. 1996. Release of alkenyl isothiocyanates and other volatiles from *Brassica rapa* seedlings during infection by *Alternaria brassicae*. Phytochemistry 43: 371–374.

Doughty, K.J., A.J.R. Porter, A.M. Morton, G. Kiddle, C.H. Bock and R. Wallsgrove. 1991. Variation in the glucosinolate content of oilseed rape (*Brassica napus* L.) leaves. Ann. Appl. Biol. 118: 469–477.

Edger, P.P., H.M. Heidel-Fischer, M. Bekaert, J. Rota, G. Glöckner, A.E. Platts et al. 2015. The butterfly plant arms-race escalated by gene and genome duplications. Proc. Natl. Acad. Sci. USA 112: 8362–8366.

Ehrlich, P.R. and P.H. Raven. 1964. Butterflies and plants: a study in coevolution. Evolution 18: 586–608.

Eichenseer, H., M.C. Mathews, J.L. Bi, J.B. Murphy and G.W. Felton. 1999. Salivary glucose oxidase: multifunctional roles for *Helicoverpa zea*? Arch. Insect Biochem. Physiol. 42: 99–109.

Endo, Y. and J. Nishitsutsuji-Uwo. 1981. Gut endocrine cells in insects: the ultrastructure of the gut endocrine cells of the lepidopterous species. Biomed. Res. 2: 270–280.

Fahey, J.W., A.T. Zalcmann and P. Talalay. 2001. The chemical diversity and distribution of glucosinolates and isothiocyanates among plants. Phytochemistry 56: 5–51.

Farmer, E.E. and C.A. Ryan. 1990. Interplant communication: airborne methyl jasmonate induces synthesis of proteinase inhibitors in plant leaves. Proc. Natl. Acad. Sci. USA 87: 7713–7716.

Frey, M., K. Schullehner, R. Dick, A. Fiesselmann and A. Gierl. 2009. Benzoxazinoid biosynthesis, a model for evolution of secondary metabolic pathways in plants. Phytochemistry 70: 1645–1651.

Felton, G.W., K. Donato, R.J. Del Vecchio and S.S. Duffey. 1989. Activation of plant foliar oxidases by insect feeding reduces nutritive quality of foliage for noctuid herbivores. J. Chem. Ecol. 15: 2667–2694.

Funk, C.J. 2001. Alkaline phosphatase activity in whitefly salivary glands and saliva. Arch. Insect Biochem. Physiol. 46: 165–174.

Gershenzon, J. and N. Dudareva. 2007. The function of terpene natural products in the natural world. Nat. Chem. Biol. 3: 408–414.

Glauser, G., G. Marti, N. Villard, G.A. Doyen, J.L. Wolfender, T.C. Turlings et al. 2011. Induction and detoxification of maize 1,4-benzoxazin-3-ones by insect herbivores. Plant J. 68: 901–911.

Greenhalgh, J.R. and N.D. Mitchell. 1976. The involvement of flavour volatiles in the resistance to downy mildew of wild and cultivated forms of *Brassica oleracea*. New Phytol. 77: 391–398.

Houseman, J.G., F. Campos, N.M.R. Thie, B.J.R. Philogene, J. Atkinson, P. Morand et al. 1992. Effect of the maize-derived compounds DIMBOA and MBOA on growth and digestive processes of European corn borer. J. Econ. Entomol. 85: 669–674.

Huffaker, A., F. Kaplan, M.M. Vaughan, N.J. Dafoe, X. Ni, J.R. Rocca et al. 2011. Novel acidic sesquiterpenoids constitute a dominant class of pathogen-induced phytoalexins in maize. Plant Physiol. 156: 2082–2097.

Huang, M., A.M. Sanchez-Moreiras, C. Abel, R. Sohrabi, S. Lee, J. Gershenzon et al. 2012. The major volatile organic compound emitted from Arabidopsis thaliana flowers, the sesquiterpene (*E*)-β-caryophyllene, is a defense against a bacterial pathogen. New Phytol. 193: 997–1008.

Jeschke, V., J. Gershenzon and D.G. Vassão. 2016. A mode of action of glucosinolate-derived isothiocyanates: detoxification depletes glutathione and cysteine levels with ramifications on protein metabolism in *Spodoptera littoralis*. Insect Biochem. Mol. Biol. 71: 37–48.

Junker, R.R. and D. Tholl. 2013. Volatile organic compound mediated interactions at the plant-microbe interface. J. Chem. Ecol. 39: 810–825.

Kawahara, A.Y. and J.W. Breinholt. 2014. Phylogenomics provides strong evidence for relationships of butterflies and moths. Proc. R. Soc. B 281: 20140970.

Köllner, T.G., M. Held, C. Lenk, I. Hiltpold, T.C. Turlings, J. Gershenzon et al. 2008. A maize (*E*)-β-caryophyllene synthase implicated in indirect defense responses against herbivores is not expressed in most American maize varieties. Plant Cell 20: 482–494.

Konno, K., C. Hirayama, H. Yasui and M. Nakamura. 1999. Enzymatic activation of oleuropein: a protein crosslinker used as a chemical defense in the privet tree. Proc. Natl. Acad. Sci. USA 96: 9159–9164.

Konno, K., C. Hirayama and H. Shinbo. 1997. Glycine in digestive juice: a strategy of herbivorous insects against chemical defense of host plants. J. Insect Physiol. 43: 217–224.

Konno, K., C. Hirayama, H. Yasui, S. Okada, M. Sugimura, F. Yukuhiro et al. 2009. GABA, β-alanine and glycine in the digestive juice of privet-specialist insects: convergent adaptive traits against plant iridoids. J. Chem. Ecol. 36: 983–991.

Koroleva, O.A., A. Davies, R. Deeken, M.R. Thorpe, A.D. Tomos and R. Hedrich. 2000. Identification of a new glucosinolate-rich cell type in Arabidopsis flower stalk. Plant Physiol. 124: 599–608.

Kutlesa, N.J. and S. Cavebey. 2001. Insecticidal activity of glufosinate through glutamine depletion in a caterpillar. Pest Manag. Sci. 57: 25–32.

Labandeira, C.C., D.L. Dilcher, D.R. Davis and D.L. Wagner. 1994. Ninety-seven million years of angiosperm-insect association: paleobiological insights into the meaning of coevolution. Proc. Natl. Acad. Sci. USA 91: 12278–12282.

Lampert, E.C. and M.D. Bowers. 2010. Host plant influences on iridoid glycoside sequestration on generalist and specialist caterpillars. J. Chem. Ecol. 36: 1101–1104.

Li, X., J. Baudry, M.R. Berenbaum and M.A. Schuler. 2004. Structural and functional divergence of insect CYP6B proteins: from specialist to generalist cytochrome P450. Proc. Natl. Acad. Sci. USA 101: 2939–2944.

Loader, C. and H. Damman. 1991. Nitrogen content of food plants and vulnerability of *Pieris rapae* to natural enemies. Ecology 72: 1586–1590.

Maag, D., M. Erb, J.S. Bernal, J.L. Wolfender, T.C. Turlings and G. Glauser. 2015. Maize domestication and anti-herbivore defences: leaf-specific dynamics during early ontogeny of Maize and its wild ancestors. PLoS ONE 10: e0135722.

Martínez-Ballesta, M.C., D.A. Moreno and M. Carvajal. 2013. The physiological importance of glucosinolates on plant response to abiotic stress in Brassica. Int. J. Mol. Sci. 14: 11607–11625.

Matsuoka, Y., Y. Vigouroux, M.M. Goodman, G.J. Sanchez, E. Buckler and J. Doebley. 2002. A single domestication for maize shown by multilocus microsatellite genotyping. Proc. Natl. Acad. Sci. USA 99: 6080–6084.

Mattiacci, L., M. Dicke and M.A. Posthumus. 1995. Beta-glucosidase—an elicitor of herbivore-induced plant odor that attracts host-searching parasitic wasps. Proc. Natl. Acad. Sci. USA 92: 2036–2040.

Meihls, L.N., V. Handrick, G. Glauser, H. Barbier, H. Kaur, M.M. Haribal et al. 2013. Natural variation in maize aphid resistance is associated with 2,4-dihydroxy-7-methoxy-1,4-benzoxazin-3-one glucoside methyltransferase activity. Plant Cell 25: 2341–2355.

Meyer, R.S., A.E. DuVal and H.R. Jensen. 2012. Patterns and processes in crop domestication: an historical review and quantitative analysis of 203 global food crops. New Phytol. 196: 29–48.

Mithen, R., R. Bennett and J. Marquez. 2010. Glucosinolate biochemical diversity and innovation in the Brassicales. Phytochemistry 71: 2074–2086.

Mithen, R.F., B.G. Lewis and G.R. Fenwick. 1986. In vitro activity of glucosinolates and their products against *Leptosphaeria maculans*. Trans. Br. Mycol. Soc. 87: 433–440.

Mithöfer, A. and W. Boland. 2008. Recognition of herbivory-associated molecular patterns. Plant Physiol. 146: 825–831.

Mori, N., H.T. Alborn, P.E.A. Teal and J.H. Tumlinson. 2001. Enzymatic decomposition of elicitors of plant volatiles in *Heliothis virescens* and *Helicoverpa zea*. J. Insect Physiol. 47: 749–757.

Mori, N., N. Yoshinaga, Y. Sawada, M. Fukui, M. Shimoda, K. Fujisaki et al. 2003. Identification of volicitin-related compounds from the regurgitant of lepidopteran caterpillars. Biosci. Biotechnol. Biochem. 67: 1168–1171.

Nagata, S., N. Morooka, K. Asaoka and H. Nagasawa. 2011. Identification of a novel hemolymph peptide that modulates silkworm feeding motivation. J. Biol. Chem. 286: 7161–7170.

Nagata, S., S. Matsumoto, A. Mizoguchi and H. Nagasawa. 2012a. Identification of cDNAs encoding allatotropin and allatotropin-like peptides from the silkworm, *Bombyx mori*. Peptides. 34: 98–105.

Nagata, S., S. Matsumoto, T. Nakane, A. Ohara, N. Morooka, T. Konuma et al. 2012b. Effects of starvation on brain short neuropeptide F-1, -2, and -3 levels and short neuropeptide F receptor expression levels of the silkworm, *Bombyx mori*. Front. Endocrinol. (Lausanne) 3: 3.

Nishida, R. 2005. Chemosensory basis of host recognition in butterflies—multi-component system of oviposition stimulants and deterrents. Chem. Senses 30: i293–i294.

Nomura, T., A. Ishihara, H. Imaishi, T.R. Endo, H. Ohkawa and H. Iwamura. 2002. Molecular characterisation and chromosomal localization of cytochrome P450 genes involved in the biosynthesis of cyclic hydroxamic acids in hexaploid wheat. Mol. Genet. Genomics 267: 210–217.

Nomura, T., A. Ishihara, R.C. Yanagita, T.R. Endo and H. Iwamura. 2005. Three genomes differentially contribute to the biosynthesis of benzoxazinones in hexaploid wheat. Proc. Natl. Acad. Sci. USA 102: 16490–16495.

Nomura, T., A. Ishihara, H. Iwamura and T.R. Endo. 2007. Molecular characterization of benzoxazinone-deficient mutation in diploid wheat. Phytochemistry 68: 1008–1016.

Nomura, T., S. Nasuda, K. Kawaura, Y. Ogihara, N. Kato, F. Sato et al. 2008. Structures of the three homoeologous loci of wheat benzoxazinone biosynthetic genes TaBx3 and TaBx4 and characterization of their promoter sequences. Theor. Appl. Genet. 116: 373–381.

Oikawa, A., A. Ishihara and H. Iwamura. 2002. Induction of HDMBOA-Glc accumulation and DIMBOA-Glc 4-*O*-methyltransferase by jasmonic acid in poaceous plants. Phytochemistry 61: 331–337.

Olsen, K.M. and J.F. Wendel. 2013. A bountiful harvest: genomic insights into crop domestication phenotypes. Annu. Rev. Plant Biol. 64: 47–70.

Ortego, F., M. Ruiiz and P. Castañera. 1998. Effect of DIMBOA on growth and digestive physiology of *Sesamia nonagrioides* (Lepidoptera: noctuidae) larvae. J. Insect Physiol. 44: 95–101.

Pankoke, H., M.D. Bowers and S. Dobler. 2010. Influence of iridoid glycoside containing host plants on midgut β-glucosidase activity in a polyphagous caterpillar, *Spilosoma virginica* Fabricius (Arctiidae). J. Insect Physiol. 56: 1907–1912.

Pankoke, H., M.D. Bowers and S. Dobler. 2012. The interplay between toxin-releasing β-glucosidase and plant iridoid glycosides impairs larval development in a generalist caterpillar, *Grammia incorrupta* (Arctiidae). Insect Biochem. Mol. Biol. 42: 426–434.

Pankoke, H., T. Buschmann and C. Müller. 2013. Role of plant β-glucosidases in the dual defense system of iridoid glycosides and their hydrolyzing enzymes in *Plantago lanceolata* and *Plantago major*. Phytochemistry 94: 99–107.

Paré, P.W., H.T. Alborn and J.H. Tumlinson. 1998. Concerted biosynthesis of an insect elicitor of plant volatiles. Proc. Natl. Acad. Sci. USA 95: 13971–13975.

Pentzold, S., M. Zagrobelny, F. Rook and S. Bak. 2014. How insects overcome two-component plant chemical defence: plant β-glucosidases as the main target for herbivore adaptation. Biol Rev. Camb. Philos. Soc. 89: 531–551.

Pereyra, P.C. and M.D. Bowers. 1988. Iridoid glycosides as oviposition stimulants for the buckeye butterfly, *Junonia coenia* (Nymphalidae). J. Chem. Ecol. 14: 917–928.

Pérez, F.J. and H.M. Niemeyer. 1989. Reaction of DIMBOA with amines. Phytochemistry 28: 1831–1834.

Peypoux, F., O. Laprévote, M. Pagadoy and J. Wallach. 2004. *N*-acyl derivatives of Asn, new bacterial *N*-acyl *D*-amino acids with surfactant activity. Amino Acids. 26: 209–214.

Ping, L., R. Büchler, A. Mithöfer, A. Svatos, D. Spiteller, K. Dettner et al. 2007. A novel Dps-type protein from insect gut bacteria catalyses hydrolysis and synthesis of *N*-acyl amino acids. Environ. Microbiol. 9: 1572–1583.

Piperno, D.R., A.J. Ranere, I. Holst, J. Iriarte and R. Dickau. 2009. Starch grain and phytolith evidence for early ninth millennium BP maize from the Central Balsas River Valley, Mexico. Proc. Natl. Acad. Sci. USA 106: 5019–5024.

Pohnert, G., V. Jung, E. Haukioja, K. Lempa and W. Boland. 1999. New fatty acid amides from regurgitant of lepidopteran (Noctuidae, Geometridae) caterpillars. Tetrahedron 55: 11275–11280.

Puttick, G.M. and M.D. Bowers. 1988. Effect of qualitative and quantitative variation in allelochemicals on a generalist insect: Iridoid glycosides and the southern armyworm. J. Chem. Ecol. 14: 335–351.

Rask, L., E. Andréasson, B. Ekbom, S. Eriksson, B. Pontoppidan and J. Meijer. 2000. Myrosinase: gene family evolution and herbivore defense in Brassicaceae. Plant Mol. Biol. 42: 93–113.

Ratzka, A., H. Vogel, D.J. Kliebenstein, T. Mitchell-Olds and J. Kroymann. 2002. Disarming the mustard oil bomb. Proc. Natl. Acad. Sci. USA 99: 11223–11228.

Rostás, M. 2007. The effects of 2,4-dihydroxy-7-methoxy-1,4-benzoxazin-3-one on two species of *Spodoptera* and the growth of *Setosphaeria turcica* in vitro. J. Pest. Sci. 80: 35–41.

Rosenthal, J.P. and R. Dirzo. 1997. Effects of life history, domestication and agronomic selection on plant defence against insects: evidence from maizes and wild relatives. Evol. Ecol. 11: 337–355.

Sasai, H., M. Ishida, K. Murakami, N. Tadokoro, A. Ishihara, R. Nishida et al. 2009. Species-specific glucosylation of DIMBOA in larvae of the rice Armyworm. Biosci. Biotechnol. Biochem. 73: 1333–1338.

Schmelz, E.A., M.J. Carroll, S. LeClere, S.M. Phipps, J. Meredith, P.S. Chourey et al. 2006. Fragments of ATP synthase mediate plant perception of insect attack. Proc. Natl. Acad. Sci. USA 103: 8894–8899.

Schmelz, E.A., S. LeClere, M.J. Carroll, H.T. Alborn and P.E.A. Teal. 2007. Cowpea chloroplastic ATP synthase is the source of multiple plant defense elicitors during insect herbivory. Plant Physiol. 144: 793–805.

Schnee, C., T.G. Köllner, M. Held, T.C.J. Turlings, J. Gershenzon and J. Degenhardt. 2006. The products of a single maize sesquiterpene synthase form a volatile defense signal that attracts natural enemies of maize herbivores. Proc. Natl. Acad. Sci. USA 103: 1129–1134.

Schramm, K., D.G. Vassão, M. Reichelt, J. Gershenzon and U. Wittstock. 2012. Metabolism of glucosinolate-derived isothiocyanates to glutathione conjugates in generalist lepidopteran herbivores. Insect Biochem. Mol. Biol. 42: 174–182.

Sehnal, F. and D. Žitňan. 1996. Midgut endocrine cells. pp. 55–85. *In*: M.J. Lehane and P.F. Billingsley [eds.]. Biology of the Insect Midgut. Chapman & Hall, London, UK.

Slansky, F. 1992. Allelochemical-nutrient interactions. pp. 135–165. *In*: G.A. Rosenthal and M.R. Berenbaum [eds.]. Herbivores: Their Interactions with Secondary Plant Metabolites. Vol. II: The Chemical Participants. Academic Press, San Diego, CA, USA.

Sotelo, T., M. Lemab, P. Soengasa, M.E. Carteaa and P. Velascoa. 2015. *In vitro* activity of glucosinolates and their degradation products against Brassica-pathogenic bacteria and fungi. Appl. Environ. Microbiol. 81: 432–440.

Spiteller, D., K. Dettner and W. Boland. 2000. Gut bacteria may be involved in interactions between plants, herbivores and their predators: microbial biosynthesis of *N*-acylglutamine surfactants as elicitors of plant volatiles. Biol. Chem. 381: 755–762.

Spiteller, D., N.J. Oldham and W. Boland. 2004. *N*-(17-phosphonooxylinolenoyl) glutamine and *N*-(17-phosphonooxylinoleoyl)glutamine from insect gut: the first backbone-phosphorylated fatty acid derivatives in nature. J. Org. Chem. 69: 1104–1109.

Takahashi, C.G., L.L. Kalns and J.S. Bernal. 2012. Plant defense against fall armyworm in micro-sympatric maize (*Zea mays* ssp. *mays*) and Balsas teosinte (*Zea mays* ssp. *parviglumis*). Entomol. Exp. Appl. 145: 191–200.

Vane-Wright, D. 2004. Butterflies at that awkward age. Nature 428: 477–478.

Wheat, C.W., H. Vogel, U. Wittstock, M.F. Braby, D. Underwood and T. Mitchell-Olds. 2007. The genetic basis of a plant–insect coevolutionary key innovation. Proc. Natl. Acad. Sci. USA 104: 20427–20431.

Winde, I. and U. Wittstock. 2011. Insect herbivore counteradaptations to the plant glucosinolate–myrosinase system. Phytochemistry 72: 1566–1575.

Wittstock, U., N. Agerbirk, E.J. Stauber, C.E. Olsen, M. Hippler, T. Mitchell-Olds et al. 2004. Successful herbivore attack due to metabolic diversion of a plant chemical defense. Proc. Natl. Acad. Sci. USA 101: 4859–4864.

Wouters, F.C., M. Reichelt, G. Glauser, E. Bauer, M. Erb, J. Gershenzon et al. 2014. Reglucosylation of the benzoxazinoid DIMBOA with inversion of stereochemical configuration is a detoxification strategy in lepidopteran herbivores. Angew. Chem. Int. Ed. 53: 11320–11324.

Wright, S.I., I.V. Bi, S.G. Schroeder, M. Yamasaki, J.F. Doebley, M.D. McMullen et al. 2005. The effects of artificial selection on the maize genome. Science 308: 1310–1314.

Yoshinaga, N., T. Aboshi, C. Ishikawa, M. Fukui, M. Shimoda, R. Nishida et al. 2007. Fatty acid amides, previously identified in caterpillars, found in the cricket *Teleogryllus taiwanemma* and fruit fly *Drosophila melanogaster* larvae. J. Chem. Ecol. 33: 1376–1381.

Yoshinaga, N., T. Aboshi, H. Abe, R. Nishida, H. Alborn, J.H. Tumlinson et al. 2008. Active role of fatty acid amino acid conjugates in nitrogen metabolism in *Spodoptera litura* larvae. Proc. Natl. Acad. Sci. USA 105: 18058–18063.

Yoshinaga, N., H.T. Alborn, T. Nakanishi, D.M. Suckling, R. Nishida, J.H. Tumlinson et al. 2010. Fatty acid-amino acid conjugates diversification in lepidopteran caterpillars. J. Chem. Ecol. 36: 319–325.

Yoshinaga, N., H. Abe, S. Morita, T. Yoshida, T. Aboshi, M. Fukui et al. 2014. Plant volatile eliciting FACs in lepidopteran caterpillars, fruit flies, and crickets: a convergent evolution or phylogenetic inheritance? Front. Physiol. 5: 121.

Yoshinaga, N. 2015. Physiological function and ecological aspects of fatty acid-amino acid conjugates in insects. Biosci. Biotechnol. Biochem. 80: 1274–1282.

Zahiri, R., J.D. Holloway, I.J. Kitching, J.D. Lafontaine, M. Mutanen and N. Wahlberg. 2012. Molecular phylogenetics of Erebidae (Lepidoptera, Noctuoidea). Syst. Entomol. 37: 102–124.

Zipfel, C. 2014. Plant pattern-recognition receptors. Trends Immunol. 35: 345–351.

CHAPTER

3

Chemically-mediated Interactions among Cucurbits, Insects and Microbes

Lori R. Shapiro[1]* and Kerry E. Mauck[2]

[1] Harvard Medical School, Harvard University, 1042A HIM, 77 Louis Pasteur, Boston, MA 02115, USA
E-mail: Lori.R.Shapiro@gmail.com

[2] University of California, Riverside Entomology Building, 900 University Ave., Riverside, CA 92521, USA
E-mail: kerry.mauck@ucr.edu

Abstract

The gourd family (Cucurbitaceae) contains many species that are ecologically, culturally, and economically important. Undomesticated cucurbit species are fast-growing annual vines and are commonly found in areas prone to disturbance such as riverbanks or roadsides. Cultivated varieties such as squash, pumpkins, gourds, cucumbers and melons rank among the highest acreage crops grown worldwide. As a result of their unique defensive chemistry and their ecological and agricultural importance, several wild and cultivated cucurbit species have emerged as models for studying chemically-mediated interactions among plants, microbes and insects. In this chapter, we provide an overview of research on the chemical ecology of the Cucurbitaceae and focus on wild and cultivated *Cucurbita pepo* (pumpkin, squash and gourds) and cultivated *Cucumis sativus* (cucumber). We explore how secondary metabolites, foliar volatiles and floral volatiles mediate interactions with antagonists (insect herbivores that damage plants and transmit pathogens) and mutualists (specialist pollinators and

*Corresponding author

herbivore natural enemies) and discuss more recent work on the influence of beneficial and pathogenic microbes on these chemical signals. Collectively, this research demonstrates the potential of wild and agricultural cucurbit systems to provide key insights into the ways that chemical signals and cues play a role in plant-insect interactions and in the epidemiology of vector transmitted pathogens. Finally, we provide a perspective on the promise of the Cucurbitaceae as ecological models for translating plant chemical ecology research from controlled laboratory settings to more diverse, ecologically relevant field environments.

1. Introduction

Cucurbit plants are among the most culturally important and widely planted food crops. culturally important and widely planted food crops. Wild crop progenitors and undomesticated species are also common and ecologically important inhabitants of many tropical and subtropical habitats. Many cucurbit species produce highly toxic secondary chemicals called 'cucurbitacins' which are among the most bitter compounds ever characterized. Because of this unique defensive chemistry, several cucurbit species have served as important models for understanding the evolution of plant defenses mediating interactions among plants and insects. Building on this research, more recent work has studied the effects of pathogenic microbes on plant-insect interactions. In this chapter, we synthesize this work to illustrate the growing importance of several Cucurbitaceae species as model systems for understanding chemically-mediated interactions among pathogens, plants and insects.

Here, we begin by describing cucurbitacins: the highly toxic, inducible secondary metabolites that have driven the evolution of specialist insect herbivores and pollinators as well as human-mediated evolution (domestication) of several Cucurbitaceae crop species (Section 2.1). Sections 2.2 and 2.3 focus on a second group of chemicals that mediate plant-insect interactions: volatile organic compounds (VOCs) which facilitate attraction or repellence of insects to or from host plants over long distances. Section 2.2 discusses the volatile chemistry of *Cucurbita* spp. inflorescences and how floral volatiles influence the behavior of highly co-evolved beneficial and antagonistic insects. Section 2.3 discusses herbivore-induced volatile emissions of the common cucumber, *Cucumis sativus*, which has served as a model for understanding volatile-mediated predator-prey interactions. Following this, Section 3 focuses on modifications to both vegetative and floral chemistry due to infection by viral (Section 3.2) and bacterial (Section 3.3) vector-borne pathogens. Finally, we conclude by laying out directions for future work on chemically-mediated plant-microbe and plant-insect interactions in Cucurbitaceae systems.

2. Chemical Mediation of Plant-insect Interactions in the Cucurbitaceae

2.1 A Bitter Bite: Cucurbitacin-based Defenses

Secondary metabolites are compounds produced by plants for defensive purposes and have no known role in primary metabolism. While humans have used plant secondary chemicals medicinally and nutritionally since prehistory, our current understanding of the ecological importance of these compounds began when G.S. Fraenkel recognized their roles in host plant choice by herbivorous insects (Fraenkel 1959). Soon after this, Ehrlich and Raven implicated plant secondary defensive compounds as drivers of plant and insect diversification through antagonistic coevolution. Plants are hypothesized to be under selection to deter herbivores by producing more toxic or entirely novel secondary metabolites, placing strong pressure on insect herbivores to evolve mechanisms to detoxify these compounds (Ehrlich and Raven 1964). In Cucurbitaceae, the class of secondary metabolites that function for herbivore defense are a group of oxygenated tetracyclic triterpenes collectively called 'cucurbitacins' (Fig. 1). Several closely related genera of leaf beetles were first reported to feed compulsively on cucurbits in 1939 (Contardi 1939), but the ecological function of cucurbitacins in mediating these feeding preferences was not recognized until 1966, despite these compounds being under active study for much of the twentieth century for medicinal applications and to prevent poisoning in livestock that graze on wild cucurbit plants (Chambliss 1966).

Cucurbitacins are produced constitutively in many undomesticated Cucurbitaceae species (Rehm and Wessels 1957) and are found in all parts of bitter plants (roots, leaves, stems, flowers, fruits) (Metcalf 1986). Cucurbitacins are among the most bitter compounds ever characterized (Chen et al. 2005) and are detectable by humans at concentrations as low as 1 ppb (Metcalf et al. 1980). Despite this toxicity, cucurbitacins have a long history of importance in traditional medicine for a variety of ailments (Yesilada 2005, Delazar et al. 2006, Shang et al. 2014). More recently, cucurbitacins have been researched as anti-cancer agents because of their extreme cytotoxicity (Chan et al. 2010), but selective activity against cancer cells *in vivo* has not yet been demonstrated (Lee et al. 2010). The cucurbitacin backbone is a tetracyclic cucurbitane nucleus skeleton with the chemical structure 19-(10→9β)-abeo-10α-lanost-5-ene (also known as 9β-methyl-19-nor lanosta-5-ene) (Chen et al. 2005) (Fig. 1). There are more than a dozen families of cucurbitacins that differ in their oxygenation and glycosylation patterns (Metcalf 1986, Chen et al. 2005) and these structural differences drive behavioral responses of co-evolved herbivores.

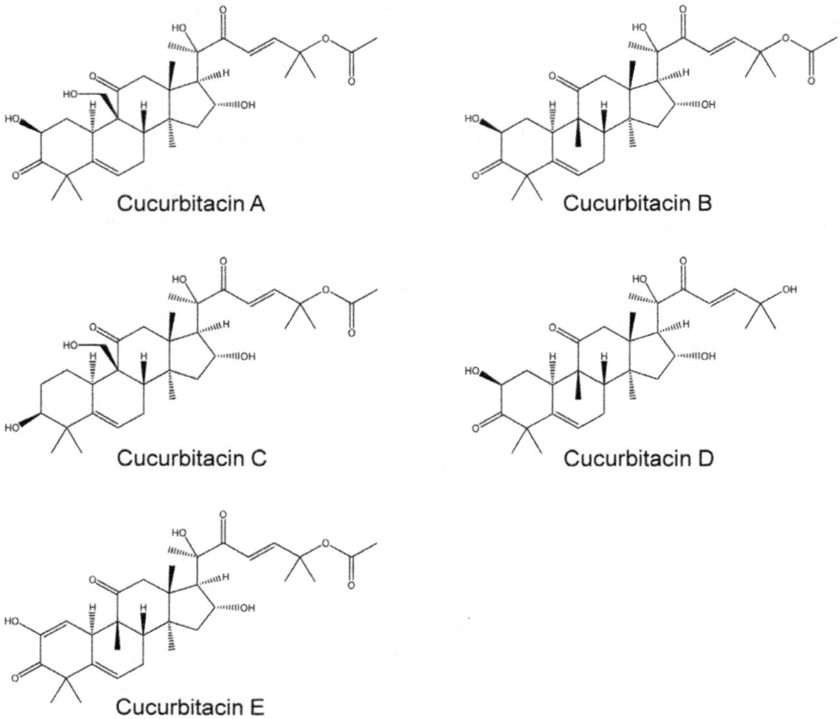

Fig. 1. Structures of cucurbitacins.

The concentration of cucurbitacins produced constitutively in wild cucurbits can be very high—between 0.03–0.4% of dry weight (Metcalf 1986)—with the highest concentrations often found in roots and fruits (Metcalf and Lampman 1989). For the vast majority of insect and mammalian herbivores (including humans) (Metcalf et al. 1980, Kistler et al. 2015), these bitter tasting cucurbitacins are highly effective herbivory and/or oviposition deterrents (Watt and Breyer-Brandwijk 1962, Da Costa and Jones 1971, Nielsen et al. 1977, Tallamy and Gorski 1997, Chen et al. 2005, Torkey et al. 2009). Cucurbitacins are highly toxic when ingested by mammals (David and Vallance 1955, Watt and Breyer-Brandwijk 1962). For mice, the intraperotineal LD_{50} value for cucurbitacin B is 1.1 mg/kg and 40 mg/kg for cucurbitacin E glycoside (David and Vallance 1955, Stoewsand et al. 1985). For most insects, cucurbitacins act as antagonists at ecdysteroid receptors and interfere with the binding of insect ecdysis hormones, interfering with molting and/or metabolism (Dinan et al. 1997, 1999).

The exceptions are several Luperini subtribes, most notably, Aulocophorina beetles from the Old World and Diabroticina beetles from the New World (Coleoptera: Chrysomelidae: Luperini) (Metcalf 1979, Gillespie et al. 2008, Eben and Espinosa 2013). For these beetles, cucurbitacins are potent arrestants and phagostimulants and induce compulsive feeding on anything containing cucurbitacins at amounts as low as 1 ng (Metcalf et al. 1980). After herbivore damage, cucurbitacins can be induced up to several fold higher than constitutive concentrations which increases the rate of specialist diabroticite beetle feeding (Tallamy 1985, McCloud et al. 1995). While several Aulocophorina species have reached pest status in the Old World (i.e., the red pumpkin beetle *Aulocophora foveicollis,* the pumpkin beetle *Aulocophora hilaris* and the plain pumpkin beetle *Aulocophora abdominalis*) (Bogawat and Pandey 1967, Metcalf and Lampman 1989, Singh et al. 2000), two closely related genera of New World Diabroticina (*Acalymma* and *Diabrotica* spp.) contain the most devastating and economically costly agricultural pests in North America (Krysan and Miller 1986).

New World Diabroticina beetles are hypothesized to have had an evolutionary association with cucurbit plants for at least 30 million years (Eben and Espinosa 2013). There are several dozen species of *Acalymma* and all are thought to be obligate cucurbit specialists at all life stages (Munroe and Smith 1980, Cabrera and Durante 2001). Only two *Acalymma* spp. have reached pest status (*Acalymma vittatum* in the Eastern US and *Acalymma trivitattum* in the Western USA). There are currently ~1,000 named *Diabrotica* spp. and perhaps as many that are still undescribed (Smith and Laurence 1967). Host records and basic natural history information are missing for the majority of *Diabrotica* spp. (Eben and Espinosa 2013), but several species are pests that cause more than a billion dollars worth of annual damage in corn, cucurbit, sweet potato and peanut crops (e.g. the spotted cucumber beetle *Diabrotica undecimpuctata,* the banded cucumber beetle *Diabrotica balteata,* the Western corn rootworm *Diabrotica virgifera virgifera,* the Northern corn rootworm *Diabrotica barberi,* the Southern corn rootworm *Diabrotica speciosa,* and *Diabrotica cristata*) (Smith 1966, Krysan and Miller 1986, Gray et al. 2009). *Diabrotica* has two economically important phylogenetic clades: The *virgifera* group is univoltine and overwinter as eggs and specialise on grasses. The *fucata* group are multivoltine, overwinter as adults and have been documented developing to maturity on eight families of non-cucurbit host plants (Branson and Krysan 1981, Krysan and Smith 1987). Because of the functional importance of cucurbitacins for host plant choice by pest *Acalymma* and for pharmacophagous forays by pest *Diabrotica* spp., the effects of cucurbitacin content and identity on host plant choice and plant-insect interactions has been an area of intensive research.

Several of the most important pest beetle species show pronounced feeding preferences for different cucurbitacins (Metcalf et al. 1980). Cucurbitacin B (91%), cucurbitacin D (69%) and cucurbitacin E (42%) are the most common cucurbitacins found in plants, with cucurbitacin

B eliciting the strongest arrestant and feeding response from most pest species from both the New World (Metcalf et al. 1980) and the Old World (Nishida et al. 1992). Cucurbitacin C is the cucurbitacin only that has been found in *Cucumis sativus* (cucumber) (Balkema-Boomstra et al. 2003). Most other cucurbitacins besides B, D and E are chemically active intermediates in cucurbitacin biosynthetic or modification pathways (Metcalf and Lampman 1989). In *Cucurbita* ssp., both cucurbitacin B and cucurbitacin E occur and they can be present as both aglycones or glycosides (Metcalf et al. 1982). *Diabrotica* spp. from the *fucata* group (*Diabrotica undecimpunctata howardi, Diabrotica speciosa)* compulsively feed on lower concentrations of cucurbitacins than the *virgifera* group and the cucurbit specialist *A. vittatum* is the least sensitive (Metcalf et al. 1980, Metcalf and Lampman 1989).

Cucurbit specialist beetles can sequester bitter cucurbitacins during both larval and adult life stages and these sequestered cucurbitacins make beetles distasteful to predators while beetles fed on cucurbitacin-free pollen are eaten readily (Ferguson and Metcalf 1985). Fitness costs associated with sequestration of cucurbitacins appear to be minimal for specialist cucurbitacin-adapted herbivores. For example, larvae of the striped cucumber beetle (*Acalymma vittatum)* grow faster when they feed on roots of bitter, cucurbitacin-producing plants vs. plants that do not produce cucurbitacin (Tallamy and Gorski 1997). This observation led to the hypothesis that diabrotacite beetles can use cucurbtacins as sources of steroid nutrition and the mechanism for this was later found to be by hydrogenation of the carbon-carbon double bond present in these compounds (Andersen et al. 1988, Halaweish et al. 1999). Significantly more *Diabrotica* males actively seek cucurbitacin-rich food sources compared to females, likely due to the transfer of cucurbitacins ingested by males to females through spermatophores (Tallamy et al. 2000). *Diabrotica undecimpunctata* females then transfer 79% of ingested or spermatophore-acquired cucurbitacins to eggs, where they protect against infection with the fungus *Metarhizium anisopliae* (Tallamy et al. 1998). The presence of high cucurbitacin concentrations in eggs may be especially advantageous given that *D. undecimpunctata,* like all other Luperine beetles, oviposits in potentially pathogen-rich soil near the base of host plants.

A wild progenitor of *Cucurbita pepo* was the first plant domesticated, i.e., brought into cultivation under human-mediated selection, in the New World more than 10,000 years ago in Southern Mexico (Smith 1997), before corn and beans were also domesticated in nearby regions in Mesoamerica (Doebley et al. 2006). *Cucurbita pepo* was independently domesticated a second time in Eastern North America (Smith 2006, Smith et al. 2007). There is evidence for at least six domestication events for *Cucurbita* spp. (Smith 1989, Sanjur et al. 2002, Piperno and Stothert 2003, Doebley et al. 2006). The high number of domestication events may be attributable to the rapid growth of wild *Cucurbita* in human disturbed habitats (Kistler et al. 2015), the non-bitter and nutritional oily seeds (Smith 1997) and the utility of dried gourd fruits (Sanjur et al. 2002, Schaefer et al. 2009). *Cucumis* crop species were each mostly likely domesticated a single time in Asia (Sebastian et al. 2010, Qi et al. 2013). During the domestication of cucurbit crops, humans selectively bred genotypes that were not able to produce cucurbitacins constitutively in leaves and at all in fruits (Qi et al. 2013). In many domesticated cucurbit varieties, cucurbitacins are still induced in foliage after herbivory (Metcalf et al. 1982, Ferguson et al. 1983) and reversion to toxic cucurbitacin production that has resulted in human poisoning from domesticated fruits can occasionally occur (Rymal et al. 1984). Control of cucurbitacin production in foliage and fruits is due to two interacting loci exhibiting Medelian inheritance (Shang et al. 2014). The dominant *Bi* and *Bt* genes control cucurbitacin C synthesis in leaves and fruits of cucumber, respectively. Bitterness in fruit require both *Bi* and *Bt*, while the recessive *bt* gene which confers non-bitterness in fruit shows evidence of human-mediated selection (Qi et al. 2013). In cultivated cucumber, the Mendelian *Bi* gene has a frameshift, rendering fruit non-bitter but still allowing for cucurbitacin production in the foliage of some varieties of cultivated cucumbers. Dominant *Bt* genes (*Btbi* or *BtBt*) result in foliar cucurbitacin synthesis, while double *bi* and *bt* recessives (*bibt*) are 'sweet' and do not produce cucurbitacins in either foliage or fruit (Da Costa and Jones 1971, Robinson et al. 1976, Qi et al. 2013, Shang et al. 2014).

Domestication of cucurbits for agriculture altered their ability to produce secondary metabolites, which was quickly followed by human-mediated geographic range expansion of domesticated cultivars. The implications for insect interactions from reduced cucurbitacin production and being grown over an expanded geographic range, and the importance for agricultural production are still unclear, but likely to be profound. The domestication and ancient indigenous methods of co-cultivating squash, beans, and corn is hypothesized to have influenced the present day feeding preferences and host plant breadth of some *Diabrotica, Acalymma* and *Ceratoma* on Cucurbitaceae, Poeaceae and Fabaceae (Whitaker and Bemis 1975), although empirical tests of this hypothesis have not been conducted. Nonetheless, there is evidence that the evolution of cucurbit specialist pollinators, such as the squash bee, *Peponapis pruinosa* (López-Uribe et al. 2016) and several species of diabroticites (Krysan 1986, Metcalf and Lampman 1989), has been influenced significantly by the co-cultivation of these plants.

Non-bitter, domesticated cucurbits that do not produce cucurbitacins constitutively have foliage that is less attractive to specialist beetle herbivores relative to foliage of bitter plants, although beetles still feed compulsively on cucurbitacins produced in cotyledons and locate host plants through floral volatile emissions (Section 2.2). However, these 'sweet' cucurbitacin-free plants are more susceptible to generalist herbivores (Da Costa and Jones 1971, Balkema-Boomstra et al. 2003) (Section 2.3). This presents a conundrum for cucurbit breeders: The presence of toxic cucurbitacins encourages damage by specialist herbivores, but the absence of cucurbitacins makes plants more susceptible to more damage by generalist herbivores. Additionally, the genetic bottleneck associated with domestication results in significantly reduced genetic diversity from which to selectively breed for herbivore resistance (Doebley et al. 2006) and this is currently an issue—a problem which is especially pronounced for cucumber (Qi et al. 2013). More broadly, the story of cucurbit domestication illustrates the need to consider how the selective elimination of key secondary metabolites unintentionally alters plant interactions with generalist and specialist herbivores (Chen 2016).

2.2 Floral Chemical Ecology of the Genus *Cucurbita*

To locate host plants, herbivorous insects often utilize volatile organic compounds (VOCs) associated with host foliage or floral resources. *Cucurbita* spp. are monoecious annual or perennial vines that are strictly New World (Whitaker and Bemis 1964) and produce large, yellow and highly scented flowers that are open and receptive once for only several hours from sunrise until late morning. A mature plant can produce dozens of flowers (one per node) in a single morning. Most flowers are male (Andersen and Metcalf 1987), and male flowers contain large quantities of pollen and release more volatiles per flower than female flowers (Mena Granera et al. 2005). The blend of volatile organic compounds released from both male and female *Cucurbita* flowers function as highly attractive, long distance olfactory host location cues for co-evolved beetle herbivores (Andersen and Metcalf 1987, Lampman and Metcalf 1988) and co-evolved pollinators (Theis and Adler 2012). A blend of more than forty individual compounds, predominantly cyclic products of the shikimate pathway, have been identified from *Cucurbita* blossoms. Nectar is the source of 1, 4-dimethoxybenzene and 1, 2, 4-methoxybenzene which makes up the largest percentage of the floral blend (Mena Granero et al. 2005), whereas emission of terpenes (linalool, indole, limonene, α-pinene and eucalyptol) is localized to petals (Mena Granero et al. 2005).

Floral volatiles act as long distance cues attracting beetles to ephemeral pollen and nectar resources. Meanwhile, cucurbitacins B and D (Fig. 1) in the flower corolla of wild species (and cultivated *Cucurbita maxima)* act as arrestants and feeding stimulants, resulting in large aggregations of specialist *Acalymma* spp. and polyphagous *Diabrotica* spp. feeding and mating in *Cucurbita* flowers in the late mornings (Andersen and Metcalf 1987). The quantity and composition of floral volatile blends differs across different *Cucurbita* species with consequences for plant-herbivore interactions. For example, in field trials comparing several *Cucurbita* species, flowers of *C. maxima* which release more than thirty compounds were more attractive to many diabroticites relative to other *Cucurbita* spp. (Andersen and Metcalf 1987, Lampman and Metcalf 1988, Metcalf and Lampman 1989). Subsequent studies found that pest diabroticite species are most attracted to a core blend of cyclic phenylpropanoid derivative compounds composed of 1, 2, 4-trimethoxybenzene, indole, and *trans*-cinnamaldehyde (Lewis et al. 1990, Metcalf and Lampman 1991). 1, 2, 4-trimethoxybenzene, and *trans*-cinnamaldehyde are both moderately attractive to several diabroticite beetle species. Indole, while not attractive as a single compound (Lewis et al. 1990), acts as an olfactory synergist with 1, 2, 4-trimethoxybenzene, and *trans*-cinnamaldehyde, increasing beetle attraction several fold (Metcalf et al. 1995). These three compounds have been tested as a synthetic blend in traps laced with carbaryl insecticides to try and specifically attract and kill diabroticite beetles in agricultural settings (Metcalf et al. 1995). While they do function effectively in this role, tri-component traps of this type are not useful in practice because of significant non-target attraction of pollinators.

Two genera of largely uncharacterized solitary bees, *Peponapis* spp. and *Xenoglossa* spp., have also co-evolved with *Cucurbita* spp. and derive almost all of their nutrition from squash blossom pollen and nectar rewards (Hurd et al. 1971). Both male and female *P. pruiniosa* are effective pollinators (Cane et al. 2011) and can be found foraging on *Cucurbita* blossoms from sunrise until mid-morning, when flowers close and nectar is reabsorbed. Females of *Peponapis pruinosa,* the most common species in the Northeastern U.S. are ground nesting while males can often be found in closed blossoms overnight. Native *Peponapis* or *Xenoglossa* populations can provide sufficient pollination for commercial *Cucurbita* plantings (Tepedino 1981, Adler and Hazzard 2009), yet pollination services for most agricultural cucurbit production often relies on introduced European honeybees (Walters and Taylor 2006). Restoring native *Peponapis* and *Xenoglossa* habitat and populations is considered essential to securing future cucurbit yields as European honeybee populations continue to decline (Kremen et al. 2002, Julier and Roulston 2009). However, dual pollinator and herbivore attraction to the same floral volatile cues presents a difficult challenge for increasing pollinator, but not diabroticite herbivore visitation to *Cucurbita* plants.

Despite the ubiquity, efficiency, and specialization of *Peponapis* and *Xenoglossa* spp. as cucurbit pollinators, the floral morphology of *Cucurbita* flowers (large, yellow, and with an open bowl shape) is suggestive of beetles being the ancestral pollinators of *Cucurbita* (Baker and Hurd Jr 1968). The ancestral state of all chrysomelid leaf beetles is thought to be pollinivory (Crowson 1960). Pollen found in the guts of dissected *Acalymma* (Samuelson 1994) and the morphological adaptations of *Acalymma* spp. mouthparts suggest these beetles are still largely pollinivorous (Cabrera and Durante 2001). While *Diabrotica* spp. develop primarily on non-cucurbit hosts, most described species have been recorded visiting flowers on wild or cultivated *Cucurbita* plants. Foliage of mature *Cucurbita* spp. plants is well defended both physically and chemically against diabroticite beetle herbivory (McCloud et al. 1995, Eben 2007), potentially making pollen and floral structures the primary sources of protein for both *Acalymma* and *Diabrotica* spp.

Conflicting selection pressures can occur when the same chemical host location cue is utilized by both herbivores and pollinators. *Cucurbita* are monoecious, but have a mixed mating system that permits both outcrossing and self-fertilization (Kohn and Biardi 1995). Outcrossed plants with high levels of floral volatiles and large floral displays could potentially produce more outcrossed offspring through both male and female functions, but may also experience greater herbivore attack as a result of specialist beetles using floral visual and volatile cues to locate hosts. For example, Theis and Adler (2012) demonstrated that *Cucurbita* plants have reduced fitness when the floral volatile blend is artificially augmented, as the enhanced blend results in higher levels of beetle herbivory. These costs can be even more significant given that these same beetle herbivores also transmit a fatal bacterial pathogen (discussed in Section 3.3). However, there can also be significant costs to reducing floral cues and opting for self-fertilization, because the increase in homozygosity can lead to inbreeding depression (Hayes et al. 2004). Inbreeding depression can impact overall plant vigor and reproductive output as well as specific plant traits that affect mutualistic and antagonistic insect interactions.

Due to their unique relationship with both pollinators and pollinivorous specialist herbivores, cucurbits have served as an important model for understanding how beneficial and antagonistic insects shape the evolution of mixed mating systems in plants. For instance, several studies with *C. pepo* ssp. *texana* have explored fitness costs associated with inbreeding in order to gauge the relative benefits of maintaining floral traits that enhance outcrossing, but expose plants to greater herbivory. These studies found that inbreeding depression significantly decreases male and female flower production (Stephenson et al. 2004), impairs pollen performance (Jóhannsson et al. 1998) and decreases fruit production (Hayes et al. 2005). Furthermore, inbred *C. pepo* ssp. *texana* also emit less total floral volatiles compared to outcrossed plants (Ferrari et al. 2006). Flowers on outcrossed plants produce significantly more 1, 4-dimethoxybenzene, linalool, and (3*E*)-4, 8-dimethyl-1, 3, 7-nonatriene than inbred plants (Ferrari et al. 2006). Both pollinators and herbivores utilize the main *Cucurbita* floral volatile 1, 4-dimethoxybenzene as a long distance host location cue (Andrews et al. 2007). This same study found differences in the quantitative and qualitative composition of the volatile blend across several maternal lines, suggesting that capacity for volatile production is a variable and heritable trait and therefore is likely under selection (Ferrari et al. 2006). Despite inbreeding-driven reductions in floral cues, inbred *Cucurbita* plants still suffer more beetle herbivore damage in the field (Hayes et al. 2004), providing evidence that induced and/or constitutive anti-herbivory defenses (which are heritable traits as well) are also impaired by inbreeding depression, negating any positive effects of reduced floral apparency. These studies highlight the numerous costs associated with inbreeding and suggests that selection will not favor genotypes that reduce floral cues to avoid herbivory because self-fertilization is likely to lead to offspring that are both less fit and less well-defended against herbivore attack.

Variation in floral phenotype can also occur due to environmental influences, including changes in plant chemistry and physical structure following herbivory. For example, foliar leaf loss or damage to roots caused by adult and larval forms of specialist diabroticite beetles, respectively, can incur metabolic costs due to induced anti-herbivore defenses and loss of photosynthetic tissue which leads to reductions in flower numbers and pollen production. In *C. pepo* ssp. *texana,* beetle foliar herbivory delayed the production of male flowers (Theis et al. 2009) and reduced pollen production (Quesada et al. 1995). Simulated foliar herbivory also increases the amount of several terpenoid volatiles emitted by male, but not female flowers (Theis et al. 2009). While the full implications of the interaction between mating system, insect attraction and induced defenses on cucurbit fitness are not known, these studies demonstrate the complexity of chemically mediated interactions between beneficial and harmful insects and the need to consider the environmental context when studying the effects of beneficial and antagonistic insects on the evolution of floral cues.

2.3 Induced Indirect Defenses in the Common Cucumber, *Cucumis sativus*

Plants release foliar volatiles as a result of normal physiological activity and in response to attack by antagonists, including herbivorous insects (Gish et al. 2015). While wild and cultivated varieties of *Cucurbita* spp. have served as important model systems for understanding plant-insect interactions mediated by floral volatiles and secondary defensive metabolites, domesticated varieties of common cucumber (*Cucumis sativus*) have served a similar function as models for studying the ecological roles of herbivore-induced foliar volatiles. *Cucumis* spp. whose native progenitors are strictly Eurasian (Decker-Walters et al. 2002, Sebastian et al. 2010), have a significantly less conspicuous floral display relative to New World *Cucurbita* spp., in terms of both visual and odor cues. However, both wild and cultivated *C. sativus* genotypes release complex blends of foliar volatiles constitutively and in response to attack by herbivorous insects (e.g., Kappers et al. 2010, 2011). A major herbivore of cultivated *C. satitvus* is the spider mite, *Tetranychus urticae*, which commonly attacks non-bitter agricultural plants (genotype *bibi*) that do not produce cucurbitacin C (Da Costa and Jones 1971, Balkema-Boomstra et al. 2003). Spider mite damage induces changes in blends of foliar volatiles released from *C. sativus* (Kappers et al. 2010, 2011). As discussed below, these volatiles convey specific information about the defense status of emitting plants to both herbivorous and predatory insects.

Typically, foliar volatiles emitted following attack by herbivores are quantitatively and qualitatively different from volatiles emitted during normal metabolism (reviewed in Rowen and Kaplan 2016). As a result, herbivore-induced plant volatiles (HIPVs) can convey information to receivers about the defense status of the emitting plant. Receivers can be distant parts of the emitter plant (Heil and Bueno 2007), other plants (Karban et al. 2014), other herbivores (De Moraes et al. 2001, Khan et al. 2016), or natural enemies of herbivores, such as predators and parasitoids (De Moraes et al. 1998, McCormick et al. 2012). It has been proposed that the induction of novel volatile emissions in response to herbivore attack is an adaptive response on the part of the plant in order to facilitate top-down control of herbivore attackers by natural enemies (discussed in Hare 2011). Indeed, natural enemies have been shown to have both innate and learned preferences for specific blends of HIPVs in numerous behavioral studies, supporting this hypothesis (McCormick et al. 2012). Some of the earliest work exploring HIPV-mediated predator-prey interactions was performed using *Cucumis sativus* (cucumber) as a model organism (Dicke and Sabelis 1987). As a result, we now have considerable information about the biosynthesis and emission of foliar HIPVs (Bouwmeester et al. 1999, Mercke et al. 2004), the putative relationship of HIPVs with direct defenses cucurbitacins (Agrawal et al. 2002) and the role of HIPVs in structuring predator-prey interactions for this agriculturally important cucurbit (Takabayashi et al. 1994, Dicke et al. 1998, Agrawal et al. 2002, De Boer et al. 2008, Kappers et al. 2010).

Cucumis sativus responds to feeding by herbivorous spider mites with emission of several new classes of volatiles not present in undamaged plant blends (e.g., monoterpenes, C11 and C16 homoterpenes, nitriles, oximes and occasionally, methyl salicylate) and reductions in emission of compounds constitutively produced by undamaged plants (5-10 carbon alcohols, aldehydes, ketones and esters) (Takabayashi et al. 1994, Dicke et al. 1998, Kappers et al. 2010, 2011). Using dual choice test assays that present odors from different host plants simultaneously, it was shown that predatory mites preferentially orient towards odor blends of *C. sativus* plants damaged by their spider mite prey over undamaged plants

(Takabayashi et al. 1994, Dicke et al. 1998). In one of the earliest studies to demonstrate learning of volatile blends in predatory arthropods, Dicke et al. (1990) showed that predatory mites reared on spider-mite-infested *C. sativus* also preferred odors of mite-damaged *C. sativus* over the odors of a different mite-damaged plant to which they did not have prior exposure (*Phaseolus lunatus*). Takabayashi et al. (1994) further showed that predatory mite responses vary depending on the age of the leaves that are being attacked. Older leaves of *C. sativus* emit a blend that is qualitatively similar to young leaves, but differs in that oximes make up a higher proportion of the old leaf blend. Old leaves are less attractive than young leaves and mixtures of old plus young leaves are also less attractive than young leaves alone, suggesting that the altered proportions of oximes in the old leaf blends possibly mask attraction to the monoterpenes and homoterpenes that dominate blends of young leaves. This variation interacts with prior experience of predatory mites—those that have prior prey-finding experience on *C. sativus* are better able to use volatiles of both young and old leaves to locate prey, while those that lack experience with *C. sativus* are not attracted to mixtures of old and young leaves (Takabayashi et al. 1994). These data suggest that variation in volatile emissions from *C. sativus* reflects the relative value of different plant tissues (with preferential recruitment of predators to young, growing tips) while predator learning can help to ensure that less valuable plant parts (older leaves) will eventually be visited by experienced predators.

The efficacy of *C. sativus* HIPVs in recruiting predatory mites and the subsequent effectiveness of predatory mites as regulators of herbivorous mite populations, suggest that HIPV emission is an adaptive strategy on the part of the plant for managing herbivore attack. Mechanistic studies of HIPV regulation in *C. sativus* in response to true herbivore damage vs. mechanical damage also support this hypothesis (Bouwmeester et al. 1999, Mercke et al. 2004). For example, the predator-attracting compounds (*E*)-β-ocimene (monoterpene); (3*E*)-4, 8-dimethyl-1, 3, 7-nonatriene (C11 homoterpene); methyl salicylate (aromatic); and (3*E*,7*E*)-4,8,12-trimethyl-1, 3, 7, 11-tridecatetraene (C16 homoterpene) are only produced in response to spider mite damage and are not produced in response to mechanical wounding with a fine abrasive that mimics damage (Bouwmeester et al. 1999). Detailed studies of precursor synthesis demonstrate that production of the homoterpenes occurs de novo upon attack by spider mites. Damage induces accumulation of nerolidol synthase, which catalyzes the formation of farnesyl diphosphate (the ubiquitous precursor of many terpenoids, including cucurbitacins) to (3*S*)-(*E*)-nerolidol. This compound is then converted to (3*E*)-4, 8-dimethyl-1, 3, 7-nonatriene by enzymes (likely peroxidases) that are present constitutively (Donath and Boland 1994, Bouwmeester et al. 1999, Mercke et al. 2004).

The compounds emitted from spider mite-damaged *C. sativus* have subsequently been shown to be common HIPVs emitted from a variety of plants (albeit in very different quantities and ratios depending on the species) and to be attractive to a variety of natural enemies (Dudareva et al. 2013). This convergence of induced defenses supports the idea that HIPVs of *C. sativus* have evolved specifically as a means of facilitating natural enemy recruitment, either via learning or innate attraction. More recent work on induced HIPVs in *C. sativus* also supports this assertion by demonstrating that there is genetic variation among cucumber varieties for the quantity and type of volatiles emitted in response to herbivore attack (Kappers et al. 2010, 2011). Blends that were attractive to predatory mites tended to have higher proportions of terpenes [(*E*)-β-ocimene, (*E*)-β-caryophyllene, (*E,E*)-α-farnesene] and homoterpenes [(3*E*)-4, 8-dimethyl-1, 3, 7-nonatriene and (3*E*, 7*E*)-4, 8, 12-trimethyl-1, 3, 7, 11-tridecatetraene] (Kappers et al. 2010, 2011). In order for a trait to evolve, there must first be heritable genetic variation for that trait on which selection can act. These studies suggest that such genetic variation exists for herbivore-induced volatile phenotypes among *C. sativus* genotypes and that predators preferentially protect plants that produce specific volatiles in response to attack by herbivorous prey (Kappers et al. 2010, 2011).

Cucumis sativus HIPVs can also vary based solely on the direct defense capability of a particular genotype with significant implications for recruitment of predators. Agrawal et al. (2002) found that a cucurbitacin-producing genotype emitted more (*E*)-β-ocimene and (3*E*)-4, 8-dimethyl-1, 3, 7-nonatriene relative to a near isoline without inducible cucurbitacins. The cucurbitacin-producing genotype also emitted a number of other compounds, notably oximes that were not present in the blend of the cucurbitacin-free genotype. The blend of the cucurbitacin-free genotype attracted more predatory mites even though these plants had reduced quantities of known attractants, possibly as a result of co-occurring, repellent compounds (oximes) in the blend of the cucurbitacin-producer (Takabayashi et al. 1994, Agrawal et al. 2002). Thus, for *C. sativus*, there may be trade-offs in direct and indirect defenses: a genotype with strong direct defenses is less attractive to predators than a genotype with weak direct defenses.

Since the completion of this work and much of the other earlier work on induced indirect and direct defenses in *Cucumis*, there has been significant progress in the development of *Cucumis* spp. as model organisms. Sequencing and partial annotation of the cucumber genome was completed in 2009 (Huang et al. 2009). Subsequent work has elucidated the regulation of cucurbitacin production (Shang et al. 2014) and the transcriptome of *Cucumis* plants under generalist herbivore attack (aphids) (Liang et al. 2015). Future research should leverage these tools to better understand the relationship between herbivore-induced direct and indirect defenses in wild and cultivated *Cucumis* species interacting with both generalist and specialist cucurbit herbivores.

3. Modification of Cucurbit Chemical Phenotypes by Pathogenic Microbes

3.1 Insect-transmitted Pathogens of Cucurbits

Insect vectors of plant pathogens, like all herbivores, collect information about the quality of their host plants by interpreting plant phenotypes through various sensory modalities. Visual, olfactory and gustatory plant quality indicators, such as VOCs, levels of nutrients, floral display, or foliar coloration, are unavoidable by-products of normal plant physiology and metabolism that can serve as cues for foraging herbivores. Pathogens often have broad-scale effects on the entire physiology of a host plant, including on cues that mediate interactions between hosts and vector and non-vector insects (Mauck et al. 2010, 2012, 2016). In addition to having direct implications for pathogen spread through effects on vector behavior, such changes can also alter plant interactions with other antagonists capable of influencing host survival and fitness (such as specialist and generalist herbivores or pollinators) as well as plant responses to abiotic stressors (Xu et al. 2008, Westwood et al. 2013).

To date, few experiments have explored how microbes influence chemical cues that mediate ecologically relevant plant-insect interactions in field environments. Cucurbits represent a promising group with which to study the chemical ecology of pathogen-host-vector interactions because of their susceptibility to a large number of pathogens and ubiquity in both wild and cultivated habitats. Despite having evolutionary centers of origin on different continents (*Cucumis* in Eurasia, *Cucurbita* in the Americas, and *Citrullus* in Africa), many species are now grown side-by-side in the same geographic areas, where they are frequently attacked by many of the same insect herbivores and the microbial pathogens that these herbivores transmit. For example, viral diseases of cucurbits are problematic in numerous agroecosystems around the globe and are also present in wild cucurbits and non-cucurbitaceous weeds (Lecoq and Katis 2014). These pathogens are transmitted by cucurbit-colonizing and non-colonizing vectors, largely aphids (*Cucumoviruses, Potyviruses* and *Poleroviruses*) and whiteflies (*Criniviruses, Ipomoviruses* and *Begomoviruses*) and in a few cases, beetles (*Squash mosaic virus*) and thrips (*Melon yellow spot virus* and *Watermelon silver mottle virus*) (Lecoq and Katis 2014). Additionally, in Eastern North America, a recently emerged diabroticite-transmitted bacterial pathogen, *Erwinia tracheiphila*, has become a significant threat to production of both *Cucurbita* spp. and *Cucumis* spp. (Rojas et al. 2015, Shapiro et al. 2016). Each of these pathogens can induce significant changes in the chemical and physical phenotypes of their host plants, including effects on odor, palatability, color, size, floral characteristics and even leaf turgor pressure. These effects (discussed below) are likely to have implications for plant interactions with vectors (via effects on vector attraction, pathogen acquisition, vector dispersal and inoculation of healthy plants) and pathogen fitness, in addition to potentially altering plant interactions with non-vector herbivores, predators and pollinators.

3.2 Plant Virus Effects on Cucurbit Chemical Phenotypes

Aphids and whiteflies are the primary vectors of plant viruses in cucurbit systems (Fig. 2). These insects are efficient vectors because they feed on plant sap using a piercing-sucking mechanism that leaves cells largely intact—a key requirement for virus acquisition and transmission (Ng and Falk 2006, Hogenhout et al. 2008). Most research on aphid and whitefly-transmitted viruses has focused on direct alleviation of the economic consequences of infection or vector feeding for fruit production and quality (Lecoq and Katis 2014). However, a growing number of studies have also sought to understand the mechanisms underlying pathogen alteration of plant physiology and the consequences of pathogen effects on plants for disease transmission by insect vectors. As a result, cucurbits have become important models for studying how phytopathogens alter host plant phenotypes in ways that affect vector behavior and disease transmission. For example, infection by *Cucumber mosaic virus* (CMV: Bromoviridae) changes volatile emissions, nutrient levels and coloration of *Cucurbita pepo* cv. 'Dixie' in ways that influence plant-insect interactions. Two cucurbit-origin isolates of CMV (CMV-Fny and CMV-KVPG2) both induce an increase in plant volatile emissions without altering the blend of compounds produced, resulting in an infected plant phenotype that effectively mimics that of a large, healthy plant (Mauck et al. 2010, 2014). Aphid vectors of CMV preferentially orient towards these enhanced odor emissions relative to emissions from healthy plant tissue (Mauck et al. 2010, 2014). CMV also induces changes in the quantities and ratios of sugars and amino acids within non-vascular plant cells (mesophyll) and vascular tissue (phloem) (Mauck et al. 2014). Thus, CMV-infected *C. pepo* tastes different and has altered quality relative to healthy *C. pepo*. Aphid vectors find the taste of CMV-infected *C. pepo* unpalatable and quickly disperse after probing infected hosts (Mauck et al. 2010, 2014). This effect is beneficial for a non-persistently transmitted virus like CMV which is only acquired and retained successfully when aphids briefly probe infected plants and then disperse before initiating long-term feeding in the phloem. A similar effect on palatability was also seen for a different cucurbit-origin isolate of CMV infecting *Cucumis sativus* (Carmo-Sousa

Fig. 2. Aphid vectors of non-persistently transmitted cucurbit-infecting viruses; *Myzus persicae* (a) and *Aphis gossypii* (b), and cultivated *Cucurbita pepo* bush variety; a plant infected with *Cucumber mosaic virus* (c) and a healthy plant (d).

et al. 2014).

Several widely prevalent, non-persistently transmitted viruses in the family Potyviridae also infect cucurbits. One of these pathogens, *Zucchini yellow mosaic virus* (ZYMV) has been the subject of studies seeking to understand virus effects on host phenotypes and vector behavior. Early work with ZYMV infecting *Cucurbita pepo* (cv. Cheffini) found similar reductions in plant quality and palatability as those seen for CMV (Blua and Perring 1992a, b). Interestingly, these changes were only present in later stages of viral disease progression: Immediately after inoculation, viral infection enhanced aphid fecundity and the production of winged aphids, but about 3-4 weeks after infection, these patterns were reversed (Blua and Perring 1992a, b). This effect is putatively linked to lower quantities of simple sugars in the phloem of older ZYMV-infected hosts (Blua et al. 1994). Aphid vectors use phloem sugar concentrations as cues during stylet penetration and initiation of long-term feeding (Douglas 2003) and reductions in sugar concentrations could make phloem location difficult. Regardless of the exact mechanism, ZYMV effects on *C. pepo* palatability and quality appear to be beneficial for virus spread, since initially large aphid populations will become increasingly restless as plant quality declines, leading to enhanced probing behavior (virus acquisition) and eventual dispersal of infective vectors due to low host plant nutrient quality (Blua and Perring 1992a, b). In a more recent study, ZYMV also induced volatile and color changes in *C. pepo* that resulted in greater aphid attraction to infected hosts vs. healthy hosts (Salvaudon et al. 2013).

Virus-induced changes in host phenotype also influence host interactions with other, non-vector organisms, with implications for host plant survival, fitness and ability to serve as an inoculum source. For example, the reductions in plant palatability induced by CMV infection in *Cucurbita pepo* (cv. Dixie) allow infected plants to escape herbivory by the squash bug, *Anasa tristis,* which spends less time feeding on infected compared to healthy plants (Mauck et al. 2015b). CMV infection has slightly beneficial effects on attraction of predators (Mauck et al. 2015b) and promotes the establishment of aphid parasitoids (Mauck et al. 2015a) which will aid the spread of non-persistently transmitted viruses like CMV by increasing vector disturbance and encouraging dispersal of infective vectors (Dáder et al. 2012). CMV infection can also affect responses to abiotic stress, such as an increase in drought resistance (Xu et al. 2008). Collectively, these studies suggest that CMV-induced changes in cucurbit phenotypes can be beneficial for the infected plant if such changes enable survival in the face of stressors (such as high herbivore pressure or drought) that would normally kill the plant and eliminate all fitness. Greater survival of infected hosts is also beneficial for the virus, since these plants will remain in the landscape longer to serve as a source of inoculum.

Infection of both wild and cultivated cucurbits by ZYMV also influences interactions with vector and non-vector insects via effects on plant chemical phenotype. In wild *Cucurbita pepo* (ssp. *texana*), phenotypic changes in floral morphology and chemistry due to ZYMV infection render hosts less attractive to specialist diabroticite beetles (Sasu et al. 2009, 2010c, Shapiro et al. 2012) (See Section 3.3 for an in-depth discussion of this interaction). As a result, *C. pepo* with ZYMV infections are less likely to contract a fatal bacterial wilt disease (*Erwinia tracheiphila*) that is transmitted by the beetles (see the next section for an in-depth discussion of this interaction). In cultivated *C. pepo* (cv. Dixie) fields, ZYMV often occurs as a co-infection with *Watermelon mosaic virus* (WMV), another non-persistently transmitted potyvirus. When ZYMV and WMV co-infect the same host plant, the co-infection induces a volatile phenotype and yellow coloration that is more attractive to foraging aphids than a single WMV infection. While WMV is present at lower titre in co-infected compared to singly-infected plants, WMV is still readily transmitted from co-infected plants, suggesting that it benefits from 'hitch-hiking' on the ZYMV-induced phenotype (Salvaudon et al. 2013).

Virus induced changes in plant quality, palatability or attractiveness to vectors have also been shown for various persistently-transmitted viruses that are common in cucurbit agroecosystems. Unlike non-persistently transmitted viruses like CMV, ZYMV and WMV, these viruses require vectors to stay and feed on a host for long periods of time in order for virus acquisition to occur. Consistent with this requirement for prolonged feeding, persistently-transmitted viruses of cucurbits tend to increase plant palatability and quality for vectors (Costa et al. 1991, Carmo-Sousa et al. 2016). This contrasts with the reductions in plant palatability reported for several non-persistently transmitted viruses (described above) and suggests that virus effects are not merely by-products of infection, but could instead be the result of adaptation on the part of a pathogen for manipulating plant chemistry in ways that are conducive to its own transmission mechanism (Mann et al. 2012, Mauck et al. 2012). These adaptations may be selected for in monocultures of cucurbit hosts, where viral pathogens have the potential for local adaptation to the cucurbit cultivar that dominates that landscape in terms of infectivity, replication rate and possibly also in terms of effects on aspects of the host phenotype that mediate interactions with vectors (e.g. Mauck et al. 2014). However, more research is needed that directly compares the effects of persistently and non-persistently transmitted viral pathogens taken from different host environments in order to more rigorously test the hypothesis that virus effects on host phenotype are adaptive.

3.3 Effects of a Bacterial Pathogen on Host Plant Volatile Phenotype

Cucumis and *Cucurbita* spp. are susceptible to infection by *Erwinia tracheiphila* (Enterobacteriaceae), an obligately beetle-transmitted bacterial pathogen that causes a fatal wilt disease in susceptible cucurbit host plants (Smith 1920, Rojas et al. 2015, Shapiro et al. 2015). *Erwinia tracheiphila* causes significant economic losses in *Cucumis sativus* (cucumber), *Cucumis melo* (muskmelon) and *Cucurbita* spp. (squash and pumpkin) crops in Northeastern and Midwestern North America (Rojas et al. 2015). Characteristic wilting symptoms and eventual plant death occur due to secretion of an exopolysaccharide matrix during *E. tracheiphila* replication in xylem vessels which impedes the flow of xylem sap (Smith 1920). Only the spotted and striped cucumber beetles (*Diabrotica undecimpunctata* and *Acalymma vittatum,* respectively; Fig. 3) are confirmed to transmit *E. tracheiphila,* and as a result, the host range of this pathogen is limited to the cucurbit hosts utilized by its vectors. Instead of being retained in the mouthparts or salivary glands like many viral pathogens, *E. tracheiphila* is consumed with infected foliage (via chewing) and passes through

Fig. 3. Luperini beetles, *Acalymma vittatum* (a) and *Diabrotica undecimpunctata* (b) bacterial wilt pathogen, *Erwinia tracheiphila*.

the beetle digestive tract along with plant material (Shapiro et al. 2014). Transmission can occur indirectly when frass from infective beetles falls on recent foliage wounds created by beetle feeding damage (Smith 1920, Rand and Cash 1920, Rand and Enlows 1920, Mitchell and Hanks 2009, Shapiro et al. 2014). More recently, *E. tracheiphila* was also shown to be able to infect *C. pepo* plants through floral nectaries (Sasu et al. 2010a) once pollinators remove anti-microbial nectar (Sasu et al. 2010b)

In the Eastern United States, *E. tracheiphila* and ZYMV epidemics co-occur annually in cucurbit agroecosystems. Thus, cucurbits provide a unique opportunity to examine the indirect (plant-mediated) impacts of two very different pathogens on each other's fitness and the ways that pathogen effects on host plant chemical phenotype might impact these dynamics. In a series of field experiments conducted over multiple seasons in Central Pennsylvania, it was found that infection rates of *E. tracheiphila* in virus-free (healthy) wild gourds (*C. pepo* ssp. *texana)* were as high as 50%, but less than 1% of ZYMV-infected plants contracted a secondary *E. tracheiphila* infection (Sasu et al. 2009, 2010c; Fig. 4). This lack of co-infection in plants with a primary virus infection could be explained by two, non-mutually exclusive mechanisms: viral infection may induce plant defenses that directly inhibit bacterial disease development; it may reduce the probability of exposure to *E. tracheiphila* by rendering plants less palatable or less attractive to foraging beetle vectors; or both mechanisms may operate simultaneously.

Teasing apart the relative contributions of these two mechanisms

Fig. 4. Wild *Cucurbita pepo* subsp. *texana* co-infected with *Zucchini yellow mosaic virus* and *Erwinia tracheiphila* (a) and virus-free wild *C. pepo* subsp. *texana* infected with *E. tracheiphila* (b).

can be accomplished by measuring susceptibility of virus-infected plants to controlled *E. tracheiphila* inoculations in conjunction with measurements of virus effects on plant palatability and attractiveness to the diabroticite beetle vectors of the bacteria. While ZYMV was found to induce accumulation of the defense signaling hormone salicylic acid (SA) which generally mediates induced anti-pathogen defenses (Vlot et al. 2011), this induction of SA did not translate into increased resistance to *E. tracheiphila* (Shapiro et al. 2013). Further, ZYMV infection did not deter vector feeding, as virus infected plants were equally as palatable to striped cucumber beetles (*Acalymma vittatum*) as healthy plants (Shapiro et al. 2012). These results indicate that the drastic reduction in bacterial wilt disease incidence among plants with primary ZYMV infections is not due to a direct protective effect of the virus on its host plant. Instead, these findings suggest that the second proposed mechanism is more likely and observed differences in wilt disease transmission among ZYMV-infected and healthy plants are indeed due to virus effects on plant cues that mediate host location by the beetle vectors.

As discussed in Section 2.3, volatiles from *Cucurbita* spp. flowers are the primary host-location cues for the cucumber beetles that transmit *E. tracheiphila*. In the field, ZYMV infection in *C. pepo* reduces the total number of flowers and the total quantity of floral volatiles emitted per flower (Shapiro et al. 2012). This ZYMV-induced change in floral chemical phenotype leads to a reduction in visitation by cucumber beetles and thus a reduction in exposure to the wilt pathogen. Curiously, *E. tracheiphila* infection also reduces attractiveness of flowers to beetle vectors, which at first appears mal-adaptive for this pathogen if it reduces vector contacts with infected hosts. However, *E. tracheiphila* infection in foliage induces production of a vegetative volatile blend that is highly attractive to beetles. Meanwhile, undamaged leaves on either healthy or ZYMV-infected *C. pepo* ssp. *texana* produce few detectable volatiles in either the field or greenhouse (Shapiro et al. 2012). These low volatile phenotypes contrast with the high quantity of unique volatiles induced by symptomatic leaves of *E. tracheiphila*-infected plants which emit a blend predominantly composed of (*E*)-2-hexenal, with 1-pentanol, hexenal and (*Z*)-3-hexenol occurring as minor components. These compounds are only emitted from wilting, symptomatic leaves and are not detected from non-symptomatic leaves of infected plants, leaves of healthy plants, or leaves of virus-infected plants (Shapiro et al. 2012). Physical defenses are also impeded by the wilting phenotype, as the blockage of the xylem tissue by the bacterial exopolysaccharides physically alters infected plants by reducing turgor pressure which reduces the efficacy of P-proteins (McCloud et al. 1995). The reduction in turgor pressure makes wilting plants physically easier for beetle herbivores to feed on compared to leaves on healthy or ZYMV infected plants (Shapiro et al. 2012) and increases the probability of vectors acquiring *E. tracheiphila* (Shapiro et al. 2014).

The overall effects of each pathogen on host chemical phenotype results in cucumber beetle vectors preferentially visiting and feeding on the foliage of *E. tracheiphila*-infected plants, then dispersing to the highly attractive and abundant flowers of healthy (virus-free) plants, where *E. tracheiphila* transmission can occur via nectaries or subsequent sampling of foliage (Sasu et al. 2010a, b, Shapiro et al. 2012, 2014). Meanwhile, the reduced floral cues of ZYMV-infected plants result in beetle vectors avoiding these hosts in favor of the more prominent floral cues associated with healthy plants. Several earlier studies provide further support for the hypothesis that the differential infection of healthy plants with *E. tracheiphila* is driven by differences in floral volatile cues between healthy and ZYMV-infected plants. For example, the highest infection rates by *E. tracheiphila* are observed for *C. pepo* ssp. *texana* genotypes that produce the most flowers (outcrossed plants) and the lowest infection rates are observed for genotypes that produce the fewest flowers (inbred plants) (Ferrari et al. 2007, Sasu et al. 2009, 2010c). During the late summer, up to 95% of flowers were found to have beetle frass containing *E. tracheiphila* in nectaries. Yet, fewer than half the plants showed wilt disease development (Sasu et al. 2010a), suggesting that attraction of large numbers of beetles to each flower is necessary to provide the cumulative exposure to *E. tracheiphila* that leads to development of bacterial wilt disease (Yao 1996). Suppression of both floral and foliar volatiles in plants prevents this critical beetle recruitment threshold from being reached and plants are not exposed to sufficient levels of *E. tracheiphila* to contract the infection. Meanwhile, flowers on healthy plants serve as a chemical magnet for beetles due to their large floral displays with high volatile emissions and are thus exposed to high *E. tracheiphila* levels which results in significantly higher rates of *E. tracheiphila* disease development.

3.4 Consideration of Pathogen Effects in Pest and Disease Management Strategies

In the wild gourd system described above, lack of co-infection is almost entirely driven by the different effects of a bacterial and a viral pathogen on plant cues that are important for attracting specialist beetle vectors. Healthy plants are significantly more likely to be exposed to the fatal *E. tracheiphila* infection because of vector attraction to their relatively higher floral cues, while ZYMV-infected plants with a reduced floral display escape vector visitation and exposure to a bacterial pathogen (and death), but suffer from reduced reproductive output. In cultivated cucurbit systems, both of these diseases are problematic for yield and fruit quality, but the co-infection dynamics indicate that controlling for the non-fatal viral disease (ZYMV) will potentially enhance the probability of plants contracting a fatal bacterial disease (*E. tracheiphila).*

Significant progress has been made in managing viral diseases in some cucurbit crops. For instance, summer squash (*Cucurbita pepo*) with protein coat-mediated resistance against several potyviruses was one of the first transgenic plants approved for field use (Tricoll et al. 1995) and many transgenic *Cucurbita* spp. cultivars are now widely planted. However, undomesticated *Cucurbita* spp. are still abundant in the New World tropics and subtropics, often growing alongside their domesticated counterparts. Thus, there are ample opportunities for the flow of coat protein resistance genes from transgenic plants into natural populations. Given that healthy (virus-free) plants are more likely to contract *E. tracheiphila*, the widespread introduction of virus resistance into wild could have significant ecological consequences for host plant populations, the *E. tracheiphila* pathogen and the beetle vectors (Decker-Walters et al. 1990, Quemada et al. 2008, Kistler et al. 2015).

This scenario has been studied directly in a series of field experiments that took advantage of the ease with which domesticated cucurbits can be crossed with wild progenitors (Spencer 2001). Cultivated summer squash carrying the virus resistant transgene (VRT) was introgressed into the wild progenitor *C. pepo* ssp. *texana* through an initial cross followed by a series of back-crosses of transgene-carrying offspring with the original wild parent genotypes. This generated progeny that were essentially wild in all traits except the transgene (Sasu et al. 2009, 2010c). When mosaic viruses are present in the field, wild plants with the VRT have higher pollen and seed production, while virus-infected plants suffer reproductive penalties due to low flower production (Sasu et al. 2009), reduced floral volatile emissions (Shapiro et al. 2012) and poor pollen quality (Harth et al. 2016). However, the positive fitness effect of resistance to viral diseases is mitigated by increases in cucumber beetle herbivory and *E. tracheiphila* exposure for healthy plants. As virus epidemics begin to spread, only the VRT plants remain healthy. These plants continue to produce large floral displays and high floral volatile emissions and therefore attract more foraging beetles than plants lacking the VRT gene (Sasu et al. 2009). Thus, spread of the VRT gene in wild *C. pepo* populations where both virus and *E. tracheiphila* co-circulate may be reduced or halted entirely if plants carrying the VRT gene die due to *E. tracheiphila* infection prior to reproduction.

The non-target effects of the VRT on susceptibility to herbivory and other co-circulating pathogens shows the complexity of implementing sustainable control strategies when multiple pathogens and herbivores are present. If *E. tracheiphila* is not present, the VRT gene confers fitness advantages under pressure from viral diseases. But, if *E. tracheiphila* is present, then the VRT plants have increased exposure and death rates from this fatal pathogen, mitigating the fitness benefits of the VRT. In cucurbit production areas where *E. tracheiphila* or its vectors are rare, use of crops with transgenic virus resistance would enhance cucurbit yields and fruit quality. However, if *E. tracheiphila* and vectors are a consistent concern across growing seasons, then alternative virus-control strategies should be considered. These may include use of genotypes that tolerate virus infection instead of genotypes that outright resist virus infection. Under this scenario, tolerant cultivars that have virus infection might still exhibit some phenotypic changes in plants that mitigate exposure to *E. tracheiphila* or vectors (e.g., reduced floral volatiles), while maintaining a sufficient level of fruit production and quality to be profitable. Tolerance to virus infection in cucurbits can also be achieved through deployment of plant-growth-promoting rhizobacteria which could also induce resistance against *E. tracheiphila* directly (Zehnder et al. 1997a, b). Despite these promising initial results using rhizobacteria-mediated resistance to herbivory and disease in cucurbit agro-ecosystems, the topic has not been revisited since the introduction of modern high-throughput microbiome sequencing technologies.

Another group of insects that respond to host plant cues and that may be influenced by pathogen effects on host phenotype, are the natural enemies of the insect vectors responsible for spreading plant viruses and *E. tracheiphila*. Natural enemies have the potential to limit the spread of pathogens via top-down control of insect vectors and could be another viable option for reducing the impact of diseases in cucurbit agroecosystems. The efficacy of natural enemies depends on their ability to locate and attack large numbers of hosts (Smyth and Hoffmann 2010). As discussed in Section 2.3, predators and parasitoids frequently use induced plant volatiles to locate their herbivorous. In cucurbits, viral and bacterial pathogens change host plant cues in ways that influence plant-herbivore interactions. It is likely that these changes in plant volatile phenotype and plant quality also influence plant-herbivore-predator interactions. For example, *Centises diabroticae* (Hymenoptera: Braconidae), a parasitoid of the striped cucumber beetle *Acalymma vittatum*, is only attracted to volatiles released due to active beetle herbivory (Ichiki et al., unpublished data). Damage by cucumber beetles normally induces elevated concentrations of monoterpenes and sesquiterpenes, including (*E*)-β-ocimene, linalool, (*E*, *E*)-α-farnesene and nerolidol. However, the quantity of induced volatiles released from plants with dual *E. tracheiphila* infections and beetle herbivore damage is suppressed compared to the total quantity of volatiles released from healthy plants with beetle damage which exhibit increases in the volatiles mentioned above. Meanwhile, undamaged *E. tracheiphila* plants have elevated levels of a very different blend of volatiles relative to healthy damaged plants (Shapiro et al. 2012). This *E. tracheiphila* stimulated foliar volatile blend attracts beetle vectors to undamaged *E. tracheiphila* plants (Shapiro et al. 2012), but because volatile induction in these plants is suppressed in response to subsequent beetle feeding, *C. diabroticae* are less attracted to *E. tracheiphila* infected plants that are being damaged by beetles.

4. Conclusions and Future Perspectives

Many of the most important ecological interactions in natural and agricultural communities are mediated by chemical cues and signals. Isolation and identification of plant chemicals that convey information between plants and insects or that serve as direct defenses, gives us insight into how different insect community members shape the evolution of plant chemical phenotypes in wild populations and how domestication has impacted these relationships. In this chapter we have shown the potential of the Cucurbitaceae family to serve as a source of ecological model species for studying chemically-mediated plant-insect and plant-pathogen interactions in laboratory and field environments. The suitability of the Cucurbitaceae for this role is due to several factors. First, there is a large amount of extant genetic diversity. Genetically compatible wild progenitors are available for several important *Cucurbita* and *Cucumis* species, enabling crosses between cultivated and wild genotypes (e.g., Sasu et al. 2009, 2010c) as well as direct comparisons of chemical-mediated signaling in natural and agricultural cucurbit systems. Second, cucurbits are chemically diverse, possessing unique secondary metabolites (cucurbitacins) and heritable variation in production of both floral and foliar volatiles (e.g., Ferrari et al. 2006, Kappers et al. 2010, 2011).

This unique cucurbit chemistry has driven the evolution of specialist herbivores and pollinators (e.g., Luperine leaf beetles and *Peponapis* and *Xenoglossa* spp. squash bees) while the modification of this chemistry due to domestication has led to new associations with generalists (e.g., spider mites and aphids). Finally, the widespread planting of both Eurasian and New World species of Cucurbitaceae in agroecosystems around the globe has facilitated the adaptation of local cucurbit pathogens to new cucurbit hosts as well as the emergence of novel pathogens (e.g., *E. tracheiphila*) (Shapiro et al. 2016). In particular, there is a large amount of diversity of vector-borne pathogens affecting wild, cultivated, native and introduced cucurbit host plant species. This species level and within-species genetic diversity in cucurbit systems will enable the exploration of the genetic basis of plant-insect and plant-microbe interactions.

The studies discussed in the prior sections show how research in cucurbit systems has enriched our understanding of the ecological consequences of changes in plant chemical phenotypes due to domestication, herbivore attack, pollinator selection pressure and pathogen manipulation. Moving forward, new research should leverage the ongoing genome sequencing and annotation projects that are bolstering the value of cucurbits as genetic models (Huang et al. 2009, Kistler et al. 2015). A greater emphasis on these tools would help us to link each of the sources of variation listed above with changes in the hormonal signaling and biosynthetic pathways responsible for production of plant chemical cues and signals. This could, in turn, lead to the development of novel cucurbit cultivars bred with the plant chemical phenotype or genetic resistance mechanisms in mind. In addition to incorporating these new genetic tools, future work should also begin to dissect complex interactions among the various members of cucurbit phytobiomes (Khalaf and Raizada 2016, Eevers et al. 2016). For example, even though high throughput sequencing has enabled detailed profiling of plant-associated microbe communities, studies have not taken advantage of this technology to explore the potential for beneficial microbes to modify cucurbit chemical phenotypes (Zehnder et al. 1997a, b, Raupach and Kloepper 1998). Similarly unexplored are the effects of pathogen infection in a host plant on predator or parasitoid foraging and the efficacy of biological control as a means of managing vectors and plant pathogens (but see Mauck et al. 2015a, b). These areas are relevant to our basic understanding of complex, chemically-mediated interactions among plants and other organisms and are also likely to inform the development of sustainable biological control methods for curbing the impacts of herbivores and pathogens in cucurbit agroecosystems.

REFERENCES

Adler, L. and R. Hazzard. 2009. Comparison of perimeter trap crop varieties: effects on herbivory, pollination, and yield in butternut squash. Environ. Entomol. 38: 207–215.

Agrawal, A.A., A. Janssen , J. Bruin, M.A. Posthumus and M.W. Sabelis. 2002. An ecological cost of plant defence: attractiveness of bitter cucumber plants to natural enemies of herbivores. Ecol. Lett. 5: 377–385.

Andersen, J.F., R.D. Plattner and D. Weisleder. 1988. Metabolic transformations of cucurbitacins by *Diabrotica virgifera virgifera* Leconte and *D. undecimpunctata howardi* Barber. Insect Biochem. 18: 71–77.

Andrews, E.S., N. Theis and L.S. Adler. 2007. Pollinator and herbivore attraction to *Cucurbita* floral volatiles. J. Chem. Ecol. 33: 1682–1691.

Andersen, J.F. and R.L. Metcalf. 1987. Factors influencing distribution of *Diabrotica* spp. in blossoms of cultivated *Cucurbita* spp. J. Chem. Ecol. 13: 681–699.

Baker, H.G. and P.D. Hurd Jr. 1968. Intrafloral ecology. Annu. Rev. Entomol. 13: 385–414.

Balkema-Boomstra, A., S. Zijlstra, F. Verstappen, H. Inggamer, P. Mercke, M. Jongsma et al. 2003. Role of cucurbitacin C in resistance to spider mite (*Tetranychus urticae*) in cucumber (*Cucumis sativus* L.). J. Chem. Ecol. 29: 225–235.

Blua, M.J. and T.M. Perring. 1992a. Alatae production and population increase of aphid vectors on virus-infected host plants. Oecologia 92: 65–70.

Blua, M.J. and T.M. Perring. 1992b. Effects of zucchini yellow mosaic virus on colonization and feeding behavior of *Aphis gossypii* (Homoptera: Aphididae) alatae. Environ. Entomol. 21: 578–585.

Blua, M.J., T.M. Perring and M.A. Madore. 1994. Plant virus-induced changes in aphid population development and temporal fluctuations in plant nutrients. J. Chem. Ecol. 20: 691–707.

Bogawat, J. and S. Pandey. 1967. Food preference in *Aulacophora* spp. Indian J. Entomol. 29: 349–352.

Bouwmeester, H.J., F.W. Verstappen, M.A. Posthumus and M. Dicke. 1999. Spider mite-induced (3*S*)-(*E*)-nerolidol synthase activity in cucumber and lima bean. The first dedicated step in acyclic C11-homoterpene biosynthesis. Plant Physiol. 121: 173–180.

Branson, T. and J. Krysan. 1981. Feeding and oviposition behavior and life cycle strategies of *Diabrotica*: an evolutionary view with implications for pest management. Environ. Entomol. 10: 826–831.

Cabrera, N. and S. Durante. 2001. Description of mouthparts of the genus *Acalymma* Barber (Coleoptera: Chrysomelidae: Galerucinae). Trans. Am. Entomol. Soc. 127: 371–379.

Cane, J.H., B.J. Sampson and S.A. Miller. 2011. Pollination value of male bees: the specialist bee *Peponapis pruinosa* (Apidae) at summer squash (*Cucurbita pepo*). Environ. Entomol. 40: 614–620.

Carmo-Sousa, M., A. Moreno, E. Garzo and A. Fereres. 2014. A non-persistently transmitted-virus induces a pull–push strategy in its aphid vector to optimize transmission and spread.Virus Res. 186: 38–46.

Carmo-Sousa, M., A. Moreno, M. Plaza, E. Garzo and A. Fereres. 2016. *Cucurbit aphid-borne yellows virus* (CABYV) modifies the alighting, settling and probing behaviour of its vector *Aphis gossypii* favouring its own spread: persistent transmitted viruses effect on vector behaviour. Ann. Appl. Bot. 169: 284–297.

Chambliss, O.L. 1966. Cucurbitacins: specific insect attractants in Cucurbitaceae. Science 153: 1392.

Chan, K.T., K. Li, S.L. Liu, K.H. Chu, M. Toh and W.D. Xie. 2010. Cucurbitacin B inhibits STAT3 and the Raf/MEK/ERK pathway in leukemia cell line K562. Cancer Lett. 289: 46–52.

Chen, J.C., M.H. Chiu, R.L. Nie, G.A. Cordell and S.X. Qiu. 2005. Cucurbitacins and cucurbitane glycosides: structures and biological activities. Nat. Prod. Rep. 22: 386–399.

Chen, Y.H. 2016. Crop domestication, global human-mediated migration, and the unresolved role of geography in pest control. Elementa 4: 000106.

Contardi, H. 1939. Estudios genéticos en *Cucurbita* y consideraciones agronómicas. Physis 18: 332–347.

Costa, H.S., J.K. Brown and D.N. Byrne. 1991. Life history traits of the whitefly, *Bemisia tabaci* (Homoptera: Aleyrodidae), on six virus-infected or healthy plant species. Environ. Entomol. 20: 1102–1107.

Crowson, R. 1960. The phylogeny of Coleoptera. Annu. Rev. Entomol. 5: 111–134.

Da Costa, C.P. and C.M. Jones. 1971. Cucumber beetle resistance and mite susceptibility controlled by the bitter gene in *Cucumis sativus* L. Science 172: 1145–1146.

Dáder, B., A. Moreno, E. Viñuela and A. Fereres. 2012. Spatio-temporal dynamics of viruses are differentially affected by parasitoids depending on the mode of transmission. Viruses 4: 3069–3089.

David, A. and D. Vallance. 1955. Bitter principles of Cucurbitaceae. J. Pharm. Pharmacol. 7: 295–296.

De Boer, J.G., C.A. Hordijk, M.A. Posthumus and M. Dicke. 2008. Prey and non-prey arthropods sharing a host plant: effects on induced volatile emission and predator attraction. J. Chem. Ecol. 34: 281–290.

De Moraes, C.M., W.J. Lewis, P.W. Pare, H.T. Alborn and J.H. Tumlinson. 1998. Herbivore-infested plants selectively attract parasitoids. Nature 393: 570–573.

De Moraes, C.M., M.C. Mescher and J.H. Tumlinson. 2001. Caterpillar-induced nocturnal plant volatiles repel conspecific females. Nature 410: 577–580.

Decker-Walters, D., T. Walters, U. Posluszny and P. Kevan. 1990. Genealogy and gene flow among annual domesticated species of *Cucurbita*. Can. J. Bot. 68: 782–789.

Decker-Walters, D., S.-M. Chung, J. Staub, H. Quemada and A. López-Sesé. 2002. The origin and genetic affinities of wild populations of melon (*Cucumis melo*, Cucurbitaceae) in North America. Plant Syst. Evol. 233: 183–197.

Delazar, A., S. Gibbons, A.R. Kosari, H. Nazemiyeh, M. Modarresi, L. Nahar et al. 2006. Flavone C-glycosides and cucurbitacin glycosides from *Citrullus colocynthis*. DARU J. Pharm. Sci. 14: 109–114.

Dicke, M. and M.W. Sabelis. 1987. How plants obtain predatory mites as bodyguards. Neth. J. Zool. 38: 148–165.

Dicke, M., K.J. van der Maas, J. Takabayashi and L. Vet. 1990. Learning affects response to volatile allelochemicals by predatory mites. pp. 31-36. *In*: Nederlandse Entomologische Vereniging [ed.]. Proceedings of Annual Meeting of the Section Experimental and Applied Entomology of the Netherlands Entomological Society. Amsterdam, Netherlands.

Dicke, M., J. Takabayashi, M.A. Posthumus, C. Schütte and O.E. Krips. 1998. Plant–phytoseiid interactions mediated by herbivore-induced plant volatiles: variation in production of cues and in responses of predatory mites. Exp. Appl. Acarol. 22: 311–333.

Dinan, L., T. Savchenko, P. Whiting and S.D. Sarker. 1999. Plant natural products as insect steroid receptor agonists and antagonists. Pestic. Sci. 55: 331–335.

Dinan, L., P. Whiting, J.-P. Girault, R. Lafont, S.T. Dhadallia, E.D. Cress et al. 1997. Cucurbitacins are insect steroid hormone antagonists acting at the ecdysteroid receptor. Biochem. J. 327: 643–650.

Doebley, J.F., B.S. Gaut and B.D. Smith. 2006. The molecular genetics of crop domestication. Cell 127: 1309–1321.

Donath, J. and W. Boland. 1994. Biosynthesis of acyclic homoterpenes in higher plants parallels steroid hormone metabolism. J. Plant Physiol. 143: 473–478.

Douglas, A.E. 2003. The nutritional physiology of aphids. Adv. Insect Physiol. 31: 73–140.

Dudareva, N., A. Klempien, J.K. Muhlemann and I. Kaplan. 2013. Biosynthesis, function and metabolic engineering of plant volatile organic compounds. New Phytol. 198: 16–32.

Eben, A. 2007. Sharing the trench: a curious feeding behavior of *Diabrotica porracea* Harold (Chrysomelidae: Galerucinae) in the presence of *Epilachna tredecimnotata* (Latreille)(Coccinellidae). Coleopt. Bull. 61: 57.

Eben, A. and A. Espinosa. 2013. Tempo and mode of evolutionary radiation in *Diabroticina* beetles (genera *Acalymma*, *Cerotoma*, and *Diabrotica*). ZooKeys 332: 207–231.

Eevers, N., B. Beckers, M. Op de Beeck, J.C. White, J. Vangronsveld and N. Weyens. 2016. Comparison between cultivated and total bacterial communities associated with *Cucurbita pepo* using cultivation-dependent techniques and 454 pyrosequencing. Syst. Appl. Microbiol. 39: 58–66.

Ehrlich, P.R. and P.H. Raven. 1964. Butterflies and plants: a study in coevolution. Evolution 18: 586–608.

Ferguson, J.E. and R.L. Metcalf. 1985. Cucurbitacins: plant derived defense compounds for Diabroticina (Coleoptera: Chrysomelidae). J. Chem. Ecol. 11: 311–318.

Ferguson, J.E., E.R. Metcalf, R.L. Metcalf and A. Rhodes. 1983. Influence of cucurbitacin content in cotyledons of Cucurbitaceae cultivars upon feeding behavior of *Diabroticina* beetles (Coleoptera: Chrysomelidae). J. Econ. Entomol. 76: 47–51.

Ferrari, M.J., A.G. Stephenson, M.C. Mescher and C.M. De Moraes. 2006. Inbreeding effects on blossom volatiles in *Cucurbita pepo* subsp. *texana* (Cucurbitaceae). Am. J. Bot. 93: 1768–1774.

Ferrari, M.J., D. Du, J.A. Winsor and A.G. Stephenson. 2007. Inbreeding depression of plant quality reduces incidence of an insect-borne pathogen in a wild gourd. Int. J. Plant Sci. 168: 603–610.

Fraenkel, G.S. 1959. The raison d'être of secondary plant substances. Science 129: 1466–1470.

Gillespie, J.J., D.W. Tallamy, E.G. Riley and A.I. Cognato. 2008. Molecular phylogeny of rootworms and related galerucine beetles (Coleoptera: Chrysomelidae). Zool. Scripta 37: 195–222.

Gish, M., C.M. De Moraes and M.C. Mescher. 2015. Herbivore-induced plant volatiles in natural and agricultural ecosystems: open questions and future prospects. Curr. Opin. Insect Sci. 9: 1–6.

Gray, M.E., T.W. Sappington, N.J. Miller, J. Moeser and M.O. Bohn. 2009. Adaptation and invasiveness of western corn rootworm: intensifying research on a worsening pest. Annu. Rev. Entomol. 54: 303–321.

Halaweish, F.T., D.W. Tallamy and E. Santana. 1999. Cucurbitacins: a role in cucumber beetle steroid nutrition? J. Chem. Ecol. 25: 2373–2383.

Hare, J.D. 2011. Ecological role of volatiles produced by plants in response to damage by herbivorous insects. Annu. Rev. Entomol. 56: 161–180.

Harth, J.E., J.A. Winsor, D.R. Weakland, K.J. Nowak, M.J. Ferrari and A.G. Stephenson. 2016. Effects of virus infection on pollen production and pollen performance: implications for the spread of resistance alleles. Am. J. Bot. 103: 577–583.

Hayes, C.N., J.A. Winsor and A.G. Stephenson. 2004. Inbreeding influences herbivory in *Cucurbita pepo* ssp. *texana* (Cucurbitaceae). Oecologia 140: 601–608.

Hayes, C.N., J.A. Winsor and A.G. Stephenson. 2005. Multigenerational effects of inbreeding in *Cucurbita pepo* ssp. *texana* (Cucurbitaceae). Evolution 59: 276–286.

Heil, M. and J.C.S. Bueno. 2007. Within-plant signaling by volatiles leads to induction and priming of an indirect plant defense in nature. Proc. Natl. Acad. Sci. USA 104: 5467–5472.

Hogenhout, S.A., E.D. Ammar, A.E. Whitfield and M.G. Redinbaugh. 2008. Insect vector interactions with persistently transmitted viruses. Annu. Rev. Phytopathol. 46: 327–359.

Huang, S., R. Li, Z. Zhang, L. Li, X. Gu, W. Fan et al. 2009. The genome of the cucumber, *Cucumis sativus* L. Nat. Genet. 41: 1275–1281.

Hurd, P., E. Linsley and T. Whitaker. 1971. Squash and gourd bees (Peponapis, Xenoglossa) and the origin of the cultivated *Cucurbita*. Evolution 25: 218–234.

Jóhannsson, M., M. Gates and A. Stephenson. 1998. Inbreeding depression affects pollen performance in *Cucurbita texana*. J. Evol. Biol. 11: 579–588.

Julier, H.E. and T.H. Roulston. 2009. Wild bee abundance and pollination service in cultivated pumpkins: farm management, nesting behavior and landscape effects. J. Econ. Entomol. 102: 563–573.

Kappers, I.F., F.W. Verstappen, L.L. Luckerhoff, H.J. Bouwmeester and M. Dicke. 2010. Genetic variation in jasmonic acid- and spider mite-induced plant volatile emission of cucumber accessions and attraction of the predator *Phytoseiulus persimilis*. J. Chem. Ecol. 36: 500–512.

Kappers, I.F., H. Hoogerbrugge, H.J. Bouwmeester and M. Dicke. 2011. Variation in herbivory-induced volatiles among cucumber (*Cucumis sativus* L.) varieties has consequences for the attraction of carnivorous natural enemies. J. Chem. Ecol. 37: 150–160.

Karban, R., L.H. Yang and K.F. Edwards. 2014. Volatile communication between plants that affects herbivory: a meta-analysis. Ecol. Lett. 17: 44–52.

Khalaf, E.M. and M.N. Raizada. 2016. Taxonomic and functional diversity of cultured seed associated microbes of the cucurbit family. BMC Microbiol. 16: 131.

Khan, Z., C.A. Midega, A. Hooper and J. Pickett. 2016. Push-pull: chemical ecology-based integrated pest management technology. J. Chem. Ecol. 42: 1–9.

Kistler, L., L.A. Newsom, T.M. Ryan, A.C. Clarke, B.D. Smith and G.H. Perry. 2015. Gourds and squashes (*Cucurbita* spp.) adapted to megafaunal extinction and ecological anachronism through domestication. Proc. Natl. Acad. Sci. USA 112: 15107–15112.

Kohn, J.R. and J.E. Biardi. 1995. Outcrossing rates and inferred levels of inbreeding depression in gynodioecious *Cucurbita foetidissima* (Cucurbitaceae). Heredity 75: 77–83.

Kremen, C., N.M. Williams and R.W. Thorp. 2002. Crop pollination from native bees at risk from agricultural intensification. Proc. Natl. Acad. Sci. USA 99: 16812–16816.

Krysan, J.L. 1986. Introduction: biology, distribution, and identification of pest *Diabrotica*. pp. 1–23. *In*: J.L. Krysan and T.A. Miller [eds.]. Methods for the Study of Pest *Diabrotica*. Springer Verlag, New York, NY, USA.

Krysan, J.L. and R. Smith. 1987. Systematics of the *virgifera* species group of *Diabrotica* (Coleoptera, Chrysomelidae, Galerucinae). Entomography 5: 375–484.

Lampman, R.L. and R.L. Metcalf. 1988. The comparative response of *Diabrotica* species (Coleoptera: Chrysomelidae) to volatile attractants. Environ. Entomol. 17: 644–648.

Lecoq, H. and N. Katis. 2014. Control of cucurbit viruses. Adv. Virus Res. 90: 255–296.

Lee, D.H., G.B. Iwanski and N.H. Thoennissen. 2010. Cucurbitacin: ancient compound shedding new light on cancer treatment. Sci. World J. 10: 413–418.

Lewis, P.A., R.L. Lampman and R.L. Metcalf. 1990. Kairomonal Attractants for *Acalymma vittatum* (Coleoptera: Chrysomelidae). Environ. Entomol. 19: 8–14.

Liang, D., M. Liu, Q. Hu, M. He, X. Qi, Q. Xu et al. 2015. Identification of differentially expressed genes related to aphid resistance in cucumber (*Cucumis sativus* L.). Sci. Rep. 5: 9645.

López-Uribe, M.M., J.H. Cane, R.L. Minckley and B.N. Danforth. 2016. Crop domestication facilitated rapid geographical expansion of a specialist pollinator, the squash bee *Peponapis pruinosa*. Proc. R. Soc. B. 283: 1833.

Mann, R.S., J.G. Ali, S.L. Hermann, S. Tiwari, K.S. Pelz-Stelinski, H.T. Alborn et al. 2012. Induced release of a plant-defense volatile 'deceptively' attracts insect vectors to plants infected with a bacterial pathogen. PLoS Pathog. 8: e1002610.

Mauck, K.E., C.M. De Moraes and M.C. Mescher. 2010. Deceptive chemical signals induced by a plant virus attract insect vectors to inferior hosts. Proc. Natl. Acad. Sci. USA 107: 3600–3605.

Mauck, K.E., N.A. Bosque-Pérez, S.D. Eigenbrode, C.M. De Moraes and M.C. Mescher. 2012. Transmission mechanisms shape pathogen effects on host-vector interactions: evidence from plant viruses. Funct. Ecol. 26: 1162–1175.

Mauck, K.E., C.M. De Moraes and M.C. Mescher. 2014. Evidence of local adaptation in plant virus effects on host–vector interactions. Integr. Comp. Biol. 54: 193–209.

Mauck, K.E., C.M. De Moraes and M.C. Mescher. 2014. Biochemical and physiological mechanisms underlying effects of cucumber mosaic virus on host-plant traits that mediate transmission by aphid vectors. Plant Cell Environ. 37: 1427–1439.

Mauck, K.E., C.M. De Moraes and M.C. Mescher. 2015a. Infection of host plants by cucumber mosaic virus increases the susceptibility of *Myzus persicae* aphids to the parasitoid *Aphidius colemani*. Sci. Rep. 5: 10963.

Mauck, K.E., E. Smyers, C.M. De Moraes and M.C. Mescher. 2015b. Virus infection influences host plant interactions with non-vector herbivores and predators. Funct. Ecol. 29: 662–673.

Mauck, K.E., C.M. De Moraes and M.C. Mescher. 2016. Effects of pathogens on sensory-mediated interactions between plants and insect vectors. Curr. Opin. Plant Biol. 32: 53–61.

McCloud, E.S., D.W. Tallamy and F.T. Halaweish. 1995. Squash beetle trenching behaviour: avoidance of cucurbitacin induction or mucilaginous plant sap? Ecol. Entomol. 20: 51–59.

McCormick, A.C., S.B. Unsicker and J. Gershenzon. 2012. The specificity of herbivore-induced plant volatiles in attracting herbivore enemies. Trend. Plant Sci. 17: 303–310.

Mena Granero, A., F.J. Egea Gonzalez, J.M. Guerra Sanz and J.L. Martínez Vidal. 2005. Analysis of biogenic volatile organic compounds in zucchini flowers: identification of scent sources. J. Chem. Ecol. 31: 2309–2322.

Mercke, P., I.F. Kappers, F.W. Verstappen, O. Vorst, M. Dicke and H.J. Bouwmeester. 2004. Combined transcript and metabolite analysis reveals genes involved in spider mite induced volatile formation in cucumber plants. Plant Physiol. 135: 2012–2024.

Metcalf, R.L. 1979. Plants, chemicals, and insects: some aspects of coevolution. Bull. Entomol. Soc. Am. 25: 30–35.

Metcalf, R.L. 1986. Coevolutionary adaptations of rootworm beetles (Coleoptera: Chrysomelidae) to cucurbitacins. J. Chem. Ecol. 12: 1109–1124.

Metcalf, R.L. and R.L. Lampman. 1989. The chemical ecology of Diabroticites and Cucurbitaceae. Experientia 45: 240–247.

Metcalf, R.L. and R.L. Lampman. 1991. Evolution of diabroticite rootworm beetle (Chrysomelidae) receptors for *Cucurbita* blossom volatiles. Proc. Natl. Acad. Sci. USA 88: 1869–1872.

Metcalf, R.L., R.A. Metcalf and A.M. Rhodes. 1980. Cucurbitacins as kairomones for diabroticite beetles. Proc. Natl. Acad. Sci. USA 77: 3769–3772.

Metcalf, R.L., A. Rhodes, R.A. Metcalf, J. Ferguson, E.R. Metcalf and P.-Y. Lu. 1982. Cucurbitacin contents and diabroticite (Coleoptera: Chrysomelidae) feeding upon *Cucurbita* spp. Environ. Entomol. 11: 931–937.

Metcalf, R.L., R.L. Lampman and L. Deem-Dickson.1995. Indole as an olfactory synergist for volatile kairomones for diabroticite beetles. J. Chem. Ecol. 21: 1149–1162.

Mitchell, R.F. and L.M. Hanks. 2009. Insect frass as a pathway for transmission of bacterial wilt of cucurbits. Environ. Entomol. 38: 395–403.

Munroe, D.D. and R.F. Smith. 1980. A revision of the systematics of *Acalymma* sensu stricto Barber (Coleoptera: Chrysomelidae) from North America including Mexico. Mem. Entomol. Soc. Can. 112: 1–92.

Ng, J.C. and B.W. Falk. 2006. Virus-vector interactions mediating nonpersistent and semipersistent transmission of plant viruses. Annu. Rev. Phytopathol. 44: 183–212.

Nielsen, J.K., L.M. Larsen and H. Søorensen. 1977. Cucurbitacin E and I in *Iberis amara*: feeding inhibitors for *Phyllotreta nemorum*. Phytochemistry 16: 1519–1522.

Nishida, R., M. Yokoyama and H. Fukami. 1992. Sequestration of cucurbitacin analogs by New and Old World chrysomelid leaf beetles in the tribe Luperini. Chemoecology 3: 19–24.

Piperno, D.R. and K.E. Stothert. 2003. Phytolith evidence for early Holocene *Cucurbita* domestication in southwest Ecuador. Science 299: 1054–1057.

Qi, J., X. Liu, D. Shen, H. Miao, B. Xie, X. Li et al. 2013. A genomic variation map provides insights into the genetic basis of cucumber domestication and diversity. Nat. Genet. 45: 1510–1515.

Quemada, H., L. Strehlow, D.S. Decker-Walters and J.E. Staub. 2008. Population size and incidence of virus infection in free-living populations of *Cucurbita pepo*. Environ. Biosafety Res. 7: 185–196.

Quesada, M., K. Bollman and A.G. Stephenson. 1995. Leaf damage decreases pollen production and hinders pollen performance in *Cucurbita texana*. Ecology 76: 437–443.

Rand, F.V. and E.M.A. Enlows. 1920. Bacterial Wilt of Cucurbits. USDA Bulletin. U.S. Department of Agriculture, Washington D.C., USA.

Rand, F.V. and L.C. Cash. 1920. Some insect relations of *Bacillus tracheiphilus* Erw. Sm. Phytopathology 10: 133–140.

Raupach, G.S. and J.W. Kloepper. 1998. Mixtures of plant growth-promoting rhizobacteria enhance biological control of multiple cucumber pathogens. Phytopathology 88: 1158–1164.

Rehm, S. and J. Wessels. 1957. Bitter principles of the cucurbitaceae. VIII. Cucurbitacins in seedlings—occurrence, biochemistry and genetical aspects. J. Sci. Food Agric. 8: 687–691.

Robinson, R.W., H.M. Munger, T.W. Whitaker and G.W. Bohn. 1976. Genes of the cucurbitaceae. Hortscience 11: 554–568.

Rojas, E.S., J.C. Batzer, G.A. Beattie, S.J. Fleischer, L.R. Shapiro, M.A. Williams et al. 2015. Bacterial wilt of cucurbits: resurrecting a classic pathosystem. Plant Dis. 99: 564–574.

Rowen, E. and I. Kaplan. 2016. Eco-evolutionary factors drive induced plant volatiles: a meta-analysis. New Phytol. 210: 284–294.

Rymal, K., O. Chambliss, M. Bond and D. Smith. 1984. Squash containing toxic cucurbitacin compounds occurring in California and Alabama. J. Food Prot. 47: 270–271.

Salvaudon, L., C.M. De Moraes and M.C. Mescher. 2013. Outcomes of co-infection by two potyviruses: implications for the evolution of manipulative strategies. Proc. R. Soc. B. 280: 20122959.

Samuelson, G.A. 1994. Pollen consumption and digestion by leaf beetles. pp. 179–183. *In*: P.H. Jolivet, M.L. Cox and E. Petitpierre [eds.]. Novel Aspects of the Biology of Chrysomelidae. Springer, Dordrecht, Netherlands.

Sanjur, O.I., D.R. Piperno, T.C. Andres and L. Wessel-Beaver. 2002. Phylogenetic relationships among domesticated and wild species of *Cucurbita* (Cucurbitaceae) inferred from a mitochondrial gene: implications for crop plant evolution and areas of origin. Proc. Natl. Acad. Sci. USA 99: 535–540.

Sasu, M.A., M.J. Ferrari, D. Du, J.A. Winsor and A.G. Stephenson. 2009. Indirect costs of a nontarget pathogen mitigate the direct benefits of a virus-resistant transgene in wild *Cucurbita*. Proc. Natl. Acad. Sci. USA 106: 19067–19071.

Sasu, M.A., I. Seidl-Adams, K. Wall, J.A. Winsor and A.G. Stephenson. 2010a. Floral transmission of *Erwinia tracheiphila* by cucumber beetles in a wild *Cucurbita pepo*. Environ. Entomol. 39: 140–148.

Sasu, M.A., K.L. Wall and A.G. Stephenson. 2010b. Antimicrobial nectar inhibits a florally transmitted pathogen of a wild *Cucurbita pepo* (Cucurbitaceae). Am. J. Bot. 97: 1025–1030.

Sasu, M.A., M.J. Ferrari and A.G. Stephenson. 2010c. Interrelationships among a virus-resistance transgene, herbivory, and a bacterial disease in a wild *Cucurbita*. Int. J. Plant Sci. 171: 1048–1058.

Schaefer, H., C. Heibl and S.S. Renner. 2009. Gourds afloat: a dated phylogeny reveals an Asian origin of the gourd family (Cucurbitaceae) and numerous oversea dispersal events. Proc. R. Soc. B. 276: 843–851.

Sebastian, P., H. Schaefer, I.R.H. Telford and S.S. Renner. 2010. Cucumber (*Cucumis sativus*) and melon (*C. melo*) have numerous wild relatives in Asia and Australia, and the sister species of melon is from Australia. Proc. Natl. Acad. Sci. USA 107: 14269–14273.

Shang, Y., Y. Ma, Y. Zhou, H. Zhang, L. Duan, H. Chen et al. 2014. Biosynthesis, regulation, and domestication of bitterness in cucumber. Science 346: 1084–1088.

Shapiro, L., C.M. De Moraes, A.G. Stephenson and M.C. Mescher. 2012. Pathogen effects on vegetative and floral odours mediate vector attraction and host exposure in a complex pathosystem. Ecol. Lett. 15: 1430–1438.

Shapiro, L.R., L. Salvaudon, K.E. Mauck, H. Pulido, C.M. De Moraes, A.G. Stephenson et al. 2013. Disease interactions in a shared host plant: effects of pre-existing viral infection on cucurbit plant defense responses and resistance to bacterial wilt disease. PLoS ONE 8: e77393.

Shapiro, L.R., I. Seidl-Adams, C.M. De Moraes, A.G. Stephenson and M.C. Mescher. 2014. Dynamics of short- and long-term association between a bacterial plant pathogen and its arthropod vector. Sci. Rep. 4: 4155.

Shapiro, L.R., E.D. Scully, D. Roberts, T.J. Straub, S.M. Geib, J. Park et al. 2015. Draft genome sequence of *Erwinia tracheiphila*, an economically important bacterial pathogen of cucurbits. GenomeA 3: e00482-00415.

Shapiro, L.R., E.D. Scully, T.J. Straub, J. Park, A.G. Stephenson, G.A. Beattie et al. 2016. Horizontal gene acquisitions, mobile element proliferation, and genome decay in the host-restricted plant pathogen *Erwinia tracheiphila*. Genome Biol. Evol. 8: 649–664.

Singh, S., A. Mishra, R. Bisen and Y. Malik. 2000. Host preference of red pumpkin beetle, *Aulacophora foveicollis* and melon fruit fly, *Dacus cucurbitae*. Indian J. Entomol. 62: 242–246.

Smith, B.D. 1989. Origins of Agriculture in Eastern North America. American Association for the Advancement of Science, Washington D.C., USA.

Smith, B.D. 1997. The initial domestication of *Cucurbita pepo* in the Americas 10,000 years ago. Science 276: 932–934.

Smith, B.D. 2006. Eastern North America as an independent center of plant domestication. Proc. Natl. Acad. Sci. USA 103: 12223–12228.

Smith, B.D., W.C. Cowan and M.P. Hoffman. 2007. Rivers of Change: Essays on Early Agriculture in Eastern North America. University of Alabama Press, Tuscaloosa, AL, USA.

Smith, E.F. 1920. An Introduction to Bacterial Diseases of Plants. W.B. Saunders Company, Philadelphia, PA, USA.

Smith, R.F. 1966. The distribution of Diabroticites in Western North America. Bull. Entomol. Soc. Am. 12: 108–110.

Smith, R.F. and J.F. Laurence. 1967. Clarification of the status of type specimens of Diabroticites (Coleoptera, Chrysomelidae, Galerucinae). Univ. Cal. Publ. Entomol. 45: 1–174.

Smyth, R.R. and M.P. Hoffmann. 2010. Seasonal incidence of two co-occurring adult parasitoids of *Acalymma vittatum* in New York State: *Centistes (Syrrhizus) diabroticae* and *Celatoria setosa*. BioControl 55: 219–228.

Spencer, L.J. 2001. Fecundity of transgenic wild-crop hybrids of *Cucurbita pepo* (Cucurbitaceae): implications for crop-to-wild gene flow. Heredity 86: 694–702.

Stephenson, A.G., B. Leyshon, S.E. Travers and C.N. Hayes. 2004. Interrelationships among inbreeding, herbivory, and disease on reproduction in a wild gourd. Ecology 85: 3023–3034.

Stoewsand, G.S., A. Jaworski, S. Shannon and R.W. Robinson. 1985. Toxicologic response in mice fed *Cucurbita* fruit. J. Food Prot. 48: 50–51.

Takabayashi, J., M. Dicke and M.A. Posthumus. 1994. Volatile herbivore-induced terpenoids in plant-mite interactions: variation caused by biotic and abiotic factors. J. Chem. Ecol. 20: 1329–1354.

Tallamy, D.W. 1985. Squash beetle feeding behavior: an adaptation against induced cucurbit defenses. Ecology 66: 1574–1579.

Tallamy, D.W. and P.M. Gorski. 1997. Long- and short-term effect of cucurbitacin consumption on *Acalymma vittatum* (Coleoptera: Chrysomelidae) fitness. Environ. Entomol. 26: 672–677.

Tallamy, D.W., D.P. Whittington, F. Defurio, D.A. Fontaine, P.M. Gorski and P.W. Gothro. 1998. Sequestered cucurbitacins and pathogenicity of *Metarhizium anisopliae* (Moniliales: Moniliaceae) on spotted cucumber beetle eggs and larvae (Coleoptera: Chrysomelidae). Environ. Entomol. 27: 366–372.

Tallamy, D.W., P.M. Gorski and J.K. Burzon. 2000. Fate of male-derived cucurbitacins in spotted cucumber beetle females. J. Chem. Ecol. 26: 413–427.

Tepedino, V.J. 1981. The pollination efficiency of the squash bee (*Peponapis pruinosa*) and the honey bee (*Apis mellifera*) on summer squash (*Cucurbita pepo*). J. Kans. Entomol. Soc. 54: 359–377.

Theis, N. and L.S. Adler. 2012. Advertising to the enemy: enhanced floral fragrance increases beetle attraction and reduces plant reproduction. Ecology 93: 430–435.

Theis, N., K. Kesler and L.S. Adler. 2009. Leaf herbivory increases floral fragrance in male but not female *Cucurbita pepo* subsp. *texana* (Cucurbitaceae) flowers. Am. J. Bot. 96: 897–903.

Torkey, H., H. Abou-Yousef, A. Abdel Azeiz and E. Hoda. 2009. Insecticidal effect of cucurbitacin E glycoside isolated from *Citrullus colocynthis* against *Aphis craccivora*. Aust. J. Basic Appl. Sci. 3: 4060–4066.

Tricoll, D.M., K.J. Carney, P.F. Russell, J.R. McMaster, D.W. Groff, K.C. Hadden et al. 1995. Field evaluation of transgenic squash containing single or multiple virus coat protein gene constructs for resistance to cucumber mosaic virus, watermelon mosaic virus 2, and zucchini yellow mosaic virus. Nat. Biotechnol. 13: 1458–1465.

Vlot, A., D. Dempsey and D. Klessig. 2011. Salicylic acid, a multifaceted hormone to combat disease. Annu. Rev. Phytopathol. 47: 177–206.

Walters, S.A. and B.H. Taylor. 2006. Effects of honey bee pollination on pumpkin fruit and seed yield. HortScience 41: 370–373.

Watt, J.M. and M.G. Breyer-Brandwijk. 1962. The Medicinal and Poisonous Plants of Southern and Eastern Africa. E. & S. Livingstone Ltd., Edinburgh, UK.

Westwood, J.H., L. Mccann, M. Naish, H. Dixon, A.M. Murphy, M.A. Stancombe et al. 2013. A viral RNA silencing suppressor interferes with abscisic acid-mediated signalling and induces drought tolerance in *Arabidopsis thaliana*. Mol. Plant Pathol. 14: 158–170.

Whitaker, T.W. and W.P. Bemis. 1964. Evolution in the genus *Cucurbit*a. Evolution 18: 553–559.

Whitaker, T.W. and W.P. Bemis. 1975. Origin and evolution of the cultivated *Cucurbita*. Bull. Torrey Bot. Club 102: 362–368.

Xu, P., F. Chen, J.P. Mannas, T. Feldman, L.W. Sumner and M.J. Roossinck. 2008. Virus infection improves drought tolerance. New Phytol. 180: 911–921.

Yao, C., G. Zehnder, E. Bauske and J. Kloepper. 1996. Relationship between cucumber beetle (Coleoptera: Chrysomelidae) density and incidence of bacterial wilt of cucurbits. J. Econ. Entomol. 89: 510–514.

Yesilada, E. 2005. Past and future contributions to traditional medicine in the health care system of the Middle-East. J. Ethnopharmacol. 100: 135–137.

Zehnder, G., J. Kloepper, S. Tuzun, C. Yao, G. Wei, O. Chambliss et al. 1997a. Insect feeding on cucumber mediated by rhizobacteria-induced plant resistance. Entomol. Exp. Appl. 83: 81–85.

Zehnder, G., J. Kloepper, C. Yao and G. Wei. 1997b. Induction of systemic resistance in cucumber against cucumber beetles (Coleoptera: Chrysomelidae) by plant growth-promoting rhizobacteria. J. Econ. Entomol. 90: 391–396.

CHAPTER

4

Chemoecology and Behavior of Parasitic Nematode—Host Interactions: Implications for Management

Denis S. Willett[1], Xavier Martini[2], and Lukasz L. Stelinski[3]*

[1] Citrus Research and Education Center, Entomology and Nematology Department, University of Florida, Lake Alfred, Florida 33850, USA
(Present Address)
Center for Medical, Agricultural, and Veterinary Entomology, Agricultural Research Service, U.S. Department of Agriculture, Gainsesville, Florida 32608, USA
E-mail: dwillettuf@gmail.com

[2] Citrus Research and Education Center, Entomology and Nematology Department, University of Florida, Lake Alfred, Florida 33850, USA
(Present Address)
North Florida Research and Education Center, Entomology and Nematology Department, University of Florida, Quincy, Florida 33850, USA
E-mail: xmartini@ufl.edu

[3] Citrus Research and Education Center, Entomology and Nematology Department, University of Florida Center, Lake Alfred, Florida 33850, USA
email: stelinski@ufl.edu

Abstract

The chemical ecology of insects, plants, vertebrates and nematodes is closely intertwined. These animals occur in complex ecosystems and in some cases become pests of humans. Nematode parasites inhabit almost

*Corresponding author

every ecosystem on the planet and infect insects, plants, and vertebrates. All nematode parasites undergo a host infection phase, reproduce, disperse and find new hosts. Transitions within and between life stages are mediated by communication among nematodes and between nematodes and their environment. This communication is conserved—similar blends of molecules regulate mate finding, life stage transitions and dispersal. Furthermore, this communication occurs in a social context and it appears nematodes can even learn from other species to adapt to dynamic environments. Knowledge of nematode communication can be used in conjunction with their learning ability to disrupt their lifecycle. This may enhance infection of insect parasitic (entomopathogenic) nematodes for biological control and/or prevent infection by vertebrate or plant parasitic nematodes.

1. Nematodes as Parasites

When addressing insect-microbe interactions, it is nearly impossible not to consider nematodes when investigating the chemical ecology of insect ecosystems. Nematodes or roundworms are numerous and ubiquitous. There are thirty thousand described species, but it is estimated that there are still one million undescribed species. Consequently, nematodes could be the second most important group beneath the Arthropods (Hugot et al. 2001). They inhabit every environment on earth and it is said that if all matter besides nematodes were removed, we would still be left with a physical representation of our planet (Cobb 1915). While 40% of the described species are free living nematodes, such as the free living bactivorous nematode, *Caenorhabditis elegans*, the majority of described nematodes are parasites of plants, vertebrates or insects (Hugot et al. 2001, Félix and Braendle 2010). Parasitism has evolved multiple times within Nematoda (Blaxter 2003). Plant, vertebrate and insect nematode parasites are instrumental to many of the critical processes we require for sustaining life on our planet and can cause and resolve problems affecting billions of people worldwide. Plant parasitic nematodes can transmit viruses (Brown et al. 1995, Lamberti et al. 2012) and through direct feeding cause agricultural crop losses upwards of 100 billion USD annually worldwide (Singh et al. 2013). Vertebrate parasitic nematodes cause devastating illnesses such as filariasis or ascariasis and losses of livestock and humans. Over three billion people are infected with vertebrate parasitic nematodes that induce symptomes ranging from lack of energy and vigor to blindness and malformations (Stoll 1999, Jasmer et al. 2003, Castillo et al. 2011). On the other hand, insect parasitic nematodes (entomopathogenic nematodes) are effective biocontrol agents and hold great promise for the study of nematode parasitic systems (Kaya and Gaugler 1993, Gaugler and Kaya 1990).

Traditionally, the study of nematode parasitology has been isolated and segregated. The free living and bacteriophagous nematode, *C. elegans*, has been studied as a model organism by geneticists; plant parasitic nematodes are often studied by plant pathologists; vertebrate parasitic nematodes are often the focus of veterinary medicine researchers; human parasitic nematodes are a common subject in medical schools; while entomopathogenic nematodes are often investigated by nematologists, entomologists and now chemical ecologists interested in pest management. While these endeavors have produced significant insights into the differences between various nematode parasites (Jasmer et al. 2003) and the evolutionary ecology of parasitism, we suggest that broader conclusions and understanding have been slow to make it across 'departmental' divides. Instead of discussing the differences between plant, vertebrate and entomopathogenic nematode parasites—what makes them unique—we hope to initially highlight the similarities in the parasitic lifestyle: the life stages, behaviors and activities that all parasitic nematodes undertake irrespective of host. In highlighting those similarities, we point to new research suggesting that such similarities might be stronger than comparable differences. Additionally, and possibly more importantly, the similarities in lifecycle, communication between nematodes, learning ability and observed aggregation can be used for developing management strategies broadly applicable to all parasitic nematodes; what works well for one species may work well or even better for others. General strategies may affect many parasitic species simultaneously and specifically without broad, non-target and undesirable environmental consequences.

1.1 Parasitic Lifecycle

The core similarity between all plant, vertebrate and entomopathogenic nematodes is their lifecycle. While there are individual differences among groups and species, those differences can be considered as variations on a similar theme: all parasitic nematodes undergo four key life stages in their quest for proliferation (Fig. 1). Principally, all parasitic nematodes must infect their host. To do so, they must enter the host or pass through some sort of physical barrier, whether it be the plant cuticle, animal skin or insect cuticle. For vertebrate and entomopathogenic nematodes, this can also occur by entering through host orifices, such as, in the case of entomopathogenic nematodes, spriracles or through ingestion in the case of vertebrate parasites (Kaya and Gaugler 1993). After breaching the barrier, to complete infection, they must overcome the host immune response. This can be done by release of toxins, overwhelming the immune response with large numbers (Wang et al. 1994) or through downregulation of the host immune system (McSorley and Maizels 2012). In addition to infection, nematodes must reproduce, usually within the host. This

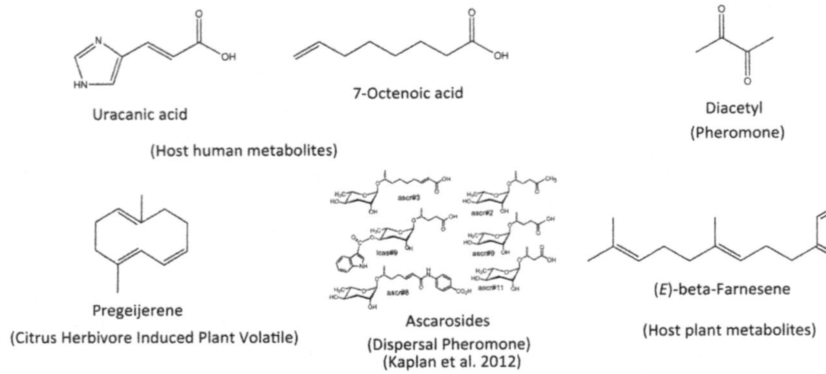

Fig. 1. Representative structures of nematode semiochemicals.

necessitates mate-finding and communication (Simon and Sternberg 2002). After infection, and often after reproduction, the nematodes must leave the host and disperse. This occurs for entomopathologenic nematodes when the host becomes unsuitable for maintaining additional parasites or becomes weakened to the point at which resources for the parasites become unavailable (Kaplan et al. 2012). This dispersal often results in a free-living phase where the nematodes reside in an external environment until finding a new host. For many parasitic nematodes, this free-living phase occurs in soil pore spaces where the nematodes must navigate a dynamic and stimulus rich environment while searching for a host. It is during this free living phase when most parasitic nematodes initiate host finding behavior. Strategies vary, some entomopathogenic nematodes are sit and wait ambush predators while others are cruisers that actively seek out hosts for parasitism (Lewis et al. 2006). Irrespective of strategy, during the host finding stage nematodes are highly attuned to host cues and signals, whether they may be derived from plants, vertebrates, insects or interactions among those three trophic levels (Lewis et al. 1993).

2. Nematode Chemical Communication

Transitions to, from and within these four primary life stages are all mediated by communication, often chemical communication, between and within nematodes, their hosts and their environment (Table 1, Fig. 1). While there are specific differences in how this communication is carried out, in many instances and indeed with many of the chemicals, communication is conserved among species and environments.

Host finding behavior is regulated by nematode-host-environment communication. Many vertebrate parasitic nematodes demonstrate thermo and chemotactic host finding behavior recruiting to warmer stimuli and using CO_2 and/or host specific chemicals to locate hosts (Granzer and Hass 1991, Castelletto et al. 2014). These chemicals may be nonvolatile and exuded by the host; the human pathogenic *Strongyloides stercoralis* responds to sodium chloride gradients (Forbes et al. 2003) and *S. ratti* responds to serum proteins (Koga and Tada 2000). In addition, vertebrate parasitic nematodes recruit to host specific chemicals; urocanic acid from skin extracts and human specific 7-octenoic acid attract *S. stercoralis* (Castelletto et al. 2014, Safer et al. 2007). Environmental cues from other species also play significant roles in host finding behavior; the ruminant parasite *Haemonchus contortus* responds to grass odors (Castelletto et al. 2014). Likewise, entomopathogenic nematodes use CO_2, host and environmental cues to locate insect larvae (Dillman et al. 2012, Lewis et al. 1992). Indeed, the role of environmental cues, particularly herbivore induced plant volatiles (HIPVs) released by plants in response to herbivory by the insect host is well established for entomopathogenic nematodes in a variety of systems. *Heterorhabditis megidis* Poinar, Jackson, and Klein responds to HIPVs released by the white cedar *Thuja occidentalis* (van Tol et al. 2001) and (*E*)-β-caryophyllene released by maize (Rasmann et al. 2005), while a wide variety of entomopathogenic nematodes including *Steinernema* and *Heterorhabditis* species are attracted to the HIPV, pregeijerene, released by citrus (Ali et al. 2010). Finally, plant parasite nematodes are also attracted by CO_2 that is emitted by roots, as well as, root exudates (Bird 1959, Robinson 1995, Zuckerman and Jansen 1984, Perry 1996, Xu et al. 2015). Additionally, more specific abiotic cues might be used by these nematodes to find their host. Interestingly, the plant parasite nematode, *Meloidogyne hapla,* is also attracted toward acidic gradients produced by growing root cells that extrude protons to acidify the region around the root elongation zone (Wang et al. 2009). Therefore, this acid gradient is a more specific cue for root finding than CO_2.

Cue specificity is of critical importance and from the examples above, CO_2 is a common attractant for most parasitic nematodes irrespective of their host (insect, plant or vertebrate). It demonstrates the ubiquitousness of CO_2 and suggests that CO_2 is not a reliable source to find a host because it is emitted by such a diversity and quantity of organisms, including non-hosts. Suitable hosts should specifically emit the most reliable cues. Therefore, it could be suggested that CO_2 consists of a response activator that alerts parasitic nematodes to the general presence of living organisms and works in association with more specific cues. Turlings et al. (2012) tested this hypothesis and demonstrated that CO_2 and induced plant volatiles acted synergistically to attract the entomopathogenic nematode, *H. megidis*. To our knowledge, synergy between CO_2 and host volatiles has not been tested for vertebrate or plant parasitic nematodes and should be the focus of further investigation.

Table 1. Examples of chemical communication in insect, plant and vertebrate parasitic nematodes. See representative structures in Fig. 1.

Nematode	Status	Host	Chemical	Source of chemical	Role	References
Caenorhabditis elegans	Free living, bacteriovorous		Diacetyl Ascarosides	Nematode	Attractant Life stage transition Dispersal pheromone Mating pheromone	Bone and Shorey 1978, Jaffe et al. 1989, Meyer and Huetell 1996, Simon and Sternberg 2002, McSorley and Maizels 2012
Steinernema spp.	Insect parasite	Beetle larvae Crickets	Ascaroside	Nematode	Dispersal pheromone	Simon and Sternberg 2012
			Pregeijerene	Herbivore-induced plant volatile	Attractant	Rasmann et al. 2005
			Host derived odorants	Host	Attractant	Safer et al. 2007
Heterorhabditis spp.	Insect parasite	Beetle larvae Caterpillars	Ascaroside	Nematode	Dispersal pheromone	Simon and Safer 2002
			Pregeijerene	Herbivore-induced plant volatile	Attractant	Rasmann et al. 2005
			(*E*)-β-Caryophyllene	Herbivore-induced plant volatile	Attractant	van Tol et al. 2001

(*Contd.*)

Meloidogyne spp.	Plant parasite	Cotton Soy Peanut Solanums Fruit	Ascaroside	Nematode	Dispersal pheromone	Simon and Safer 2002
			Root exudate	Host	Attractant	Ali et al. 2011
			CO_2	Host	Attractant	Bird 1959
Tylenchulus semipenetrans	Plant parasite	Citrus	Pregeijerene	Host	Attractant	Ali et al. 2010
Ditylenchus destructor	Plant parasite	Potatoes	Root exudates	Host	Attractant	Xu et al. 2015
Ancylostoma caninum	Vertebrate parasite	Dogs	CO_2	Host	Attractant	Lewis et al. 1993
Strongyloides stercoralis	Vertebrate parasite	Humans	Uracanic acid	Host	Attractant	Koga and Tada 2000
			7-Octenoic acid	Host	Attractant	Granzer and Haas 1991
Strongyloides ratti	Vertebrate parasite	Rat	Rat serum albumin	Host	Attractant	Forbes et al. 2003
Haemonchus contortus	Vertebrate parasite	Ruminant	Grass odor	Food of hosts	Attractant	Granzer and Haas 1991

Host infection also follows similar patterns for insect, plant and vertebrate parasitic nematodes and is primarily regulated by host-nematode communication. The utility of this communication, from the nematode perspective, is regulation of the host immune system to facilitate successful infection and proliferation. Much of this communication occurs at the nematode surface coat (Davies and Curtis 2011). Vertebrate parasitic nematode surface coat secretions can combat oxyradical attacks, suppress immune cell proliferation (both Th1 and Th2 cells in humans) and modulate Toll-like receptor signaling and interactions (Allen and MacDonald 1998, Davies and Curtis 2011, McSorley et al. 2013, Reynolds et al. 2015). Similarly, entomopathogenic nematode surface coat secretions suppress immune response of host insects preventing encapsulation and regulating enzymatic cascades such as the prophenoloxidase-activating system in *Galleria mellonella* (Wang and Gaugler 1999, Brivio et al. 2002, Li et al. 2007). Likewise, plant parasitic nematodes use similar strategies to facilitate successful long term infection; surface coat secretions regulate enzymatic pathways and contain antioxidative enzymes (Davies and Curtis 2011, Goverse and Smant 2014).

In addition to relying on surface coat secretions to mediate nematode-host interactions, nematodes may utilize host-endosymbiotic bacteria to interface with the host immune system. Entomopathogenic nematodes, for example, use highly specific endosymbionts to regulate host immune response and kill their insect larval hosts (McSorley and Maizels 2012). The role of bacteria in mediating interactions between plant or vertebrate parasitic nematodes and their host is little explored, but bears investigation (See Outstanding Questions: Section 5). Given the close phylogenetic relationship between *Strongyloides* and the entomopathogenic nematodes in *Steinernema*, the dual origin of bacterial endosymbiosis in insect parasite clades and the critical role of *Wolbachia* in development of verterbrate parasitic nematodes, bacteria may play a large role in mediating nematode parasite host infection (Dorris et al. 1999).

Following host infection, reproduction becomes a priority for parasitic nematodes. In this stage, nematode-nematode communication via pheromones becomes paramount for regulating mate finding and life stage transitions. A wide range of vertebrate, plant and entomopathogenic nematodes have demonstrated pheromone mediated attraction (Bone and Shorey 1978). While some nematode sex pheromones have been found to be acid or fatty-acid based (Huettel 1986, Jaffe et al. 1989, Meyer and Huettel 1996), blends of dideoxy sugar derived ascarosides play a prominent role as pheromones regulating mate-finding and life stage development (Butcher et al. 2007, Srinivasan et al. 2008, Edison 2009). Secretion of these ascaroside blends depends on diet and varies over the life of a nematode; older nematodes, adults in particular, tend to produce larger ascaroside quantities (Kaplan et al. 2011). The use of ascarosides as nematode pheromones is widely conserved; many vertebrate and entomopathogenic nematodes utilize ascaroside blends for signaling (Choe et al. 2012). Plant parasitic nematodes also likely use ascaroside blends to regulate life stage transitions (Bird et al. 2009).

Indeed, equally as important as mate finding for nematodes is life stage transition, principally into the infective stage in which the nematodes disperse to seek new hosts. This stage is analogous to the dauer stage in free living nematodes such as *C. elegans* and is the third larval stage in vertebrate parasitic nematodes. Similarly, nematodes in this stage are called infective juveniles in entomopathogenic nematodes and second stage juveniles in plant parasitic nematodes. While ascaroside blends regulate entry into this lifestage, they also regulate behavior, inducing dispersal away from the host at the commencement of the free living stage of the lifecycle (Kaplan et al. 2012). Indeed, this nematode-nematode pheromone communication may also play a role after nematode dispersal as infective juveniles enter the host finding phase and aggregate to overcome host immune defenses where nematodes follow one another to identify and infect hosts (Fushing et al. 2008, El-Borai et al. 2012, Shapiro-Ilan et al. 2014). Similar to the conserved nature of ascaroside signaling for life cycle development, aggregative signaling may also be widely conserved. There is evidence of interspecific following behavior in entomopathogenic nematodes; *Heterorhabditis indica* tend to follow *Steinernema diaprepesi* responding to learned volatile preferences (Willett et al. 2015). This following behavior may result from conserved signaling among nematodes and could have evolved to allow multiple species to take advantage of available subterranean resources. This pattern of conserved communication is likely to be found among other parasitic nematodes as well and could be adapted for management purposes.

The use of diverse ascarosides assembled from glycoside building blocks in nematodes emphasizes the important role of small molecule signaling in mediating communication (Schroeder 2015). The modular nature of these molecules engenders the ability to encode a vast array of information via simple, readily constructed changes to functional groups (Srinivasan et al. 2012). Given the conserved nature of these signaling mechanisms across nematode genera (Choe et al. 2012) and the ability for other organisms to perceive these signals (Manosalva et al. 2015), further research into the roles and mechanisms behind nematode derived modular metabolite signaling will yield a plethora of insights into managing parasitic nematode behavior.

2.1 Nematode Behavior: from Passive Aggregation to Learning

Aggregative behavior has long been observed in vertebrate, plant and insect parasitic nematodes. Vertebrate parasitic nematodes aggregate within and among hosts (Conway et al. 1995, Grenfell et al. 1995, Shaw et al. 1998), plant parasitic nematodes demonstrate aggregation in the field (Boag and Topham 1984, Noe and Campbell 1985) and entomopathogenic nematodes demonstrate aggregative distributions in the field and within hosts (Stuart and Gaugler 1994, Westermann 1999, Stuart et al. 2006). Indeed, aggregative behavior is a defining characteristic of many parasitic nematodes (Poulin and Morand 2000). Aggregation is particularly advantageous for parasitic nematodes during host infestation, as increasing numbers of nematodes attacking the same host reduces the defenses of the host more efficiently (Morill and Forbes 2015). This advantage is likely to overcome the costs of aggregation such as suboptimal hosts and/or higher host mortality (Morill and Forbes 2015).

Nematodes can also learn to respond to signals in their environment and from each other. Despite having a limited number of neurons [*C. elegans* possesses only 302 (Hobert 2003)], nematodes demonstrate a remarkable capability for behavioral plasticity in response to thermal, mechanical and chemical cues (Hobert 2003, Sasakura and Mori 2013). *C. elegans* will return to its rearing temperature when presented with a thermal gradient (Hedgecock and Russell 1975) and habituate to tapping (Rose and Rankin 2001). In addition to learning responses to a single stimulus, nematodes are adept at associative learning between paired stimuli and migrate to specific concentrations of cAMP, lysine and sodium chloride previously associated with food or rearing conditions (Bargmann and Horvitz 1991, Nuttley et al. 2002, Sasakura and Mori 2013). *C. elegans* recruits across temperature and salt gradients presented in conjunction with an *Escherichia coli* food source (Sasakura and Mori 2013). While nematode learning has been primarily explored using *C. elegans* as a model system, entomopathogenic nematodes also demonstrate a remarkable ability for behavioral plasticity in response to volatile cues and even demonstrate a capability for long and short term memory, recognising ephemeral cues in their environment (Willett et al. 2015). These abilities are not merely innate responses to environmental stimuli; control nematode cohorts reared under the same conditions do not exhibit the same behavioral responses as those exposed to a conditioning stimulus (Willett et al. 2015).

2.2 Social Behavior

Learning can also occur in a social context, nematodes can learn from each other and from members of a different species (Willett et al. 2015).

Aggregative social behavior has long been observed in vertebrate, plant and insect parasitic nematodes. Indeed, aggregative behavior is a defining characteristic of many parasitic nematodes (Hobert 2003). Whether because of variation in environmental resources (Hobert 2003, Wang et al. 2009) or nematode-nematode communication conferring infection advantages (Gaugler and Kaya 1990, Choe et al. 2012), the end result is that nematodes spend substantial portions of their lifecycle in close proximity to one another inter or intra-specifically.

The consequences of this aggregative behavior are just beginning to be understood. No longer can parasites be treated as individual pre-programmed entities with innate preferences; rather, nematode parasites exist in a dynamic social network that influences all aspects of their behavior and lifecycle, from feeding (de Bono et al. 2002) to reproduction, dispersal and host finding. Communication and behavioral plasticity in a social context allows parasitic nematodes to rapidly adapt to dynamic environments, surviving environmental stressors (de Bono et al. 2002, Artyukhin et al. 2015) and taking advantage of ephemeral opportunities, such as swarming to resource rich environs (Hollis 1962, McBride and Hollis 1966) in a manner impossible to achieve through genetic evolutionary adaptation alone (Galef and Laland 2005).

2.3 Case Study I: Enhancing Biological Control of Root-feeding Insect Pests of Citrus and Blueberry by Manipulating Native Communities of Entomopathogenic Nematodes with Semiochemicals

Ali et al. (2010, 2011, 2012) postulated that biological control activity of a broad range of native entomopathogenic nematode species may be enhanced through application of HIPVs to subterranean root zones of citrus and blueberry to control belowground feeding by insect herbivores of economic importance. HIPV emissions benefit plants by recruiting natural enemies of herbivorous insects and such multitrophic interactions have been thoroughly examined in the above ground terrestrial environment (Tumlinson et al. 1992, Turlings and Ton 2006). It has become increasingly evident that similar multitrophic interactions occur below ground (Rasmann et al. 2005, Turlings et al. 2012). The larvae of the root weevil, *Diaprepes abbreviatus*, are serious pests of citrus (Simpson et al. 2000). Entomopathogenic nematodes have varying and unpredictable, efficacy in controlling the weevil and interactions between the plant, insect and nematode have been poorly understood (Duncan et al. 1999). In a root zone bioassay, root weevil infested rootstock (Swingle citrumelo) recruited significantly more entomopathogenic nematodes (*S. diaprepesi*) than non-infested or mechanically damaged roots or larvae alone (Ali et al. 2010). By dynamic *in situ* collection and gas chromatography-mass spectroscopy (GC-MS) analysis of volatiles from soil, in combination with a two choice sand-column bioassay, it was found that Swingle citrus roots release induced volatiles in response to herbivore feeding and that some of these induced volatiles function as attractants for entomopathogenic nematodes (Ali et al. 2011).

Subsequently, Ali et al. (2011) examined the extent to which below ground recruitment cues modify behavior of nematode species representing various foraging strategies and trophic levels. They compared attraction to extracts of infested roots and non-infested roots from hybrid, Swingle citrus rootstock and a parent line of the hybrid, *Poncirus trifoliata*. Swingle roots infested by weevils attracted more nematodes than non-infested roots irrespective of nematode foraging strategy and trophic status. The parental line of the Swingle rootstock, *P. trifoliata*, attracted all nematode species irrespective of insect herbivory. Dynamic *in situ* collection and GC-MS analysis of soil volatiles revealed that *P. trifoliata* roots released recruitment signals constitutively, regardless of weevil feeding. A different non-hybrid citrus species (Sour orange, *Citrus* × *aurantium*) released nematode recruitment cues only in response to larval feeding. Volatile collections from above/below ground portions of citrus plants revealed that above ground feeding does not induce production of nematode recruitment cues analogous to that induced by root damage nor does damage by larvae below ground induce a similar cue above ground. These results suggested that constitutive release of nematode attractants by citrus roots occurs broadly and can be constant or herbivore-induced. The major constituent of this indirect cue is produced by roots, not shoots, in response to below ground, not above ground herbivory. These findings suggest that this recruitment cue acts on nematode species broadly, attracting both entomopathogenic and plant parasitic nematodes.

Finally, Ali et al. (2012) identified pregeijerene (1,5-dimethylcyclodeca-1, 5, 7-triene) as the main constituent released upon weevil damage to roots. This entomopathogenic nematode attracting HIPV released by citrus roots peaked 9-12 hours after initiation of larval root feeding. In field assays, lab-collected pregeijerene attracted native Florida entomopathogenic nematode species and increased mortality of beetle larvae compared with controls. In addition, field applications of 8 ng/μl of pregijerene isolated from the common rue *Ruta graveolens* caused similar results. Field collections of root volatiles from mature citrus trees proved that pregeijerene is released in the field in response to herbivore damage. Using species-specific probes designed to identify entomopathogenic nematode species, Ali et al. (2012) determined by quantitative real-time polymerase chain reaction (PCR) that field application of pregeijerene increased pest mortality by attracting four species of entomopathogenic nematodes native to Florida, as well as, native entomopathogenic nematode species in blueberry in New Jersey. Combined, these investigations identified in real-time a specific belowground 'cry for help' released by citrus roots that attracts natural enemies of polyphagous root pests. The unusual 12-carbon terpene, pregeijerene or its thermal breakdown product geijerene, has been found in root extracts of several plants, but an ecological role of this natural plant product had not been described previously. This and similar chemicals may have broad application for controlling agriculturally significant insect root pests by enhancing the activity of entomopathogenic nematodes.

2.4 Case Study II: Increasing the Efficacy of Entomopathogenic Nematodes through Pre-conditioning to HIPV Cues Induced by Subterranean Insect Herbivory

Using two-choice olfactometers, Willett et al. (2015) assayed preferences of entomopathogenic nematodes (*S. diaprepesi* and *H. indica*) to volatiles commonly present in the citrus rhizosphere both before and after exposure to those volatiles. Prior to exposure, both species did not display preferences for (*E*)-β-caryophyllene or (+)-limonene, but did innately prefer pregeijerene. Exposure to volatiles altered their behavior (Willett et al. 2015). Exposure to (*E*)-β caryophyllene resulted in preference for this HIPV for both species. Also, exposure to (+)-limonene resulted in preference for (+)-limonene for both species. In nematode species that displayed a high intrinsic preference for a volatile, exposure to that volatile had little effect on behavior. However, if nematodes were intrinsically repelled by a volatile (e.g. α-pinene), exposure to that volatile resulted in a large positive behavioral change (Willett et al. 2015). Furthermore, exposure altered preference for other volatiles to which the nematodes had not been exposed. Specifically, preference for exposed volatiles increased, while preference for volatiles to which they had not been exposed decreased (Willet et al. 2015).

While the ability to demonstrate specific behavioral plasticity in response to volatile exposure should benefit nematodes for adapting to dynamic environments, the rapidity with which nematodes learn and the degree to which they remember may explain their ability to respond to persistent and ephemeral signals in their environment. To quantify the effect of the interaction between exposure time (learning time) and post exposure time (memory) on volatile preference, Willett et al. (2015) used a D-optimal exposure x post exposure response surface design sufficient to detect quadratic curvature for both factors. *S. diaprepesi* infective juveniles were exposed from 1 to 48 hours (exposure time) to 1 μg/ml of α-pinene, then washed and placed in water for an additional 1 to 48 hours (post exposure time) prior to use in two choice behavior assays with α-pinene vs. a blank control. A preference index (I_P; the ratio between number of infective juvenile nematodes responding to α-pinene and the number responding to a blank control; repellence: $I_p < 1$; preference $I_p > 1$) was constructed for each assay to facilitate comparisons between assays. Both exposure time, post exposure time and their interaction affected behavioral plasticity (Willett et al. 2015). At low levels of exposure, α-pinene was repellent ($I_p < 1$) and not affected by increasing post exposure time. At high levels of exposure, α-pinene was highly attractive ($I_p > 1$) at low post exposure time, but preference declined with increasing post exposure time (Willett et al. 2015). The ability to both gain and lose preference for exposed volatiles over a period of 48 hours suggests that nematodes respond to relatively ephemeral cues in their environment. However, the interaction between exposure and post exposure time suggests an ability to differentiate certain (perhaps important) signals with persistent exposure causing a 'longer-term' memory.

Furthermore, to determine the extent of social behavioral plasticity in entomopathogenic nematodes, Willett et al (2015) initially assayed α-pinene preferences for both non exposed and exposed (to 1 µg/ml α-pinene for 48 hours) *S. diaprepesi* and *H. indica* independently in two choice olfactometers. Thereafter, they combined either 100 non exposed or 100 exposed *S. diaprepesi* infective juveniles simultaneously with 2500 non exposed *H. indica*. In independent trials, α-pinene was initially repellent. Exposure to α-pinene reversed repellency and increased preference for the exposed volatile. In combined simultaneous trials, non exposed *S. diaprepesi* and non exposed *H. indica* were repelled by α-pinene. However, when exposed *S. diaprepesi* were paired with non exposed *H. indica*, both species exhibited a preference for α-pinene. Non exposed *H. indica* altered their preference for α-pinene when paired with exposed *S. diaprepesi* suggesting a 'social transference' of behavioral plasticity from exposed to non exposed cohorts of nematode species.

The social transference of behavioral plasticity not only from exposed to non exposed cohorts, but also across entomopathogenic nematode species, suggests a broad role for social behavioral plasticity in below ground communities. Individual experiences can affect future behavior on the part of an individual. As individuals learn to prefer certain cues at the expense of others while retaining and losing that preference over time, the ability to transfer that experience interspecifically suggests a mechanism for rapid adaptation of communities to dynamic below ground environments. This ability to rapidly learn based not only on past experience, but also from individuals of other species may allow communities an adaptive response not available genetically alone. This rapidly adaptive behavior may suggest selection for such traits in an environment permeated with ephemeral cues and signals.

Previously, it was suggested that semiochemical cues that attract entomopathogenic nematodes may be exploited for management of root-zone pests of agricultural crops with biological control (Rasmann et al. 2005). The investigation by Willet et al (2015) suggests that this tactic could be more effective than previously thought given the social transfer of learned behavior between multiple nematode species. In other words, an attractant for one entomopathogenic nematode species may work quite broadly for general attraction of commensal species. This could possibly improve infection and associated management of insect pests with biological control in the subterranean environment. Initial testing of social transference of learned (pre-conditioned) behavior indeed indicates that infestation of insect herbivores by entomopathogenic nematodes increases due to interspecific transference of learned behavior among a broad range of entomopathogenic nematode species (Stelinski et al. unpublished data).

3. Opportunities for Practical Applications and Future Research

Vertebrate and plant parasitic nematodes cause significant economic damage worldwide and endanger human, livestock and plant health while entomopathogenic nematodes are effective biocontrol agents that can enhance human, livestock and plant production through their beneficial effects of regulating insect pest populations (Gaugler and Kaya 1990, Kaya and Gaugler 1993). The dynamic learning ability and observed social behavioral plasticity of parasitic nematodes can be exploited for practical management (Fig. 2). Indeed, possible management tactics for vertebrate, plant and insect parasitic nematodes may be considered essentially complementary. Once we understand the social communication system inherent in the parasitic nematode lifecycle, we can then enhance or disrupt certain aspects as desired (Fig. 2).

For entomopathogenic nematodes, enhancing their efficiency and efficacy as biocontrol agents is the goal. Entomopathogenic nematodes are used to control a wide variety of below ground agricultural pests (Kaya and Guagler 1993). While effective, applications can be costly and treatment effects can be short-lived (Gaugler and Kaya 1990). Previous work has demonstrated that manipulating environmental cues through application of herbivore induced plant volatiles (Ali et al. 2012) or genetic engineering (Degenhardt et al. 2009) can enhance the ability of entomopathogenic nematodes to manage subterranean pests such as the black vine weevil *Otiorhynchus sulcatus* in ornamentals, the citrus root weevil *D. abbreviatus* in citrus and the Western corn rootworm, *Diabrotica virgifera virgifera*, in corn.

The aggregative behavior of nematodes and intra and inter-specific, chemically mediated (Fig. 3) nematode-nematode and plant-nematode-

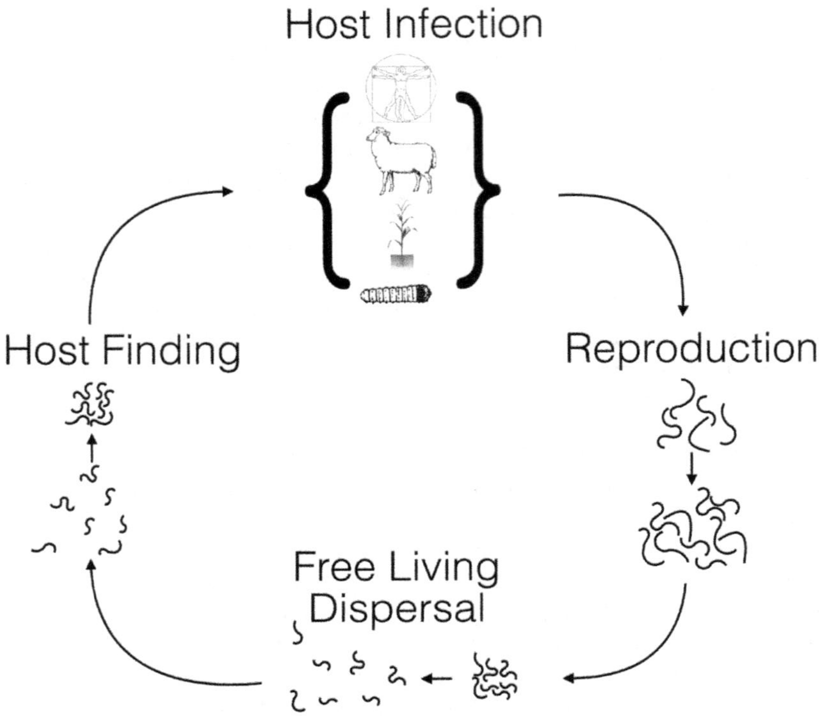

Fig. 2. Nematode Parasitic Lifecycle. All parasitic nematodes, whether they be parasites of humans, vertebrates, plants or insects, undergo four primary lifestages during which they exhibit characteristically similar behavior. All parasitic nematodes must infect their host, reproduce, disperse from their host and find new hosts.

insect interactions suggest that disrupting host-finding behavior may be an effective means of controlling parasitic nematode pests. Indeed, the application of repellents or attractants would not only affect the behavior of those nematodes directly contacted, but also of other nematodes that would 'follow the leader' subjected to behavioral manipulation. Host finding, mate recognition and aggregative signals can be disrupted or jammed to prevent completion of the lifecycle externally or potentially even with the host (Bone and Shorey 1977). Host specific odorants and attractants could be applied in areas of parasite activity to both monitor populations and disrupt host finding ability. Application of nematode pheromones can and has been used similarly to disrupt the lifecycles of parasitic nematodes (Bone and Shorey 1977). Likewise, given the conserved nature of nematode signaling and communication, resistant varieties can be developed expressing such disruptive signals to prevent and reduce infection by plant parasites.

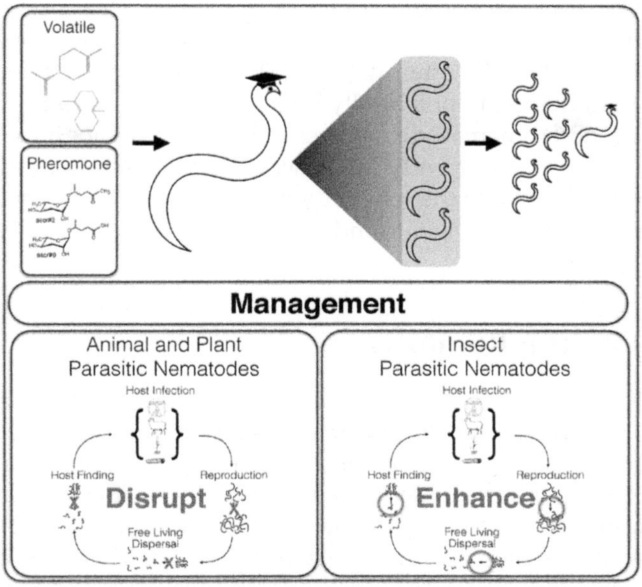

Fig. 3. Practical Management of Parasitic Nematodes. Vertebrate, plant and insect parasitic nematodes respond to volatiles in their environment and pheromones. We can use these volatiles and pheromones to communicate with and educate parasitic nematodes. The communication ability inherent in the social behavior of parasitic nematodes allows educated nematodes to recruit and influence follower nematodes. By influencing a small subset of leader parasitic nematodes through volatile and pheromone communication, we can have a disproportionate effect on the followers that are influenced by the educated nematodes. This system can be used in management strategies to disrupt the lifecycles of vertebrate and plant parasitic nematodes and enhance the lifecycles of insect parasitic nematodes for biological control.

Finally, the social and learning abilities demonstrated by these parasites could be used to train entomopathogenic nematodes to target specific pests in specific environments and create intelligent, guided roundworm 'missiles' to enhance classical or augmentation biological control. For instance, it could be envisaged that rearing entomopathogenic nematodes in the presence of specific volatiles and subsequently applying them in management programs in conjunction with those volatiles would stimulate greater infectivity or attraction to desired areas than without pre-conditioning. These efforts would likely to have compounded effects as indigenous nematodes present in the system may follow cues or signals released by trained nematodes to the food source (Willett et al. 2015). This application could be a cost effective alternative to current methods of enhancing biological control with entomopathogenic nematodes. Genetic transformation of plants to release signals that recruit entomopathogenic nematodes is costly and often unpalatable to the public. Targeting with trained nematodes could potentially be accomplished through addition of stimulant blends to nematode cultures and fit seamlessly into nematode production systems currently in place. Such targeting with trained nematodes may increase the longevity and efficacy of entomopathogenic nematodes as biocontrol agents and provide a cost effective alternative to chemical pesticides.

4. Concluding Remarks

Nematodes play a vastly important role in regulating populations of pests affecting humans and are often themselves ubitiquess pests of vertebrates and plants. The above opportunities to manipulate insect, vertebrate and plant parasitic nematodes based on understanding of the dynamic social communication present in their lifecycle provide an alternative to traditional pest management using chemical pesticides. Given increasing resistance to anthelmintics and pesticides, management techniques based on disrupting nematode communication may provide a viable, long-lasting alternative. Indeed, using nematode communication in a social context will potentially allow us to influence nematode parasite behavior to a degree not previously possible with insecticide and nematicide based techniques alone. By using 'Pied Piper' like strategies to call and influence parasitic nematodes with volatile cues, nematode-nematode communication and follow the leader behavior of parasitic nematodes may enhance management of both nematode and arthropod pests. For entomopathogenic nematodes, judicious lifecycle intervention will engender effective biological control and potentially obviate the need for chemical pesticides. Instead of combating infection of vertebrate and plant parasitic nematodes with nematicides, through judicious lifecycle intervention, we may be able to prevent infection entirely.

Acknowledgments

This work was supported by University of Florida Research Foundation Professorship to L.L.S.

REFERENCES

Ali, J.G., H.T. Alborn and L.L. Stelinski. 2010. Subterranean herbivore-induced volatiles released by citrus roots upon feeding by *Diaprepes abbreviatus* recruit entomopathogenic nematodes. J. Chem. Ecol. 36: 361–368.

Ali, J.G., H.T. Alborn and L.L. Stelinski 2011. Constitutive and induced subterranean plant volatiles attract both entomopathogenic and plant parasitic nematodes. J. Ecol. 99: 26–35.

Ali, J.G., H.T. Alborn, R. Campos-Herrera, F. Kaplan, L.W. Duncan, C. Rodriguez-Saona et al. 2012. Subterranean, herbivore-induced plant volatile increases biological control activity of multiple beneficial nematode species in distinct habitats. PLoS ONE 7: e38146.

Allen, J.E. and A.S. MacDonald. 1998. Profound suppression of cellular proliferation mediated by the secretions of nematodes. Parasite Immunol. 20: 241–247.

Artyukhin, A.B., J.J. Yim, M.C. Choeng and L. Avery. 2015. Starvation-induced collective behavior in *C. elegans*. Sci. Rep. 5: 10647.

Bargmann, C.I. and H.R. Horvitz. 1991. Chemosensory neurons with overlapping functions direct chemotaxis to multiple chemicals in *C. elegans*. Neuron 7: 729–742.

Blaxter, M.L. 2003. Nematoda: genes, genomes and the evolution of parasitism. Adv. Parasitol. 54: 101–195.

Bird, A.F. 1959. The attractiveness of roots to the plant parasitic nematodes *Meloidogyne javanica* and *M. hapla*. Nematologica 4: 322–335.

Bird, D.M., C.H. Opperman and V.M. Williamson. 2009. Plant infection by root-knot nematode. pp. 1–13. *In*: R.H. Berg and C.Taylor [eds.]. Cell Biology of Plant Nematode Parasitism. Springer-Verlag, Berlin, Germany.

Boag, B. and P. Topham. 1984. Aggregation of plant parasitic nematodes and taylor's power law. Nematologica 30: 348–357.

Bone, L.H. and H. Shorey. 1977. Disruption of sex pheromone communication in a nematode. Science 197: 694–695.

Bone, L.H. and H. Shorey. 1978. Nematode sex pheromones. J. Chem. Ecol. 4: 595–612.

Brivio, M.F., M. Pagani and S. Restelli. 2002. Immune suppression of *Galleria mellonella* (insecta, lepidoptera) humoural defenses induced by *Steinernema feltiae* (nematoda, rhabditida): involvement of the parasite cuticle. Exp. Parasitol. 101: 149–156.

Brown, D.J.F., W.M. Robertson and D.L. Trudgill. 1995. Transmission of viruses by plant nematodes. Annu. Rev. Phytopathol. 33: 223–249.

Butcher, R.A., M. Fujita, F.C. Schroeder and J. Clardy. 2007. Small-molecule pheromones that control dauer development in *Caenorhabditis elegans*. Nat. Chem. Biol. 3: 420–422.

Castillo, J.C., S.E. Reynolds and I. Eleftherionos. 2011. Insect immune responses to nematode parasites. Trends Parasitol. 27: 537–547.

Castelletto, M.L., S.S. Gang, R.P. Okubo, A.A. Tselikova, T.J. Nolan, E.D. Platzer et al. 2014. Diverse host-seeking behaviours of skin-penetrating nematodes. PloS Pathog. 10: e1004305.

Choe, A., S.H. von Reuss, D. Kogan, R.B. Gassar, E.G. Platzer, F.C. Shroeder et al. 2012. Ascaroside signaling is widely conserved among nematodes. Curr. Biol. 22: 772–780.

Cobb, N.A. 1915. Nematodes and their Relationships. U.S. Government Printing Office, Washington D.C., USA.

Conway, D.J., A. Hall, K.S. Anwar, M.L. Rahman and D.A.P. Bundy. 1995. Household aggregation of *Strongyloides stercoralis* infection in Bangladesh. Trans. R. Soc. Trop. Med. Hyg. 89: 258–261.

Davies, K.G. and R.H. Curtis. 2011. Cuticle surface coat of plant-parasitic nematodes. Annu. Rev. Phytopathol. 49: 135–156.

de Bono M., D.M. Tobin, M.W. Davis, L. Avery and C.I. Bargmann. 2002. Social feeding in *Caenorhabditis elegans* is induced by neurons that detect aversive stimuli. Nature 419: 899–903.

Degenhardt, J., I. Hiltpold, T.G. Koellner, M. Frey, A. Gieri, J. Gershenzon et al. 2009. Restoring a maize root signal that attracts insect-killing nematodes to control a major pest. Proc. Natl. Acad. Sci. USA 106: 13213–13218.

Dillman, A.R., M.L. Guilermin, J.H. Lee, B. Kim, P.W. Sternberg and E.A. Hallem. 2012. Olfaction shapes host–parasite interactions in parasitic nematodes. Proc. Natl. Acad. Sci. USA 109: E2324–E2333.

Dorris, M., P. De Ley and M.L. Blaxter. 1999. Molecular analysis of nematode diversity and the evolution of parasitism. Parasitol. Today 15: 188–193.

Duncan, L.W., I.D. Shapiro, C.W. McCoy and J.H. Graham. 1999. Entomopathogenic nematodes as a component of citrus root weevil IPM. pp. 69–78. *In*: S. Polavarapu [ed.]. Optimal Use of Insecticidal Nematodes in Pest Management. Rutgers University Press, New Brunswick, Canada.

Edison, A.S. 2009. *Caenorhabditis elegans* pheromones regulate multiple complex behaviors. Curr. Opin. Neurobiol. 19: 378–388.

El–Borai, F.E., R. Campos–Herrera, R.J. Stuart and L.W. Duncan. 2011. Substrate modulation, group effects and the behavioural responses of entomopathogenic nematodes to nematophagous fungi. J. Invertebr. Pathol. 106: 347–356.

Félix, M.A. and C. Braendle. 2010. The natural history of *Caenorhabditis elegans*. Curr. Biol. 20: R965–R969.

Forbes, W., F.T. Ashton, R.C. Boston and G.A. Schad. 2003. Chemotactic behaviour of *Strongyloides stercoralis* infective larvae on a sodium chloride gradient. Parasitology 127: 189–197.

Fushing, H., L. Zhu, D.I. Shapiro-Ilan, J.F. Campbell and E.E. Lewis. 2008. State-space based mass event-history model 1: Many decision-making agents with one target. Ann. Appl. Stat. 2: 1503–1522.

Galef, B.G. and K.N. Laland. 2005. Social learning in animals: empirical studies and theoretical models. Bioscience 55: 489–499.

Gaugler, R. and H.K. Kaya. 1990. Entomopathogenic Nematodes in Biological Control. CRC Press, Boca Raton, USA.

Goverse, A. and G. Smant. 2014. The activation and suppression of plant innate immunity by parasitic nematodes. Annu. Rev. Phytopathol. 52: 243–265.

Granzer, M. and W. Haas. 1991. Host-finding and host recognition of infective *Ancylostoma caninum* larvae. Int. J. Parasitol. 21: 429–440.

Grenfell, B.T., K. Wilson, V.S. Isham, H.E.G. Boyd and K. Dietz 1995. Modelling patterns of parasite aggregation in natural populations: trichostrongylid nematode–ruminant interactions as a case study. Parasitology 111: S135–S151.

Hedgecock, E.M. and R.L. Russell. 1975. Normal and mutant thermotaxis in the nematode *Caenorhabditis elegans*. Proc. Natl. Acad. Sci. USA 72: 4061–4065.

Hobert, O. 2003. Behavioural plasticity in *C. elegans*: paradigms, circuits, genes. J. Neurobiol. 54: 203–223.

Hollis, J.P. 1962. Nature of swarming in nematodes. Nature 193: 798–799.

Huettel, R. 1986. Chemical communicators in nematodes. J. Nematol. 18: 3–8.

Hugot, J.P., P. Baujard and S. Morand. 2001. Biodiversity in helminths and nematodes as a field of study: an overview. Nematology 3: 199–208.

Jaffe, H., R.N. Heuttel, A.B. Demilo, D.K. Hayes and R.V. Rebois. 1989. Isolation and identification of a compound from soybean cyst nematode, *Heterodera glycines*, with sex pheromone activity. J. Chem. Ecol. 15: 2031–2043.

Jasmer, D.P., A. Goverse and G. Smant. 2003. Parasitic nematode interactions with mammals and plants. Annu. Rev. Phytopathol. 41: 245–270.

Kaplan, F., J. Srinivasan, P. Mahanti, R. Ajredini, O. Durak, R. Nimalendran et al. 2011. Ascaroside expression in *Caenorhabditis elegans* is strongly dependent on diet and developmental stage. PLoS ONE 6: e17804.

Kaplan, F., H.T. Alborn, S.H. von Ruess, R. Ajredini, J.G. Ali, F. Akyazi et al. 2012. Interspecific nematode signals regulate dispersal behaviour. PLoS ONE 7: e38735.

Kaya, H.K. and R. Gaugler. 1993. Entomopathogenic nematodes. Annu. Rev. Entomol. 38: 181–206.

Koga, M. and I. Tada. 2000. *Strongyloides ratti*: chemotactic responses of third-stage larvae to selected serum proteins and albumins. J. Helminth. 74: 247–252.

Lamberti, F., C.V. Taylor and J.W. Seinhorst. 2012. Nematode Vectors of Plant Viruses. Volume 2, Series A, Life Sciences. Springer, New York, NY, USA.

Lewis, E.E., R. Gaugler and R. Harrison. 1992. Entomopathogenic nematode host finding: response to host contact cues by cruise and ambush foragers. Parasitology 105: 309–315.

Lewis, E.E., R. Gaugler and R. Harrison. 1993. Response of cruiser and ambusher entomopathogenic nematodes (steinernematidae) to host volatile cues. Can. J. Zool. 71: 765–769.

Lewis, E.E., J. Campbell, C. Griffin, H. Kaya and A. Peters. 2006. Behavioural ecology of entomopathogenic nematodes. Biol. Control 38: 66–79.

Li, X.Y., R.S. Cowles, E.A. Cowles, R. Gaugler and D.L. Cox-Foster. 2007. Relationship between the successful infection by entomopathogenic nematodes and the host immune response. Int. J. Parasitol. 37: 365–374.

Manosalva, P., M. Monohar, S.H. von Reuss, S. Chen, A. Koch, F. Kaplan et al. 2015. Conserved nematode signalling molecules elicit plant defenses and pathogen resistance. Nat. Comm. 6: 7795.

McBride, J.M. and J.P. Hollis. 1966. Phenomenon of swarming in nematodes. Nature 211: 545–546.

McSorley, H.J. and R.M. Maizels. 2012. Helminth infections and host immune regulation. Clin. Microb. Rev. 25: 585–608.

McSorley, H.J., J.P. Hewitson and R.M. Maziels 2013. Immunomodulation by helminth parasites: defining mechanisms and mediators. Int. J. Parasitol. 43: 301–310.

Meyer, S. and R. Huettel. 1996. Application of a sex pheromone, pheromone analogs, and *Verticillium lecanii* for management of *Heterodera glycines*. J. Nematol. 28: 36–42.

Morrill, A. and M.R. Forbes. 2015. Aggregation of infective stages of parasites as an adaptation and its implications for the study of parasite-host interactions. Am. Nat. 187: 225–235.

Noe, J. and C. Campbell. 1985. Spatial pattern analysis of plant-parasitic nematodes. J. Nematol. 17: 86–93.

Nuttley, W.M., K.P. Atkinson-Leadbeater and D. van der Kooy. 2002. Serotonin mediates food-odour associative learning in the nematode *Caenorhabditis elegans*. Proc. Natl. Acad. Sci. USA 99: 12449–12454.

Poulin, R. and S. Morand. 2000. Parasite body size and interspecific variation in levels of aggregation among nematodes. J. Parasitol. 86: 642–647.

Perry, R.N. 1996. Chemoreception in plant parasitic nematodes. Annu. Rev. Phytopathol. 34: 181–199.

Rasmann, S., T.G. Koellner, J. Degenhardt, I. Hiltpold, S. Toepfer, U. Kuhlmann et al. 2005. Recruitment of entomopathogenic nematodes by insect-damaged maize roots. Nature 434: 732–737.

Reynolds, L.A., B.B. Finlay and R.M. Maziels. 2015. Cohabitation in the intestine: interactions among helminth parasites, bacterial microbiota, and host immunity. J. Immunol. 195: 4059–4066.

Robinson, A. 1995. Optimal release rates for attracting *Meloidogyne incognita, Rotylenchulus reniformis*, and other nematodes to carbon dioxide in sand. J. Nematol. 27: 422–450.

Rose, J.K. and C.H. Rankin. 2001. Analyses of habituation in *Caenorhabditis elegans*. Learn. Mem. 8: 63–69.

Safer, D., M. Brenes, S. Dunipace and G. Schad. 2007. Urocanic acid is a major chemoattractant for the skin-penetrating parasitic nematode *Strongyloides stercoralis*. Proc. Natl. Acad. Sci. USA 104: 1627–1630.

Sasakura, H. and I. Mori. 2013. Behavioural plasticity, learning, and memory in *C. elegans*. Curr. Opin. Neurobiol. 23: 92–99.

Shapiro-Ilan, D.I., E.E. Lewis and P. Schliekelman. 2014. Aggregative group behaviour in insect parasitic nematode dispersal. Int. J. Parasitol. 44: 49–54.

Shaw, D., B.T. Grenfell and A.P. Dobson. 1998. Patterns of macroparasite aggregation in wildlife host populations. Parasitology 117: 597–610.

Schroeder, F.C. 2015. Modular assembly of primary metabolic building blocks: a chemical language in *C. elegans*. Chem. Biol. 22: 7–16.

Simon, J.M. and P.W. Sternberg. 2002. Evidence of a mate-finding cue in the hermaphrodite nematode *Caenorhabditis elegans*. Proc. Natl. Acad. Sci. USA 99: 1598–1603.

Simpson, S.E., H.N. Nigg and J.L. Napp. 2000. Host plants of Diaprepes root weevil and their implications to the regulatory process. pp. 19-37. *In*: Diaprepes Short Course Proceedings. Citrus Research and Education Center, Lake Alfred, USA.

Singh, S., M. Hodda and G.J. Ash. 2013. Plant-parasitic nematodes of potential phytosanitary importance, their main hosts and reported yield losses. Bull. OEPP/EPPO 43: 334–374.

Srinivasan, J., F. Kaplan, R. Ajredini, C. Zaharia, H.T. Alborn, P.A. Teal et al. 2008. A blend of small molecules regulates both mating and development in *Caenorhabditis elegans*. Nature 454: 1115–1118.

Srinivasan, J., S.H. von Reuss, N. Bose, A. Zaslaver, P. Mahatni, M.C. Ho et al. 2012. A modular library of small molecule signals regulates social behaviours in *Caenorhabditis elegans*. PLoS Biol. 10: e1001237.

Stoll, N.R. 1999. This wormy world. J. Parasitol. 85: 392–396.

Stuart, R. and R. Gaugler. 1994. Patchiness in populations of entomopathogenic nematodes. J. Invertebr. Pathol. 64: 39–45.

Stuart, R.J., M. Barbercheck, P.S. Grewal, R.A.J. Talor and C. Hoy. 2006. Population biology of entomopathogenic nematodes: concepts, issues, and models. Biol. Control 38: 80–102.

Tumlinson, J.H., T.C.J. Turlings and W.J. Lewis. 1992. The semiochemical complexes that mediate insect parasitoid foraging. Agric. Zool. Rev. 5: 221–252.

Turlings, T.C.J. and J. Ton. 2006. Exploiting scents of distress: the prospect of manipulating herbivore-induced plant odurs to enhance the control of agricultural pests. Curr. Opin. Plant Biol. 9: 421–427.

Turlings, T.C.J., I. Hiltpold and S. Rasmann. 2012. The importance of root-produced volatiles as foraging cues for entomopathogenic nematodes. Plant Soil: 358: 51–60.

van Tol, R.W.H.M., A.T.C. van der Sommen, M.I.C. Boff, J. van Bezooijen, M.W. Sabelis and P.H. Smits. 2001. Plants protect their roots by alerting the enemies of grubs. Ecol. Lett. 4: 292–294.

Wang, C., G. Bruening and V.M. Williamson. 2009. Determination of preferred pH for root-knot nematode aggregation using pluronic F-127 gel. J. Chem. Ecol. 35: 1242–1251.

Wang, Y., R. Gaugler and L. Cui. 1994. Variations in immune response of *Popillia japonica* and *Acheta domesticus* to *Heterorhabditis bacteriophora* and *Steinernema* species. J. Nematol. 26: 11–18.

Wang, Y. and R. Gaugler. 1999. *Steinernema glaseri* surface coat protein suppresses the immune response of *Popillia japonica* (Coleoptera: Scarabaeidae) larvae. Biol. Control 14: 45–50.

Westerman, P. 1999. Aggregation of entomopathogenic nematodes, *Heterorhabditis* spp. and *Steinernema* spp., among host insects at 9 and 20°C and effects on efficacy. J. Invertebr. Pathol. 73: 206–213.

Willett, D.S., H.T. Alborn, L.W. Duncan and L.L. Stelinski. 2015. Social networks of educated nematodes. Sci. Rep. 5: 14388.

Xu, Z., Y.Q. Zhao, D.J. Yang, H.J. Sun, C.L. Zhang and Y.P. Xie. 2015. Attractant and repellent effects of sweet potato root exudates on the potato rot nematode, *Ditylenchus destructor*. Nematology 17: 117–124.

Zuckerman, B.M. and H. Jansson. 1984. Nematode chemotaxis and possible mechanisms of host/prey recognition. Annu. Rev. Phytopathol. 22: 95–11.

CHAPTER

5

Microbial Endosymbionts and Chemical Ecology

Daisuke Kageyama

National Agriculture and Food Research Organization
1-2 Owashi, Tsukuba-city, Ibaraki 305-0851, Japan
E-mail: kagymad@affrc.go.jp

Abstract

Insects provide rich and diverse arenas for various microbes. A considerable proportion of insect species harbor endosymbionts—bacteria, protists, and viruses—in their gut or cells. Here, I review various case studies in which endosymbionts were shown to have profound effects on their hosts, with special emphasis on chemical ecology. Gut bacteria in locusts and cockroaches play important roles in the production of fecal aggregation pheromones. Gut bacteria in *Drosophila* flies have critical effects on the biosynthesis of hydrocarbons, which act as contact sex pheromones, thereby having a potentially important role for speciation. Some bacteria exert their effects as chemical defense against predators, parasites, or pathogens in a variety of insects. Besides this, there are numerous cases where direct or indirect contribution of microorganisms to insect chemical ecology is suspected. The utility and benefit of the hologenomic concept in the light of microbe-associated chemical ecology is also discussed.

1. Introduction

Recent studies have revealed that symbiotic microorganisms have much more diverse and profound effects on the biology of eukaryotes than previously appreciated (Mayer 2016, Yong 2016). They may feminize the male woodlice (Rigaud 1997, Rigaud et al. 1997), change the host plant preference of aphids (Tsuchida et al. 2004), and illuminate squids and fishes (Ruby 1996) to name but a few. They may even affect emotional behavior and related brain systems in humans (Mayer et al. 2014, Mayer 2016).

Chemical ecology is a discipline that aims to understand how the distribution and abundance of organisms, as well as the complex interactions between them, are mediated by chemical agents (Raguso et al. 2015). In this sense, almost every aspect of biological phenomena can be viewed from the context of chemical ecology. In this chapter, I summarize the diverse interactions between insect hosts and their endosymbionts, particularly emphasizing their outcomes on chemical ecology.

2. Endosymbiotic Microorganisms

Insects provide a rich and diverse environment for many species of microorganism. By definition, every organism living inside a host organism is called endosymbiont. However, conventionally, the term endosymbiont is often used for bacteria, protists and, less commonly, viruses. A considerable proportion of insect species harbor symbiotic microorganisms in their bodies—be they intracellular or extracellular—that play important roles in metabolism and reproduction (Kageyama et al. 2012, Zug and Hammerstein 2012, Douglas 2015).

Wolbachia pipientis, a bacterial species belonging to the alpha subdivision of Proteobacteria, often simply referred to as *Wolbachia*, is undoubtedly the most prevalent—more than 40% of insect species are estimated to harbor *Wolbachia* (Zug and Hammerstein 2012). *Wolbachia* basically relies on vertical transmission from mothers to offspring, although molecular phylogenetic data suggests that it has moved horizontally between distant hosts multiple times in its evolutionary past (Werren et al. 2008). *Wolbachia* usually resides in the intracellular environment of almost all the insect tissues (Dobson et al. 1999, Ijichi et al. 2002, Narita et al. 2007). Although it is an evolutionary dead-end, somatic tissue tropism is considered to be an important aspect in *Wolbachia* biology (Faria and Sucena 2013, Pietri et al. 2016). Somatic tropism can be viewed as a self-sacrificing behavior of *Wolbachia* that allows sister members to be transmitted to subsequent host generations. Among various endosymbionts, *Wolbachia* is prominent in the fact that it can manipulate reproductive system of a wide array of arthropods in various ways for their own benefit, which will be explained in the next section.

Insects that rely on restricted diet for their entire lives (such as aphids that feed on plant sap, tsetse flies that feed on animal blood, and longicorn beetles that feed on woody tissue) tend to have special organs called bacteriocytes (or mycetocytes), which is a cellular structure of insects that house a numerous amount of certain species of bacteria or fungi (Buchner 1965, Douglas 1989). Such microbes provide their hosts with nutrition such as vitamins and amino acids, which can supplement the critical shortage in the dietary intake. Most of the phytophagous hemipteran bugs develop a number of sac- or tube-like outgrowths in a posterior region of the midgut, wherein specific bacteria are harbored. Readers who are interested in the exciting discoveries that have been made for bacteriocytes-associated endosymbioses may refer to Baumann 2005, Moran and Bennett 2014, Douglas 2014 and Wilson and Duncan 2015.

The tissue tropism of other intracellular symbionts is less well-known. Being transmitted from mother to offspring, such microbes must at least reside in the female germ tissues, wherein their presence has been observed repeatedly (Watanabe et al. 2014, Hayashi et al. 2016). In *Drosophila* flies, *Spiroplasma poulsonii*, a member of the Mollicutes, is inherited maternally and kills only male embryos (Williamson and Poulson 1979). Although they are present silently in oocytes, they are actively motile and more abundantly observed in hemolymph (Williamson and Poulson 1979, Anbutsu and Fukatsu 2006, Goto et al. 2006). On the other hand, a maternally transmitted microsporidian protist that feminizes males of the amphipod shrimp (*Gammarus duebeni*), is known to be restricted to female reproductive tissues (Terry et al. 1997).

For some microbes, the insect gut is favorable for colonization because of the ease of access to nutrients and protection from various stresses from the external environment (e.g., desiccation, ultraviolet radiation). These gut microbes are not inherited through the egg cytoplasm, but establish a new infection from the environment every generation (Dillon and Dillon 2004, Douglas 2015).

3. Effects on Reproduction

Wolbachia and other intracellular symbionts, such as *Rickettsia* (α-Proteobacteria), *Arsenophonus* (γ-Proteobacteria), *Spiroplasma* (Mollicutes), and *Cardinium* (Bacteroidetes), can have drastic effects on insect reproduction (Werren et al. 2008, Kageyama et al. 2012).

3.1 Cytoplasmic Incompatibility: Microbe-Mediated Disruption of Pheromonal Communication?

Among microbe-mediated manipulation of insect reproduction, one of the oldest known cases is cytoplasmic incompatibility (CI), wherein mating between infected males and uninfected females is incompatible (Werren 1997, Werren et al. 2008, Stouthamer et al. 1999, Poinsot et al. 2003) (Fig. 1a). Thus, uninfected females can reproduce only if they mate with uninfected males, whereas infected females can reproduce regardless of the infection status of their partners. Because the symbionts can only be transmitted from females to offspring, the infection is expected to spread rapidly in such a population (Turelli and Hoffmann 1991, Turelli et al. 1992). CI is often sensitive to the strain of the symbiont; e.g., females with a strain A symbiont are incompatible with males with a strain B symbiont, but females co-infected with strain A and strain B symbionts are compatible with males with a strain A symbiont (Fig. 1b). Thus far, CI is known to be caused only by two bacterial species, *Wolbachia pipientis* (α-Proteobacteria) and *Cardinium hertigii* (Bacteroides), which are taxonomically distantly related (Kageyama et al. 2012).

Various data on genetics and cytology suggested a modification-rescue model of CI, wherein *Wolbachia* induces reversible modifications in sperm DNA or chromatin so as to kill the eggs through abnormal mitosis. The same strain of *Wolbachia* in eggs is considered to eliminate or neutralize the modification (Poinsot et al. 2003). Recently, a groundbreaking discovery was made according to the mechanism of CI. By comparative genomics and transcriptomics of various *Wolbachia* strains in *Drosophila* flies, including

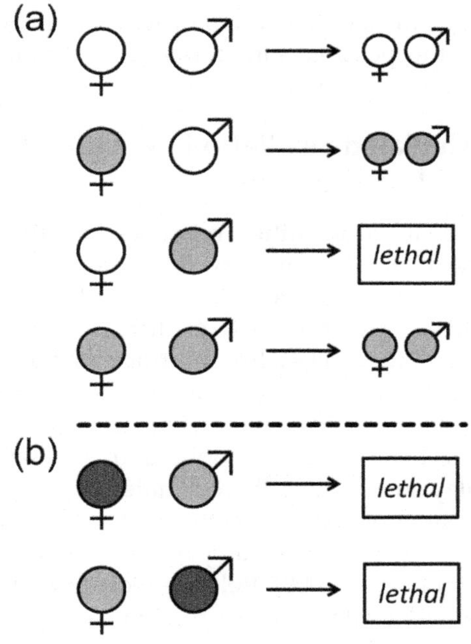

Fig. 1. Cytoplasmic incompatibility. (a) Among the four combinations between *Wolbachia*-infected (depicted in light grey) and uninfected (depicted in white) males and females, a combination with an uninfected female and infected male is incompatible—embryos die during early development (unidirectional incompatibility). (b) When different strains of *Wolbachia* are present in males and females, they are incompatible (bidirectional incompatibility).

CI-inducing and non-CI-inducing strains, two genes were identified as candidates of CI-inducing genes (LePage et al. 2017). Convincingly, one of the two genes had previously been identified as a candidate in mosquitoes (Beckmann and Fallon 2013). Although each of the two genes, transgenically expressed in *Drosophila*, did not induce CI, a simultaneous expression of the two genes induced CI (LePage et al. 2017). It was further revealed that two proteins encoded by these genes interact with each other and one of them is deubiquitinase—an enzyme that removes ubiquitin moieties from proteins (Beckmann et al. 2017). Although the targets of the enzyme are not yet unknown, the deubiquitination seems to be the mechanism for the CI phenotype.

On the basis of controlled experiments using *Drosophila*, however, Pontier and Schweisguth challenged this model by suggesting that male and female pupae communicate via pheromones for their proper development, and that CI is the consequence of the disruption of this communication by *Wolbachia* (Pontier and Schweisguth 2015). However, these results were not supported by follow-up experiments by another team, thus making it controversial (Jacquet et al. 2017, Pontier and Schweisguth 2017). Further analyses to prove or disprove the pheromone model of CI and the possible integration into the modification-rescue model of CI would be challenging.

3.2 Microbe-Mediated Sex Ratio Distortion: Potential to Alter Sexual Behavior

A variety of maternally transmitted microbes bias the sex ratio of their host insects toward females, by way of feminization, male-killing, or parthenogenesis induction (Kageyama et al. 2012). The spread of such microbes among host populations will result in a female-biased population sex ratio (Dyson and Hurst 2004). In conditions with a strongly biased sex ratio, sexual behavior is expected to change (Abe and Kamimura 2015).

In a butterfly, *Acraea encedon*, the prevalence of male-killing *Wolbachia* has resulted in a female-biased population sex ratio. Lekking is a competitive display made by gathered males, which may lure visiting females who are searching for prospective mates. However, things appear to be reversed in *A. encedon*—lekking behavior is observed in females but not males (Jiggins et al. 2000). Release–recapture of virgin and mated females at the lekking site showed that the recapture rate of virgin females (48%) was significantly higher than that of mated females (17%), supporting the view that the lekking behavior of *A. encedon* females is for mating (Jiggins et al. 2000).

Hypolimnas bolina butterflies provided an ideal opportunity to investigate the effect of population sex ratio on sexual behavior because the frequencies of individuals infected with male-killing *Wolbachia* vary greatly between island populations in the Pacific and Southeast Asia (Charlat et al. 2007). In contrast to the naive notion that female mating frequency decreases as the population becomes female-biased, female bias leads to an increase in female mating frequency, up to a point where male mating capacity becomes limiting. Increased female mating frequency can be explained as a facultative response to the depleted male mating resources in female-biased populations (Charlat et al. 2007).

Parasitic nematodes can also influence the mating behavior of insects. According to an intriguing study by Vance, more than 10% of wild-caught adults of the mayfly *Baetis bicaudatus* (Ephemeroptera; Baetidae) are parasitized by nematode *Gasteromermis* sp. (Nematoda; Mermithidae) (Vance 1996). None of the parasitized individuals (n = 126) contained visible eggs, ovaries, or testes. Among them, the external morphology of 82 individuals (65%) was indistinguishable from that of normal females. The remaining 44 parasitized individuals showed an array of intersexual morphologies. Moreover, measurement of DNA content by flow cytometry suggested that the parasitized individuals with intersexual morphologies ("parasitized intersexes") were genetically male individuals, whereas parasitized individuals visibly indistinguishable from normal females ("parasitized females") were composed of both genetically female and genetically male individuals (Vance 1996). The behavior of the mayfly were also changed by the nematode. Unparasitized males formed swarms near rivers and do not return to the water after they have emerged. Vance found that all 418 swarming individuals were unparasitized males and that the parasitized individuals showed ovipositing behavior, which was never seen in unparasitized males (Vance 1996). Laboratory studies demonstrated that parasitized individuals (both parasitized females and parasitized intersexes) became very agitated shortly after emergence as adults. Within 3–6 hours, all the parasitized mayflies had crawled into the water down the side of rocks. The nematodes could then be seen escaping through a puncture wound in the mayfly's abdomen (Vance 1996). Therefore, the *Gasteromermis* nematode can transmit horizontally to a new host. The mayflies were killed by the emergence of the nematodes. The feminization in this case seems to be an adaptive strategy for *Gasteromermis* nematodes, which do not have a vertical route of transmission.

Thus, microbe-mediated distortion of population sex ratio may have profound effects on sexual behavior. Such effects, if they lasted for long enough, may cause some evolutionary changes in various reproductive traits, such as sperm production and sex pheromone production, which may well be connected to chemical ecology.

4. Effects on Aggregation Pheromones

Some insects aggregate. This aggregation can be viewed as either a defense behavior against enemies or a fluctuating environment, or an efficient arena for finding mating partners. Of note, some gut microbes—or microbial community in the intestine—seem to play an important role in the aggregating behavior of some insects.

In desert locusts, *Schistocerca gregaria*, guaiacol (2-methoxyphenol) and phenol are volatile compounds that are both released from the feces of normal locusts. Among these, guaiacol was not released from the feces of sterile locusts whose eggs were surface-sterilized and fed with a γ-irradiated freeze-dried diet (Dillon et al. 2000). Inoculation with *Pantoea agglomerans*, an abundant bacteria from the intestine of *S. gregaria*, restored the production of guaiacol. In combination of other experiments, it was demonstrated that guaiacol originates from gut bacteria, including *P. agglomerans* (Dillon et al. 2000).

Similarly, in cockroaches, their feces contained chemical compounds that act as aggregation pheromones. However, inconsistency and variability of the candidate compounds made the identification of cockroach aggregation pheromones difficult and controversial. A recent study demonstrated that the gut microbial community contributes to the production of volatile carboxylic acids (VCAs) in the German cockroach *Blattella germanica* (Wada-Katsumata et al. 2015). Chemical analysis of the fecal extract of *B. germanica* revealed as many as 40 VCAs. Feces from sterile cockroaches—generated by sterilizing eggs and rearing nymphs on sterile food and water—lacked 12 major fecal VCAs, and 24 of the remaining VCAs were present at only extremely low amounts. Olfactory and aggregation bioassays demonstrated that nymphs strongly preferred the extract of control feces to that of sterile ones. Inoculation of sterile cockroaches with individual bacterial isolates triggered the aggregation response to the fecal extract, and inoculation with a mix of six bacterial isolates was more effective than with single isolates.

The above studies were unexpected because it has generally been anticipated that aggregation pheromones are synthesized by the insects themselves. Now we are tempted to ask—how common are microbe-associated aggregation pheromones among insects?

5. Effects on Sex Pheromone—Do They Trigger Speciation?

More surprisingly, mate choice can be governed by gut microbes (Sharon et al. 2010). A *D. melanogaster* population was separated into two groups, and one group reared on a standard medium containing cornmeal and molasses (molasses medium) and the other on a medium containing starch (starch medium). When the isolated groups were mixed, they preferred to mate with the same group members. This effect vanished when the flies were reared on antibiotic-containing media, suggesting that endosymbiotic bacteria were responsible for the phenomenon. Mating preference was regained by inoculating with bacteria (those collected from either molasses medium or starch medium upon which the flies had grown) into antibiotic-treated flies. Cloning and sequencing of bacterial 16S ribosomal RNA gene showed that flies from the molasses medium harbored at least 10 distinct strains of bacteria (belonging to six species; *Acetobacter pomorum* and *Bacillus firmus* being the most abundant) in addition to *Wolbachia*, whereas flies from the starch medium harbored a much less diverse range of bacteria—only *Lactobacillus plantarum* was detected in addition to *Wolbachia*. This was also confirmed by inoculation with a pure culture of *L. plantarum* isolated from starch flies. Chemical analyses using gas chromatography suggested that symbiotic bacteria can influence mating preference by changing the levels of cuticular hydrocarbons, which act as sex pheromones (Sharon et al. 2010).

6. Effect on Self-defense—Do They Produce Chemical Weapons?

Growing attention has been paid to the protective effects of symbionts against various enemies such as parasites, pathogens, and predators. In retrospect, we may wonder why such effects of the symbionts have not been found earlier because it was generally assumed that heritable symbionts tend to be beneficial to their hosts. The reason for this could be the lack of a viewpoint from the natural context, wherein various organisms interact with each other. Therefore, ideally simplified conditions, which are often achieved in the laboratory, would not allow us to detect these effects. Below are examples of tripartite interactions between host insects, symbiotic microbes, and parasites/pathogens/predators. Future studies may reveal the involvement of more players in these interactions, which make them more complex and exciting.

6.1 Protection against Predators

Beetles belonging to the genus *Paederus* harbor bacterial symbionts closely related to *Pseudomonas aeruginosa* (γ-Proteobacteria). These bacteria produce a toxic compound, pederin. Beetle larvae that hatched from pederin-containing eggs experienced reduced predation from wolf spiders as compared with pederin-free larvae (Kellner and Dettner 1996).

Unexpectedly, a complete gene cluster for the synthesis of diaphorin, a toxin that is structurally similar to pederin, was found in the highly reduced genome of the bacterial symbiont '*Candidatus* Profftella armatura', which is harbored by the Asian cirus psyllid, *Diaphorina citri* (Nakabachi et al. 2013). It was suggested that *Profftella* confers hosts with resistance against predators or parasites by providing diaphorin. The highly reduced genome size of *Profftella* (464,857 bp) suggested a long symbiotic relationship between *Profftella* and *D. citri*.

Protection against predators can also be achieved by producing semiochemicals by endosymbiotic bacteria. In the scent glands of the hemipteran bug *Thasus neocalifornicus*, where defensive chemicals and alarm pheromones are produced, the density of *Wolbachia* was higher than in other tissues. Decreasing the *Wolbachia* population using antibiotics led to a reduction of defensive compounds and alarm pheromones, suggesting that this symbiotic bacterium might be involved in the formation of these chemicals (Becerra et al. 2015).

6.2 Protection against Parasitic Wasps

The pea aphid, *Acyrthosiphon pisum*, is probably the most intensively studied organism in insect endosymbiont research. In addition to the indispensable *Buchnera* (γ-Proteobacteria) symbiont (called a primary symbiont), multiple species of facultative symbiont (called secondary symbionts) have been discovered, and each of the effects has been investigated using a transinfection technique and selective elimination of the bacterium by antibiotic therapies.

The parasitic wasp *Aphidius ervi* (Hymenoptera; Braconidae) is a natural enemy for *A. pisum*. *A. ervi* develops within the body of the aphids until pupation, leading to the death of the aphids. Oliver and colleagues first discovered that *A. pisum* infected with one of its secondary symbionts, *Hamiltonella defensa* (γ-Proteobacteria), survived in the face of infestation by the wasps by killing the developing wasp larvae (Oliver et al. 2003). They also revealed that the variable levels of resistance against the wasps can be explained by the genotype of *H. defensa*, not by the genotype of the aphids (Oliver et al. 2005). Furthermore, Oliver and colleagues revealed that a toxin-encoding bacteriophage (bacterial virus) that is harbored by *H. defensa* was responsible for the wasp resistance phenotype (Oliver et al. 2009).

A similar phenomenon has been found in the fruit fly, *Drosophila hydei*, wherein 23–66% of the individuals harbor maternally transmitted *Spiroplasma*, which does not cause any distinct phenotype, such as male-killing or CI, on its host (Ota et al. 1979, Kageyama et al. 2006, Osaka et al. 2010). Why is *Spiroplasma* maintained in such high frequencies was an enigma until Xie and colleagues revealed that *Spiroplasma* enhanced survival of *D. hydei* that were attacked by a parasitic wasp (Xie et al. 2010, Xie et al. 2011). Recently, *Spiroplasma* was shown to have spread rapidly among *D. hydei* populations (Xie et al. 2015).

Recent findings demonstrated that the competition for host lipids between parasitic wasps and *Spiroplasma* is the mechanism underlying *Spiroplasma*-mediated protection of *D. melanogaster* against parasitic wasps by providing three important results (Paredes et al. 2016). First, lipid quantification showed that both *Spiroplasma* and parasitoid wasps depleted *D. melanogaster* hemolymph lipids. Second, the depletion of hemolymphatic lipids by RNA interference (RNAi) reduced wasp success in larvae that were not infected with *Spiroplasma* and blocked *Spiroplasma* growth. Third, the growth of *Spiroplasma* bacteria was not affected by the presence of the wasps, indicating that when *Spiroplasma* is present, larval wasps are exposed to a lipid-depleted environment.

6.3 Protection against Parasitic Nematodes

For the fruitfly *Drosophila neotestacea*, a nematode *Howardula aoronymphium* (Allantonematidae, Tylenchida) is a dreadful parasite, because female flies parasitized by the nematode becomes completely sterile (Jaenike and Perlman 2002). Jaenike et al. discovered that maternally transmitted *Spiroplasma* is sometimes found in *D. neotestacea* and protects its host from the effects of *H. aoronymphium* (Jaenike et al. 2010). Nematode-parasitized flies that were infected experimentally with *Spiroplasma* restored their fertility. Similarly, among wild *D. neotestacea* parasitized by the nematodes, those infected with *Spiroplasma* had more eggs in their ovaries. These results well explained the field data that show *Spiroplasma* was spreading rapidly through populations of *D. neotestacea* in North America.

Spiroplasma encodes a ribosome-inactivating protein (RIP), which is related to Shiga-like toxins produced by human toxigenic strains of *Escherichia coli* and *Shigella dysenteriae*. Hamilton et al. found that the 28S rRNA of *H. aoronymphium*, but not the host *D. neotestacea*, was specifically depurinated under the presence of *Spiroplasma* in *D. neotestacea* (Hamilton et al. 2016). The recombinant *Spiroplasma* RIP catalyzed depurination of 28S rRNAs in a cell-free assay, as well as the nematode rRNA in vitro.

6.4 Protection against Pathogenic Fungi

Females of the European beewolf, *Philanthus triangulum* (Hymenoptera; Crabronidae), cultivate bacteria of the genus *Streptomyces* in specialized antennal glands and apply them to their brood cells prior to oviposition. The bacteria are taken up by the larva and occur on the walls of the cocoon. Bioassays indicated that *Streptomyces* ('*Candidatus* Streptomyces philanthi') protected the cocoon from fungal infestation and significantly enhanced the survivorship of the larva (Kaltenpoth et al. 2005). Subsequently,

Kaltenpoth and colleagues demonstrated that the bacteria produced a "cocktail" of nine antibiotic substances (streptochlorin and eight piericidin derivatives), which were distributed all over the surface of the cocoon and protect immature wasps against opportunistic pathogens (Kroiss et al. 2010, Koehler and Kaltenpoth 2013, Koehler et al. 2013, Flórez et al. 2015).

Similar to beewolves, fungus-growing ants harbor vertically and occasionally horizontally transmitted *Pseudonocardia* sp. (Actinobacteria), which defends the fungus garden against the *Escovopsis* fungal pathogens by producing multiple antibiotics (Oh et al. 2009, Barke et al. 2010).

6.5 Protection against Pathogenic Viruses

In *D. melanogaster*, *Wolbachia* (a strain *w*Mel) causes only weak level of CI, which cannot explain the persistence of *Wolbachia* in natural populations (Hoffmann et al. 1998). Although Hoffmann and colleagues examined various fitness-related factors, these data did not sufficiently explain the *Wolbachia* prevalence in *D. melanogaster* (Olsen et al. 2001, Harcombe and Hoffmann 2004). Later, two research groups independently found a phenomenon that can explain the persistence of *Wolbachia* in *D. melanogaster*: namely they confer protection against pathogenic viruses (Hedges et al. 2008, Teixeira et al. 2008).

Drosophila C virus (DCV), a member of the Dicistroviridae family, is a natural pathogen of *D. melanogaster*. Infection of adult *D. melanogaster* with DCV by injection can result in 100% mortality within 3–4 days. Although variation in the susceptibility of fly strains to DCV-induced mortality has been recorded, the underlying basis for this variation has not been determined. In *Wolbachia*-positive flies, DCV-induced mortality was significantly delayed compared with flies cured of *Wolbachia* infection. The delay in mortality corresponded with a delay in virus accumulation in *Wolbachia*-positive flies. These results indicate that *Wolbachia* have an antiviral effect in *Drosophila*. *Wolbachia* also had similar effects on other RNA viruses, such as cricket paralysis virus (Dicistroviridae), Flock House virus (Nodaviridae), and Nora virus (Picornaviridae) (Hedges et al. 2008, Teixeira et al. 2008), but had no effect on a DNA virus (Insect Iridescent Virus 6) (Teixeira et al. 2008).

Now that wider attention has been paid to this *Wolbachia* strain, *w*Mel (and its relative, *w*MelPop) because these *Wolbachia* strains transferred to *Aedes aegypti* mosquito vectors had similar effects to suppress human pathogens such as dengue virus and chikungunya virus, as well as *Plasmodium gallinaceum*, a malaria parasite of poultry (Moreira et al. 2009). Because these *Wolbachia* strains caused strong CI in mosquitoes, *Aedes aegypti* as well as *Aedes albopictus*, transinfected individuals autonomously and rapidly increase their frequencies in mosquito populations (Hoffmann et al. 2011, Walker et al. 2011, Blagrove et al. 2012). After the successful initial trial made in North Queensland, Australia, *Wolbachia*-infected *A. aegypti* are being released into various places all over the world (e.g., Brazil, Colombia, Vietnam, Indonesia and India) where dengue fever is prevalent (Eliminate Dengue Program; http://www.eliminatedengue.com/).

7. Hologenomic Concept

Many organisms are associated with numerous microbial symbionts that can to greater or lesser extent affect their physiology, development, and evolution (Rosenberg and Zilber-Rosenberg 2011). The recognition of this organismal complexity led to the hologenomic concept that integrates the host organism and all its associated microbes into a single hologenome (Margulis and Fester 1991). At present, however, this concept is subjected to controversy. Some experts claim that this new concept is necessary and beneficial because it emphasizes the pluralistic nature of any plants and animals, whose phenotype is the outcome of multilevel selection acting on the host genome, the microbial genome, and the hologenome (Bordenstein and Theis 2015, Theis et al. 2016). Others argue that the term hologenome and the concepts associated with it can lead to misunderstanding of the evolutionary concepts in host-associated microbes, which can simply be viewed from classical population genetics theory (Moran and Sloan 2015, Douglas and Werren 2016). Although it is unnecessary to show simple phenomena in a complex way, I would like to embrace the hologenomic view because it is an interesting and stimulating way to look at nature, which inspires us to deeper thinking about the complexities of life.

Overall, it has become increasingly evident that insects are complex ecosystems involving numerous species of microbes and are undoubtedly ideal materials to investigate microbe–microbe and host–microbe interactions. I believe that many aspects of these interactions can be exciting research targets in the context of chemical ecology.

8. Conclusion

Growing data have demonstrated that microbes can play important roles in the chemical ecology of various species of insects. I believe that, by using insects as models, many exciting discoveries will be made in the interface of microbiology, evolutionary biology, and chemical ecology in years to come.

Acknowledgments

I deeply thank Dr. Jun Tabata, the editor of this book, for his continuous encouragement and patience.

REFERENCES

Abe, J. and Y. Kamimura. 2015. Sperm economy between female mating frequency and male ejaculate allocation. Am. Nat. 185: 406–416.

Anbutsu, H. and T. Fukatsu. 2006. Tissue-specific infection dynamics of male-killing and nonmale-killing spiroplasmas in *Drosophila melanogaster*. FEMS Microbiol. Ecol. 57: 40–46.

Barke, J., R.F. Seipke, S. Grüschow, D. Heavens, N. Drou, M.J. Bibb et al. 2010. A mixed community of actinomycetes produce multiple antibiotics for the fungus farming ant *Acromyrmex octospinosus*. BMC Biol. 8: 109.

Baumann, P. 2005. Biology bacteriocyte-associated endosymbionts of plant sap-sucking insects. Annu. Rev. Microbiol. 59: 155–189.

Becerra, J.X., G.X. Venable and V. Saeidi. 2015. *Wolbachia*-free heteropterans do not produce defensive chemicals or alarm pheromones. J. Chem. Ecol. 41: 593–601.

Beckmann, J.F. and A.M. Fallon. 2013. Detection of the *Wolbachia* protein WPIP0282 in mosquito spermathecae: implications for cytoplasmic incompatibility. Insect Biochem. Mol. Biol. 43: 867–878.

Beckmann, J.F., J.A. Ronau and M. Hochstrasser. 2017. A *Wolbachia* deubiquitylating enzyme induces cytoplasmic incompatibility. Nature Microbiol. 2: 17007.

Blagrove, M.S.C., C. Arias-Goeta, A.-B. Failloux and S.P. Sinkinset. 2012. *Wolbachia* strain *w*Mel induces cytoplasmic incompatibility and blocks dengue transmission in *Aedes albopictus*. Proc. Natl. Acad. Sci. USA 109: 255–260.

Bordenstein, S.R. and K.R. Theis. 2015. Host biology in light of the microbiome: ten principles of holobionts and hologenomes. PLoS Biol. 13: 8.

Buchner, P. 1965. Endosymbiosis of Animals with Plant Microorganisms. InterScience Publishers, New York, NY.

Charlat, S., M. Reuter, E.A. Dyson, E.A. Hornett, A. Duplouy, N. Davies et al. 2007. Male-killing bacteria trigger a cycle of increasing male fatigue and female promiscuity. Curr. Biol. 17: 273–277.

Dillon, R.J. and V. Dillon. 2004. The gut bacteria of insects: nonpathogenic interactions. Annu. Rev. Entomol. 49: 71–92.

Dillon, R.J., C.T. Vennard and A.K. Charnley. 2000. Pheromones: exploitation of gut bacteria in the locust. Nature 403: 851.

Dobson, S.L., K. Bourtzis, H.R. Braig, B.F. Jones, W. Zhou and F. Rousset. 1999. *Wolbachia* infections are distributed throughout insect somatic and germ line tissues. Insect Biochem. Mol. Biol. 29: 153–160.

Douglas, A.E. 1989. Mycetocyte symbiosis in insects. Biol. Rev. 64: 409–434.

Douglas, A.E. 2014. The molecular basis of bacterial-insect symbiosis. J. Mol. Biol. 426: 3830–3837.

Douglas, A.E. 2015. Multiorganismal insects: diversity and function of resident microorganisms. Annu. Rev. Entomol. 60: 17–34.

Douglas, A.E. and J.H. Werren. 2016. Holes in the hologenome: why host-microbe symbioses are not holobionts. mBio 7: e02099-15.

Dyson, E.A. and G.D.D. Hurst. 2004. Persistence of an extreme sex-ratio bias in a natural population. Proc. Natl. Acad. Sci. USA 101: 6520–6523.

Faria, V.G. and É. Sucena. 2013. *Wolbachia* in the Malpighian tubules: evolutionary dead-end or adaptation. J. Exp. Zool. B. Mol. Dev. Evol. 320: 195–199.

Flórez, L.V., P.H. Biedermann, T. Engl and M. Kaltenpoth. 2015. Defensive symbioses of animals with prokaryotic and eukaryotic microorganisms. Nat. Prod. Rep. 32: 904–936.

Goto, S., H. Anbutsu and T. Fukatsu. 2006. Asymmetrical interactions between *Wolbachia* and *Spiroplasma* endosymbionts coexisting in the same insect host. Appl. Environ. Microbiol. 72: 4805–4810.

Hamilton, P.T., F. Peng, M.J. Boulanger and S.J. Perlman. 2016. A ribosome-inactivating protein in a *Drosophila* defensive symbiont. Proc. Natl. Acad. Sci. USA 113: 350–355.

Harcombe, W. and A.A. Hoffmann. 2004. *Wolbachia* effects in *Drosophila melanogaster*: in search of fitness benefits. J. Invertebr. Pathol. 87: 45–50.

Hayashi, M., M. Watanabe, F. Yukuhiro, M. Nomura and D. Kageyama. 2016. A nightmare for males? A maternally transmitted male-killing bacterium and strong female bias in a green lacewing population. PLoS ONE 11: e0155794.

Hedges, L.M., J.C. Brownlie, S.L. O'Neill and K.N. Johnson. 2008. *Wolbachia* and virus protection in insects. Science 322: 702.

Hoffmann, A.A., B.L. Montgomery, J. Popovici, I. Iturbe-Ormaetxe, P.H. Johnson, F. Muzzi et al. 2011. Successful establishment of *Wolbachia* in *Aedes* populations to suppress dengue transmission. Nature 476: 454–457.

Hoffmann, A.A., M. Hercus and H. Dagher. 1998. Population dynamics of the *Wolbachia* infection causing cytoplasmic incompatibility in *Drosophila melanogaster*. Genetics 148: 221–231.

Ijichi, N., N. Kondo, R. Matsumoto, M. Shimada, H. Ishikawa and T. Fukatsu. 2002. Internal spatiotemporal population dynamics of infection with three *Wolbachia* strains in the adzuki bean beetle, *Callosobruchus chinensis* (Coleoptera: Bruchidae). Appl. Environ. Microbiol. 68: 4074–4080.

Jacquet, A., B. Horard and B. Loppin. 2017. Does pupal communication influence *Wolbachia*-mediated cytoplasmic incompatibility? Curr. Biol. 27: R53–R55.

Jaenike, J., R. Unckless, S.M. Cockburn, L.M. Boelio and S.J. Perlman. 2010. Adaptation via symbiosis: recent spread of a *Drosophila* defensive symbiont. Science 329: 212–215.

Jaenike, J. and S.J. Perlman. 2002. Ecology and evolution of host-parasite associations: mycophagous *Drosophila* and their parasitic nematodes. Am. Nat. 160: S23–S39.

Jiggins, F.M., G.D. Hurst and M.E. Majerus. 2000. Sex-ratio-distorting *Wolbachia* causes sex-role reversal in its butterfly host. Proc. R. Soc. Lond. B 267: 69–73.

Kageyama, D., H. Anbutsu, M. Watada, T. Hosokawa, M. Shimada and T. Fukatsu. 2006. Prevalence of a non-male-killing spiroplasma in natural populations of *Drosophila hydei*. Appl. Environ. Microbiol. 72: 6667–6673.

Kageyama, D., S. Narita and M. Watanabe. 2012. Insect sex determination manipulated by their endosymbionts: incidences, mechanisms and implications. Insects 3: 161–199.

Kaltenpoth, M., W. Göttler, G. Herzner and E. Strohm. 2005. Symbiotic bacteria protect wasp larvae from fungal infestation. Curr. Biol. 15: 475–479.

Kellner, R.L.L. and K. Dettner. 1996. Differential efficacy of toxic pederin in deterring potential arthropod predators of *Paederus* (Coleoptera: Staphylinidae) offspring. Oecologia 107: 293–300.

Koehler, S., J. Doubský and M. Kaltenpoth. 2013. Dynamics of symbiont-mediated antibiotic production reveal efficient long-term protection for beewolf offspring. Front. Zool. 10: 3.

Koehler, S. and M. Kaltenpoth. 2013. Maternal and environmental effects on symbiont-mediated antimicrobial defense. J. Chem. Ecol. 39: 978–988.

Kroiss, J., M. Kaltenpoth, B. Schneider, M.-G. Schwinger, C. Hertweck, R.K. Maddula et al. 2010. Symbiotic *Streptomycetes* provide antibiotic combination prophylaxis for wasp offspring. Nat. Chem. Biol. 6: 261–263.

LePage, D.P., J.A. Metcalf, S.R. Bordenstein, J. On, J.I. Perlmutter, J.D. Shropshire et al. 2017. Prophage WO genes recapitulate and enhance *Wolbachia*-induced cytoplasmic incompatibility. Nature 543: 243–247.

Margulis, L. and R. Fester. 1991. Symbiosis as a Source of Evolutionary Innovation: Speciation and Morphogenesis. MIT Press, Cambridge, MA, USA.

Mayer, E. 2016. The Mind-Gut Connection: How the Hidden Conversation within Our Bodies Impacts Our Mood, Our Choices, and Our Overall Health. HarperCollins Publishers, New York, USA.

Mayer, E., R. Knight, S.K. Mazmanian, J.F. Cryan and K. Tillisch. 2014. Gut microbes and the brain: paradigm shift in neuroscience. J. Neurosci. 34: 15490–15496.

Moran, N.A. and G.M. Bennett. 2014. The tiniest tiny genomes. Annu. Rev. Microbiol. 68: 195–215.

Moran, N.A. and D.B. Sloan. 2015. The hologenome concept: helpful or hollow? PLoS Biol. 13: e1002311.

Moreira, L.A., I. Iturbe-Ormaetxe, J.A. Jeffery, G. Lu, A.T. Pyke, L.M. Hedges et al. 2009. A *Wolbachia* symbiont in *Aedes aegypti* limits infection with dengue, chikungunya, and *Plasmodium*. Cell 139: 1268–1278.

Nakabachi, A., R. Ueoka, K. Oshima, R. Teta, A. Mangoni, M. Gurgui et al. 2013. Defensive bacteriome symbiont with a drastically reduced genome. Curr. Biol. 23: 1478–1484.

Narita, S., M. Nomura and D. Kageyama. 2007. Naturally occurring single and double infection with Wolbachia strains in the butterfly *Eurema hecabe*: transmission efficiencies and population density dynamics of each *Wolbachia* strain. FEMS Microbiol. Ecol. 61: 235–245.

Oh, D.-C., M. Poulsen, C.R. Currie and J. Clardy. 2009. Dentigerumycin: a bacterial mediator of an ant-fungus symbiosis. Nat. Chem. Biol. 5: 391–393.

Oliver, K.M., J.A. Russell, N.A. Moran and M.S. Hunter. 2003. Facultative bacterial symbionts in aphids confer resistance to parasitic wasps. Proc. Natl. Acad. Sci. USA 100: 1803–1807.

Oliver, K.M., N.A. Moran and M.S. Hunter. 2005. Variation in resistance to parasitism in aphids is due to symbionts not host genotype. Proc. Natl. Acad. Sci. USA 102: 12795–12800.

Oliver, K.M., P.H. Degnan, M.S. Hunter and N.A. Moran. 2009. Bacteriophages encode factors required for protection in a symbiotic mutualism. Science 325: 992–994.

Olsen, K., K.T. Reynolds and A.A. Hoffmann. 2001. A field cage test of the effects of the endosymbiont *Wolbachia* on *Drosophila melanogaster*. Heredity 86: 731–737.

Osaka, R., M. Watada, D. Kageyama and M. Nomura. 2010. Population dynamics of a maternally-transmitted *Spiroplasma* infection in *Drosophila hydei*. Symbiosis 52: 41–45.

Ota, T., M. Kawabe, K. Oishi and D.F. Poulson. 1979. Non-male-killing spiroplasmas in *Drosophila hydei*. J. Hered. 70: 211–213.

Paredes, J.C., J.K. Herren, F. Schüpfer and B. Lemaitre. 2016. The role of lipid competition for endosymbiont-mediated protection against parasitoid wasps in *Drosophila*. mBio 7: e01006-16.

Pietri, J.E., H. Debruhl and W. Sullivan. 2016. The rich somatic life of *Wolbachia*. Microbiology Open 5: 923–936.

Poinsot, D., S. Charlat and H. Merçot. 2003. On the mechanism of *Wolbachia*-induced cytoplasmic incompatibility: confronting the models with the facts. BioEssays 25: 259–265.

Pontier, S.M. and F. Schweisguth. 2015. A *Wolbachia*-sensitive communication between male and female pupae controls gamete compatibility in *Drosophila*. Curr. Biol. 25: 2339–2348.

Pontier, S.M. and F. Schweisguth. 2017. Response to "Does pupal communication influence *Wolbachia*-mediated cytoplasmic incompatibility?" Curr. Biol. R55–R56.

Raguso, R.A., A.A. Agrawal, A.E. Douglas, G. Jander, A. Kessler, K. Poveda et al. 2015. The raison d'etre of chemical ecology. Ecology 96: 617–630.

Rigaud, T. 1997. Inherited microorganisms and sex determination of arthropod hosts. pp. 81–101. *In*: S.L. O'Neill, A.A. Hoffmann and J.H. Werren [eds.]. Influential Passengers. Oxford University Press, Oxford, UK.

Rigaud, T., P. Juchault and J.-P. Mocquard. 1997. The evolution of sex determination in isopod crustaceans. BioEssays 19: 409–416.

Rosenberg, E. and I. Zilber-Rosenberg. 2011. Symbiosis and development: the hologenome concept. Birth Defects Res. C Embryo Today 93: 56–66.

Ruby, E.G., 1996. Lessons from a cooperative, bacterial-animal association: the *Vibrio fischeri–Euprymna scolopes* light organ symbiosis. Annu. Rev. Microbiol. 50: 591–624.

Sharon, G., D. Segal, J.M. Ringo, A. Hefetz, I. Zilber-Rosenberg and E. Rosenberg. 2010. Commensal bacteria play a role in mating preference of *Drosophila melanogaster*. Proc. Natl. Acad. Sci. USA 107: 20051–20056.

Stouthamer, R., J.A.J. Breeuwer and G.D.D. Hurst. 1999. Microbial manipulator of arthropod reproduction. Annu. Rev. Microbiol. 53: 71–102.

Teixeira, L., A. Ferreira and M. Ashburner. 2008. The bacterial symbiont *Wolbachia* induces resistance to RNA viral infections in *Drosophila melanogaster*. PLoS Biol. 6: e1000002.

Terry, R.S., A.M. Dunn and J.E. Smith. 1997. Cellular distribution of a feminizing microsporidian parasite: a strategy for transovarial transmission. Parasitology 115: 157–163.

Theis, K.R., N.M. Dheilly, J.L. Klassen, R.M. Brucker, J.F. Baines, T.C.G. Bosch et al. 2016. Getting the hologenome concept right: an eco-evolutionary framework for hosts and their microbiomes. mSystems 1: e00028-16.

Tsuchida, T., R. Koga and T. Fukatsu. 2004. Host plant specialization governed by facultative symbiont. Science 303: 1989.

Turelli, M. and A.A. Hoffmann. 1991. Rapid spread of an inherited incompatibility factor in California *Drosophila*. Nature 353: 440–442.

Turelli, M., A.A. Hoffmann and S.W. McKechnie. 1992. Dynamics of cytoplasmic incompatibility and mtDNA variation in natural *Drosophila simulans* populations. Genetics 132: 713–723.

Vance, S.A. 1996. Morphological and behavioural sex reversal in mermithid-infected mayflies. Proc. R. Soc. Lond. B 263: 907–912.

Wada-Katsumata, A., L. Zurek, G. Nalyanya, W.L. Roelofs, A. Zhang and Coby Schal. 2015. Gut bacteria mediate aggregation in the German cockroach. Proc. Natl. Acad. Sci. USA 112: 15678–15683.

Walker, T., P.H. Johnson, L.A. Moreira, I. Iturbe-Ormaetxe, F.D. Frentiu, C.J. McMeniman et al. 2011. The *w*Mel *Wolbachia* strain blocks dengue and invades caged *Aedes aegypti* populations. Nature 476: 450–453.

Watanabe, M., F. Yukuhiro, T. Maeda, K. Miura and D. Kageyama. 2014. Novel strain of *Spiroplasma* found in flower bugs of the genus *Orius* (Hemiptera: Anthocoridae): transovarial transmission, coexistence with *Wolbachia* and varied population density. Microb. Ecol. 67: 219–228.

Werren, J.H. 1997. Biology of *Wolbachia*. Annu. Rev. Entomol. 42: 587–609.

Werren, J.H., L. Baldo and M.E. Clark. 2008. *Wolbachia*: master manipulators of invertebrate biology. Nat. Rev. Microbiol. 6: 741–751.

Williamson, D.L. and D.F. Poulson. 1979. Sex ratio organisms (spiroplasmas) of *Drosophila*. pp. 175–208. *In*: R.F. Whitcomb and J.G. Tully [eds.]. The Mycoplasmas. Academic Press, New York, USA.

Wilson, A.C. and R.P. Duncan. 2015. Signatures of host/symbiont genome coevolution in insect nutritional endosymbioses. Proc. Natl. Acad. Sci. USA 112: 10255–10261.

Xie, J., I. Vilchez and M. Mateos. 2010. *Spiroplasma* bacteria enhance survival of *Drosophila hydei* attacked by the parasitic wasp *Leptopilina heterotoma*. PLoS ONE 5: e12149.

Xie, J., B. Tiner, I. Vilchez and M. Mateos. 2011. Effect of the *Drosophila* endosymbiont *Spiroplasma* on parasitoid wasp development and on the reproductive fitness of wasp-attacked fly survivors. Evol. Ecol. 25: 1065–1079.

Xie, J., C. Winter, L. Winter and M. Mateos. 2015. Rapid spread of the defensive endosymbiont *Spiroplasma* in *Drosophila hydei* under high parasitoid wasp pressure. FEMS Microbiol. Ecol. 91: 1–11.

Yong, E.D. 2016. I Contain Multitudes: The Microbes within Us and a Grander View of Life. The Bodley Head, London, UK.

Zug, R. and P. Hammerstein. 2012. Still a host of hosts for *Wolbachia*: analysis of recent data suggests that 40% of terrestrial arthropod species are infected. PLoS ONE 7: e38544.

CHAPTER

6

Chemical Ecology of Yeasts Associated with Insects

Jun Tabata* **and Hiroko Kitamoto**

National Agriculture and Food Research Organization
3-1-3 Kannondai, Tsukuba-city, Ibaraki 305-8604, Japan
E-mail: jtabata@affrc.go.jp (JT)
E-mail: kitamoto@affrc.go.jp (HK)

Abstract

Yeasts are ubiquitously found in various environments. Recent studies on yeast biodiversity have given more attention to insect ecosystems, because most insects have intimate mutualistic symbiosis with microorganisms including yeasts. This means insects enjoy the benefits of symbionts and their functions, which implies such associated microbes may have properties also useful for human use. Thus, understanding of the diversity and features of yeasts associated with insects is not only essential to elucidate the ecology and evolution of insect-microbial symbiosis but also has great potential for the development of innovative biotechnological tools. In this chapter, we review some cases that display characteristic biochemical or chemical ecological interactions among insects and associated yeasts, particularly those that are found in the phyllosphere and offer significant benefits for insect partners.

1. Introduction

In a broad sense, yeasts are eukaryotic, single-celled microorganisms belonging to the fungus kingdom. They have a cell wall but do not

*Corresponding author

have photosynthetic ability or motility. Thus, they take-up nutrition by decomposing and absorbing external organic matter. Yeast species can use their own carbon sources, and their trophic selectivity is intimately associated with their niche, which leads to great diversity among yeast species. In fact, yeasts exhibit vast species diversity with habitat specialization (Phaff et al. 1978). Yeasts include microbes of two different phyla, Ascomycota and Basidiomycota (Kurtzman and Fell 2006). Most yeasts belong to the former and, of these, members of the *Saccharomycetes* are called true yeasts.

Saccharomyces cerevisiae is one of the microorganisms most appreciated by humans because of its utility in the production of food and drinks (Stefanini et al. 2012). It owes this distinction to a single trait – its ability to produce alcohol from sugar (Goddard and Greig 2015). Ancient *S. cerevisiae* DNA has been found in prehistoric Chinese pots (7,000–5,500 B.C.) (McGovern et al. 2004) and in jars of the King Scorpion tomb in Abydos, Egypt (3,150 B.C.) (Cavalleri et al. 2003), indicating that humans have long used this yeast for wine fermentation. Nevertheless, despite its long-standing acquaintance with humans, its origin, natural habitats, ecology, and evolution are still being debated. *S. cerevisiae* preferentially produces alcohol from sugar by anaerobic fermentation, which is thought to be an adaptation to high sugar environments, and can lead to its predominance over competing microbes in crushed fruits including grapes, but *S. cerevisiae* is very scarce on natural healthy fruits; metagenome sequences suggest that members of the *Saccharomyces* comprise less than 0.005% of the fungi on ripe grapes in vineyards (Taylor et al. 2014). Intriguingly, insects are shown to play a key role as a vector and natural reservoir of *S. cerevisiae* during all seasons (Stefanini et al. 2012, 2016, Blackwell and Kurtzman 2016). Overwintering adult wasps can harbor yeast cells from autumn to spring and transmit them to their progeny (Stefanini et al. 2012). Moreover, *Saccharomyces* species have been demonstrated to be common in the intestinal tracts of social wasps including *Polistes* spp., which provide a "mating nest" for outcrossing of *S. cerevisiae* and *S. paradoxus*, indicating that insects are an important environmental niche for the evolution of yeasts as well as for their dispersion and maintenance (Stefanini et al. 2016).

Besides true yeasts, insects often harbor "yeast-like symbionts (YLSs)", which belong to the Ascomycota subclass Pezizomycotina (Vega and Dowd 2005). YLSs are obligatory insect-intracellular symbionts and are transmitted vertically from adult females to their offspring. Most of YLSs are not capable of free-living outside of host insect cells, and endosymbiotic systems between YLSs and insects seem to be much more significant to the insect than to the yeast, and as such it is hard to discern the mutualism (Vega and Dowd 2005). More equal associations are expected in facultative symbiosis of environmental yeasts and insects.

Insects can offer a dispersal opportunity and a favorable environment for yeasts (Cooke 1977). In order to take these benefits by manipulating insects, environmental yeasts often modify their habitats and resources. Such yeast-mediated biochemical processes provide great potential for studies to understand the profound interactions among insects and microbes and to apply insights for agricultural and/or industrial innovations. In this chapter, we review biochemical and ecological studies to illustrate the associations of yeasts and insects in various environments.

2. Yeasts on Fruits and Flowers

Numerous kinds and amounts of yeasts have been found on fruits that include plenty of saccharides for their nutrition resource. *S. cerevisiae* on grapes, which induce ethanol fermentation for wine, is a typical example. Historically, many strains of *S. cerevisiae* have been isolated and characterized for use in producing foods and beverages. Flowers are also well-studied sources of yeast flora; most flowers possess nectar that also supplies nutritional resources for yeasts, although relatively limited groups of yeasts are likely to be found from flower nectar (Phaff et al. 1978). For example, yeast communities in ephemeral flowers in the Neotropical, Nearctic, and Australian biogeographic regions were mostly from four clades centered around the genera *Metschnikowia*, *Kodamaea*, *Wickerhamiella*, and *Starmerella*, which are vectored by beetles, drosophilids, and bees that visit flowers (Lachance et al. 2001).

2.1 *Drosophila*

Drosophilid flies, in particular members of the genus *Drosophila*, are one of the most studied insects that have an intimate association with yeasts. *Drosophila* currently encompass approximately 1,600 species, and nearly all are attracted to fruit baits (McEvey 2008). They mostly breed on decaying plant, flowers, rotting fruits, mushrooms, leaf litter, tree bark, exudates, and fluxes, where a rich flora of yeasts and other microbes are present (Ganter 2006). Thus, *Drosophila* flies inevitably carry yeast, and numerous of yeast species have been isolated from their bodies. Yeasts are also provided as food for both adults and larvae of *Drosophila*. Flies generally prefer to feed on particular yeast species when offered a choice between pure cultures of yeasts (Cooper 1960), indicating that flies can discriminate yeast species probably due to chemical cues from the cultures.

It has long been known that *Drosophila* adults search for yeast-fermented fruits that are provided as foods and breeding sites using olfactory cues, and a recent study demonstrated experimentally the individual contribution of fruit and yeast in host finding and reproductive success of flies (Becher et al. 2012). Of note, *S. cerevisiae* on its own is sufficient for

D. melanogaster attraction, oviposition, and larval development, whereas attraction and oviposition were significantly lower if non-fermented grape juice or growth media were used, and yeast-free grapes did not support larval development either. Odors emitted by fruits were only of secondary importance because fermenting yeast without fruits induced the same fly behavior as yeast fermenting on fruit. The odorant components of the fermenting *S. cerevisiae* were identified as ethanol, acetic acid, acetoin, 2-phenylethanol, and 3-methyl-1-butanol, the synthetic mixture of which was as attractive for flies as fermenting grape juice or fermenting yeast minimal medium (Fig. 1).

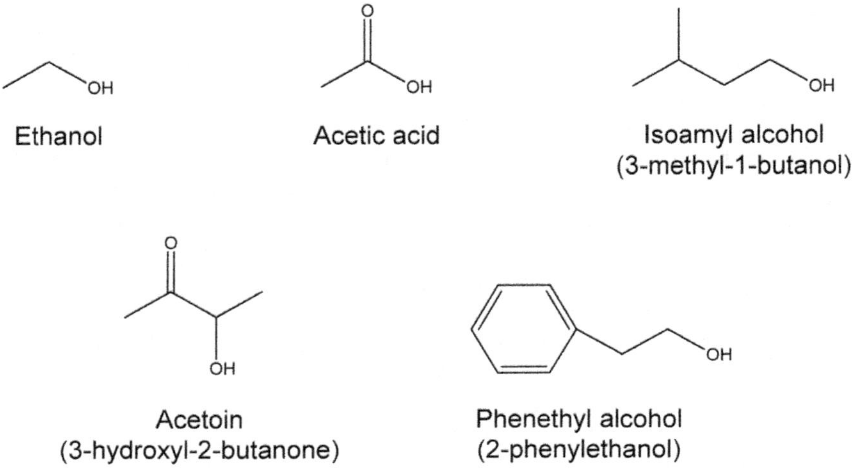

Fig. 1. Typical odorants emitted from fermenting *Saccharomyces cerevisiae*.

Moreover, each *Drosophila* species is likely to recognize its cognate yeasts based on their odorant profiles. For example, *D. suzukii*, which prefers to feed on fresh fruit rather than rotten fruit that are preferable for other *Drosophila* (Cini et al. 2012), appears to have evolved a strong mutualistic relationship with certain yeast species, including *Hanseniaspora uvarum* (Hamby et al. 2012). On the other hand, *D. melanogaster* has favorable associations with *H. uvarum, Candida zemplinina, C. californica,* and *Pichia kluyveri* (Stamps et al. 2012). Scheidler et al. (2015) performed behavioral preference assays with *D. melanogaster* and *D. suzukii* using six yeast species: *Pichia terricola, P. kluyveri, H. uvarum, C. californica, C. zemplinina,* and *S. cerevisiae*. *D. suzukii* showed the strongest response to *H. uvarum,* whereas *D. melanogaster* responded to *C. zemplinina* and *P. terricola* as well as *H. uvarum*. Chemical profiles of the yeast odor components, including isoamyl propionate and isoamyl acetate, were substantially different among the six species according to principal component analysis.

Yeast-odor attractiveness to *Drosophila* varies even among genotypes of the same yeast species. Buser et al. (2014) reported high levels of variation among strains of *S. cerevisiae* in their ability to modify fruits and attract *D. simulans*; flies showed significantly different attraction to 97 different genotypes of *S. cerevisiae* in a T-maze choice bioassay. Moreover, the more attractive strains were demonstrated to be dispersed more frequently by flies, and the flies associated with the more attractive strains were likely to have higher fecundity. These experimental results indicated mutualism between *D. simulans* and *S. cerevisiae*. The differences in attractiveness of the yeast strains could be attributed to the differences of intensities and patterns of metabolism to modify the environment and to release volatile organic compounds; actually, the volatile profiles of the attractive and repulsive strains were clearly different (Günther et al. 2015). Such "niche constructions" have been proposed to be a general mechanism driving the evolution of mutualism (Buser et al. 2014). Attraction to *S. cerevisiae* in *D. simulans* was promoted primarily by isoamyl acetate and repulsed by acetic acid, although these responses were strongly influenced by compound concentration and the environmental background chemical context (Günther et al. 2015). Chemical communication between yeasts and flies is complex, and is not driven simply by the presence of single volatiles, but modulated by compound interactions (Günther et al. 2015).

2.2 Pollinators

Trilateral mutualistic relationships among players including plants as well as insects and yeasts are often found in pollination systems. Nearly 400 strains of honeybee-associated yeasts have been found in the guts of both pollen- and nectar-feeding bees in California (Phaff and Starmer 1987). The relationships of yeasts and bees were suggested to be in a functional association, not a product of chance (Lachance et al. 2001, Stratford et al. 2002); exclusion experiments indicated that insects are obligate vectors for flower yeasts, although the degree to which yeasts and bees are associated remains uncertain (Ganter 2006). For bees, yeasts are considered to serve the fermentation process of transforming pollen into bee bread, a nutritious and preserved diet for larvae, and yeasts are also suspected of producing antibiotic substances that can keep hives healthy (Rosa et al. 2003, Gilliam 1997). Most of the yeast species associated with pollinator bees are members of *Candida, Cryptococcus, Starmerella,* and *Metschnikowia* (Rosa et al. 2003).

Torto et al. (2007a) shows an interesting example of a honeybee–yeast interaction in a different context from mutualism. The yeast *Kodamaea ohmeri* is vectored by a hive-attacking parasitoid, the small hive beetle (*Aethina tumida*) (Coleoptera: Nitidulidae), and it mimics the production of an alarm pheromone (isopentyl acetate) of the honeybee

Apis mellifera when grows on pollen in the hive. Alarm pheromones are released by honeybees when their colony is disturbed in order to induce colony defense behaviors, including stinging and mass attack, but these chemicals are often perceived by natural enemies as a kairomone, which is an attractive cue for recruiting predators and/or parasitoids. In this case, small hive beetles are strongly attracted to the honeybee alarm pheromone to parasitize hives (Torto et al. 2005, 2007b). The yeast *K. ohmeri* manipulates its vector beetle behavior by using an "imitation" of the pheromone of host bees to increase the opportunity for it to be carried by the beetles. It is noteworthy that the beetle attacks European honeybee colonies intensively but causes no serious damage to African colonies, because African honeybee subspecies have evolved effective methods to mitigate beetle infestation. Torto et al. (2007a) pointed out three key elements that appear responsible for the highly invasive attacks of small hive beetle on European honeybees: (i) the pest and disease management stresses faced by the European colonies, which are different from those experienced by the African colonies; (ii) the behavioral physiology of European honeybees resulting from domestication over many hundreds of years, for example a lower sensitivity to alarm pheromones than their African cousins (Breed et al. 2004); and (iii) the sophisticated chemical mimicry system associated with the mutualistic relationship between the beetle and yeast and based on the honeybee alarm pheromone and fermentation-related products, leading to an ideal situation for invasion and survival of the beetle because of significant stores of resources for the beetle and the failure of honeybees with a high threshold for perception of the alarm pheromones to recognize chemical signs of invasion.

Coleopteran insects are also often involved in pollination systems. Yeasts including species of the genera *Metschnikowia*, *Kodamaea*, *Wickerhamiella*, *Starmerella*, and *Candida* are found in association with flowers and floricolous beetles (Ganter 2006). Most of these beetles belong to the family Nitidulidae (sap beetles). Yeasts occur quite frequently in floral nectar, and, accordingly, yeast metabolism is supposed to contribute significantly to floral scent, which could influence the possibility for recruiting pollinators and may consequently have an impact for the distribution of nectar yeasts and the pollination success of flowers (Pozo et al. 2009, Herrera et al. 2009). Some flowers produce yeast-fermentation-like odorants even in their own scent (Jürgens 2009); for example, flowers of *Asimina* and *Deeringothamnus* (Annonaceae) smell "yeasty" and have been demonstrated to emit typical yeast fermentation volatiles including ethanol, ethyl acetate, acetic acid, and 3-methyl-1-butanol. This suggests that they mimic rotting fruit or sap, which is a likely food substrate and/or breeding site for saprophilous insects of temperate forests (Goodrich et al. 2006, Goodrich and Raguso 2009). Sap beetle pollinations are considered to be a modern convergence with similar pollination systems, and sap beetles could have used rotting plant material as food and breeding sites long before the evolutionary shifts of food habitat (Johnson and Schiestl 2016). Thus it is possible that the floral mimicry of the scent of microbial fermentation (Procheş and Johnson 2009) is an evolutionary consequence of a response to a pre-existing and ancient bias in the sensory systems of beetles (Schiestl and Dötterl 2012).

3. Yeast in Phylloplane and Caulosphere

The phyllosphere, which is the above-ground part of plants, is known as a key habitat for many microorganisms including both phytopathogens and nonphytopathogens (Lindow and Leveau 2002). In the phyllosphere, the phylloplane (leaves) and caulosphere (stems) particularly occupy a relatively large proportion of the aerial surface area of a plant. However, the aerial surfaces of plants are considered as a hostile environment for microorganisms, because the availability of nutrients is limited, exposure to sun irradiation is strong, and water availability varies (Lindow and Leveau 2002, Berlec 2012). The outer side of the phyllosphere is covered with a waterproof waxy film called the cuticle, which plays an essential role in the mechanical defense of plants against microbial invasion, herbivore infestation, and other external environmental stresses. Plant cuticles are composite structures composed of a covalently linked macromolecular scaffold of cutin and a variety of organic solvent-soluble lipids that are collectively termed waxes (Yeats and Rose 2013). The cuticle structure is often divided into two domains based on histochemical staining and its presumed chemical composition: a cutin-rich domain with embedded polysaccharides, which is referred to as the "cuticular layer," and an overlying layer that has a lower abundance of polysaccharides but is enriched in waxes, referred to as the "cuticle proper" (Yeats and Rose 2013). To colonize plant surfaces, some phyllosphere microorganisms are likely to obtain nutrients from the plant surface by degrading the cuticle itself; for example, a nonpathogenic yeast isolated from rice husks, *Pseudozyma antarctica*, characteristically produces many enzymes with fatty acid assimilation ability including a cutinase-like esterase as well as several lipases (Kitamoto et al. 2011). *In vitro* enzymatic assays clearly showed these enzymes can degrade lipid components of cuticle layers of plants, indicating they are secreted extracellularly to extract fatty acids from the plant surface and to metabolize them for the growth of yeasts (Ueda et al. 2015).

An alternative strategy of phyllosphere microorganisms to overcome the cuticle barriers of plants is to colonize a site of injury provided by herbivores. Two modes of feeding are found in herbivorous insects; (i) biting off and chewing plant tissues and (ii) imbibing liquid nourishment

(Schoonhoven et al. 2010). The former insects, known as mandibulates, possess the ancestral and more general type of mouthparts. Their mandibles serve to cut or scrape off plant epiphytic structures. Saps oozes from the wounds and provides propagation sites for microorganisms, although plants generally produce secondary metabolites with anti-microbial properties to prevent invasion by pathogens. In some insect phyla, the primitive mandibulate mouthparts borne on separate mouthpart regions have been converted to mouthpart types consisting of functionally integrated ensembles of fused elements, which are called haustellate mouthparts, and serve to pierce plant tissues and imbibe sap from inside them. The wounds inflicted by insects with haustellate mouthparts are relatively small, but the impact on plants is often more severe than that of mandibulates. Moreover, these sap-sucking insects such as aphids generally produce a large amount of excreta, which is a sugar-rich sticky liquid also called honeydew, and this serves as preferable media for microbial propagation (Leroy et al. 2012). Thus, associations among herbivorous insects and phyllosphere-indigenous microorganisms, including yeasts, may have been evolved frequently and ubiquitously.

3.1 Cactophilic *Drosophila*

One of the best studied models for insect–yeast interactions in the phyllosphere is the system of cactophilic *Drosophila* flies and their host cacti with unique yeasts around the Sonoran desert (Starmer and Fogleman 1986). This system shows solid species specificity among flies, cacti, and yeasts. In essence, each cactophilic *Drosophila* species uses a decaying cactus of its own preference as a host plant (Heed and Mangan 1986). Specific selections for host plants are found even among geographically isolated populations within the species; for example, four distinct populations of *D. mojavensis* using different cactus species across their range (Heed and Mangan 1986). Because *Drosophila* adults generally feed and breed on the same site of plants, assortative mating with respect to their host-plant preference can easily occur and lead to incipient speciation. Behavioral preferences for host plants have been demonstrated to be mediated by olfaction (Date et al. 2013). The chemical profiles of volatiles emitted by cacti are clearly different among species, and the electrophysiological as well as behavioral responses of *Drosophila* flies correspond to them. For example, odors from decaying barrel cactus *Ferocactus cylindraceus* include relatively high proportions of aromatic compounds, and barrel cactus-specialized *D. mojavensis* is especially responsive to aromatic compounds but not to esters of short-chain fatty acids, which are typical constituents of other cactus odors.

Of note, yeast communities in necrotic tissues are also likely linked with their host cactus species. Chemical constituents of each cactus plant have been shown to have direct influences on the distributions of yeast species through studies on yeast growth on media (Starmer and Fogleman 1986). For example, triterpene glycosides, which are characteristically present in tissues of *Stenocereus* cacti, strongly inhibit the growth of some yeast taxa including *Pichia heedii*, which is rare in *Stenocereus* tissues but prevalent in other cacti. Yeasts in decaying cacti appear to provide a detoxifying effect for *Drosophila* larvae; medium-chain fatty acids, which largely accumulate as ester forms in *Stenocereus* cacti, can slow development, reduce adult size, or are even lethal (Starmer 1982, Fogleman and Starmer 1985) for *D. mojavensis*. However, lipolytic yeasts including *Candida ingens* (= *Dipodascus starmeri*) and *Pichia mexicana* can decompose these compounds by hydrolytic processes by other microorganisms. The presence of lipolytic yeasts is, therefore, beneficial for *D. mojavensis* feeding on *Stenocereus* cacti. The better growth of *D. mojavensis* with faster development time and larger body size is in turn beneficial for yeasts because of the increasing opportunities and ability for dispersal. Thus cactophilic *Drosophila* and associated yeasts could have evolved as part of coadaptation processes (Starmer and Fogleman 1986). Odors of decaying cacti that can be modulated by yeast fermentation and are strongly associated with *Drosophila* host-preferences could have greatly impacted such an evolutionary scenario (Dobzhansky et al. 1956).

3.2 Exudates from Deciduous Trees

Nutrients flow from roots to leaves in the xylem and from leaves to roots in the phloem during foliation periods of deciduous trees, and exudates build up when the vascular system is injured. These exudates are generally sugar-rich and nitrogen-deficient and typically include sugars, small-molecule organic acids, amino acids, proteins, and minerals. Thus tree exudates (sap) serve as a preferred site for colonization by many microorganisms, including yeasts (Phaff and Starmer 1987). For example, yeasts of *Komagatalla*, *Ogataea*, or *Candida* have been isolated from tree exudates (Péter et al. 2003, 2006, Morais et al. 2004, Limtong et al. 2004, Kurtzman 2012). Some strains of these yeasts biosynthesize various flavor compounds and have been selected and used for producing a foods/ beverages with characteristic flavors (e.g. Suzuki et al. 2016).

Tree exudates colonized by microorganisms are also a preferred site for feeding and breeding of many insects (Phaff and Starmer 1987). These insects can be significant vectors for microbes including yeasts (Gilbert 1980). Accordingly, ecosystems with associations of yeasts and insects on tree exudates appear to be prevalent in temperate forests. However, trees generally have defense responses against mechanical damage to treat wounds from which sap is flowing (Knoblauch and van Bel 1998, van Bel 2003). Microbial patches and their associations with insects may, therefore, be temporary when trees promptly restore the wounds to stop sap efflux. However, long-term and significant interactions among multiple insect species on deciduous *Quercus* trees that produce sap, and inevitably with microbial colonies, have been found in Japan (Fig. 2). According to an extensive field survey, ≈5% of *Q. acutissima* trees showed long-term sap exudation, typically from May to November (Ichikawa and Ueda 2010). Moreover, ≈94% of these sap-exuding trees harbored larvae of the oriental carpenter moth, *Cossus jezoensis* (Lepidoptera: Cossidae). Larvae of cossid moths are live tree-borers and infest trunk tissues under the bark, where the vascular system is present, of various deciduous trees. Exudates from *Quercus* trees seem to ooze continuously with the feeding of *C. jezoensis* larvae. Yoshimoto and Nishida (2007) performed comparative experiments using trees with and without *C. jezoensis* larvae and confirmed the amount of exuded sap increased in trees infested with *C. jezoensis* larvae. Other kinds of insects, such as beetles and their

Fig. 2. *Quercus* tree exudates colonized by microorganisms including yeasts and offered as patches for feeding and mating of various insects: (a) a larva of the oriental carpenter moth (*Cossus jezoensis*), which promotes continuous exudate-oozing from a tree and prey insects attracted to odors from fermented exudates with hiding under its bark; (b) a nymphalid butterfly (*Hestina persimilis japonica*); (c) the drone beetle (*Pseudotorynorrhina japonica*); (d) the Japanese rhinoceros beetle (*Trypoxylus dichotomus*). Arrows indicate spots of tree exudates. Photos courtesy of Dr. Yukinobu Nakatani.

larvae of the family Cerambycidae, also feed on *Quercus* live trees and are potentially causative agents for exudates, but their effects appear to be relatively small, because the amount as well as the period of sap exudation was very limited in wounds damaged by these insects (Ichikawa and Ueda 2010).

Exudates from *Q. acutissima* trees include glucose (0.2–1.8%) and fructose (0.2–2.4%) as major constituents, which are converted to ethanol (1.0–2.6%), acetic acid (0.2–0.7%), or lactic acid (0.1–0.2%), via fermentation processes in microorganisms (Tanaka 2016). These microbes include yeasts of *Saccharomyces paradoxus*, *Pichia membranifaciens*, and *Kregervanrija fluxum* (Okabe 2010). Some volatile compounds released from fermented tree sap are attractive for sap feeding insects; for example, ethanol and acetic acid elicit attraction responses in a broad range of insects, including adult butterflies of Nymphalidae (Ômura et al. 2000, Ômura and Honda 2003), *Drosophila* (Devineni and Heberlein 2009), and vespine hornets (Day and Jeanne 2001). Thus tree exudates, which are promoted by larvae of the carpenter moth, are offered as patches for insect communities, including flies, hornets, ants, beetles, butterflies, and moths, with significantly higher species-richness, total abundance, and Simpson's index of diversity (Yoshimoto et al. 2005, Yoshimoto and Nishida 2008). Most notably, larvae of the carpenter moth often prey on these insects attracted to fermented sap (Ichikawa and Ueda 2010). The carpenter moth larvae have also been observed to feed on mites, such as *Hericia sanukiensis*, which propagate in sap fermented by microorganisms (Hayashi et al. 2010). Such a carnivorous feature of the carpenter moth is noteworthy, because more than 99% of moths and butterflies are phytophagous (Pierce 1995). Although members of the cossid moths are considered to be wood feeders, *C. jezoensis* is likely to have evolved to prey on other insects and mites by manipulating host trees to promote sap exudation and to lure prey insects to them. Microorganisms including yeasts and their fermentation processes produce attractive odors and preferable nutritional resources, thereby mediating interactions among these insects and constructions of community structures.

4. Yeast on Woods

Wood is a major component of forest biomass and is assumed to occupy approximately 90% of the whole biomass on earth (Tanahashi 1983, Kuwahara 1991). Wood is, therefore, the most abundant organic resource, but their tissues, which are constituted of cellulose and hemicellulose and are protected by lignin, are indigestible for animals. Most insects do not produce the enzymes necessary to degrade plant polysaccharides and lignin. Nevertheless, some phyla of insects, including beetles, moths, termites, and cockroaches, feed on wood at least in some part of their life history. Most of these insects are associated with microorganisms including yeasts, which are present inside or sometimes outside of the insect bodies and support degrading wood by supplying enzymes such as β-glucanases, β-glucosidases, and β-xylanase (Klemm et al. 2002). Microbes could benefit from insect infestation, which change the physical and physiological status of tree/wood tissues, as well as from vectoring along with insect dispersal.

4.1 Bark and Ambrosia Beetles

Bark beetles and related ambrosia beetles are currently included into two subfamilies of the Curculionidae, Scolytinae, and Platypodinae (Knížek and Beaver, 2004). Members of these groups inhabit living or decaying wood both in their adult and larval stages and include some important forestry pests (Goto 2009). These beetles possess mycangia, external pockets that hold microbes including fungi and yeasts (Whitney and Harris 1970). The structures of mycangia in bark and ambrosia beetles include complex cuticular invaginations and are well-diverged; bark beetles typically have many small pits on the surface of their body, whereas ambrosia beetles, which feed completely on their symbiotic fungi, have specialized mandibular or mesothorax pouches (Hulcr and Cognato 2010). These mycangia are often secretory, producing waxes, fatty acids, and amino acids, to hold and nourish spores and cells during transport (Ganter 2006).

Some bark beetles attack living trees, which sometimes leads to plant death. The impact on trees largely depends on the numbers of beetles, and their aggregation behaviors are often mediated by yeasts. For example, adult females of the mountain pine beetle, *Dendroctonus ponderosae*, produce *trans*-verbenol as an aggregation pheromone to attract conspesific individuals (Pitman et al. 1968), and two yeasts associated with this species, *Hansenula capsulata* and *Pichia pini* (= *Ogataea pini*), have been demonstrated to convert *trans*-verbenol efficiently into verbenone (Fig. 3). Verbenone, which Hunt and Borden (1989) indicated that microbe-reduced *D. ponderosae* are incapable of producing, had anti-aggregation properties (Ryker and Yandell 1983). Thus, the bark beetle appears to rely primarily on microbial symbionts for terminating aggregation and mass attack on host trees. A highly persistent association between *D. ponderosae* and these two yeasts are observed; no populations of *D. ponderosae* were found to be free of these yeasts in an extensive survey, indicating a strong mutualistic association (Whitney 1971). Other yeasts capable of oxidizing *cis*- and *trans*-verbenols to verbenone (Leufvén et al. 1984) have been frequently isolated from *Dendroctonus* and *Ips* bark beetles, many of which use verbenols for aggregation and verbenone for anti-aggregation. Hence

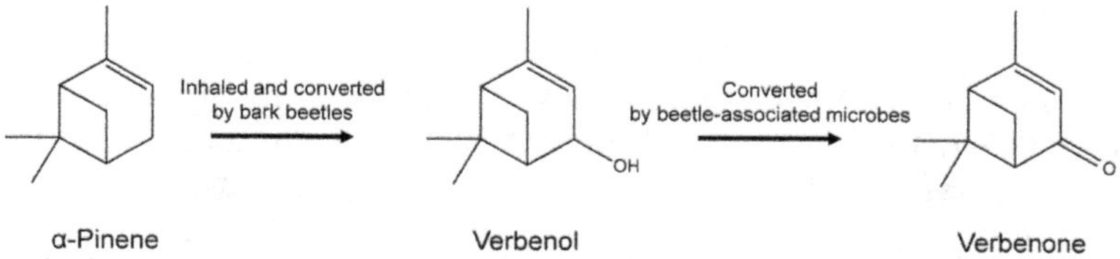

Fig. 3. Biosynthesis of aggregation and anti-aggregation pheromones of the mountain pine beetle, *Dendroctonus ponderosae*. Microbes, including two yeasts (*Hansenula capsulata* and *Pichia pini*), associated with bark beetles are attributed to the oxidative conversion of verbenol to verbenone (Hunt and Borden 1990).

yeast-mediated behavioral shifts from aggregation to dispersal may be widespread in bark beetles (Hunt and Borden 1990).

Ambrosia beetles are mycetophagous and feed on ambrosial fungi, which grow in tunnels and galleries in woods. These beetles are obligate mutualists with vertically transmitted ambrosial fungi, which are currently placed in polyphyletic genera including *Ambrosiella*, *Rafaelea*, and *Dryadomyces* (Ascomycetes: Ophiostomatales) (Mueller et al. 2005). Ambrosial associations are likely to have evolved multiple times in the lineages of Scolytinae and Platypodinae. Ambrosia beetles carry ambrosial fungi in their mycangia and cultivate them in tunnels and galleries in host trees, and they are often referred to as insect agriculturalists (Mueller et al. 2005). In addition to ambrosial fungi, most of ambrosia beetles harbor other euascomycete fungi and yeasts in their mycangia (Ganter 2006), and these auxiliary microbes have been demonstrated to be able to support the growth and development of ambrosia beetles (Batra 1966). It appears that agricultural monoculture with primary ambrosia fungi is potentially harmful to the beetles (Ganter 2006). Ambrosia beetles can discriminate the scent of a partner fungus; *Xyleborus* and *Xylosandrus* beetles have been shown to be attracted to the odors of their own symbiotic fungal species but never to non-symbiotic fungus (Hulcr et al. 2011). Moreover, most ambrosia beetles respond to ethanol, a chemical indicator of plant stress (Rangar et al. 2010, 2012, 2013, 2015, 2016), probably as a signal for decaying and fermenting plant tissue during propagations of microorganisms including yeasts.

4.2 Stag Beetles and Bess Beetles

Stag beetles (Coleoptera: Lucanidae) are well-diverged in their morphology as well as their biology. Adults of many species feed on plant sap or fruit, whereas adults of some species are carnivorous and prey on other insects (Tanaka 2016). However, the larvae of lucanid beetles commonly inhabit decaying wood and feed on wood tissues and/or fungi growing on wood (Araya 1993, Tanahashi et al. 2009). Some lucanids appear to show preferences for wood-decaying types, which are related to the faunae of microorganisms on woods (Kirk and Cowling 1984, Takahashi 1986); for example, *Aesalus asiaticus* and *Platycerus acuticollis* are more often found in brown-rot and soft-rot woods, respectively; nevertheless the great majority of decaying woods are white-rot (76%), in an extensive survey in Japanese forests (Araya 1993). These findings indicated significant associations among stag beetles and microorganisms.

Recent studies discovered that adult females of most of stag beetles possess mycangia in their abdomens (Tanahashi et al. 2010, Tanahashi and Hawes 2016). Notably, these organs for holding microbes are absent in adult males (Tanahashi et al. 2010). From female-specific mycangia of stag beetles yeasts have been isolated that are closely related to *Pichia stipites* and *Pichia segobiensis* with xylose-fermenting abilities (Tanahashi et al. 2010). A homologous mycangium is present in the European horned stag beetle, *Sinodendron cylindricum*, which belongs to an ancestral clade of the family Lucanidae (Tanahashi and Hawes 2016). Yeasts in the mycangia of *S. cylindricum* have, however, been shown to be *Scheffersomyces insectosa*, *Sugiyamaella* spp., and an undescribed taxonomically novel yeast, all of which were confirmed to metabolize xylose. *Sugiyamaella* yeasts are common in the gut of wood-feeding bess beetles (Coleoptera: Passalidae) (Houseknecht 2011), a sister phyla of Lucanidae, implying the evolutionary origin of symbiosis among yeasts and stag beetles (Tanahashi and Hawes 2016). Individual larvae of *S. cylindricum* harbor 10^4–10^6 living yeast in each gut, indicating that yeasts are associated with the digestion of wood (Tanahashi and Hawes 2016). Suzuki et al. (2013) discovered another living yeast strain, which was very closely related to *Cryptococcus magnus*, from the larval gut of *Aegus subnitidus* (= *A. laevicollis subnitidus*). These yeasts produce extensive amounts of cutinase-like esterase and appear to be beneficial for stag beetles to digest wood tissues. Because various strains of *C. magnus* have been isolated from trees, including the wattle tree, Portuguese oak, Montpellier maple, lemon tree, and from shrubs (Fonseca et al. 2011), *C. magnus*-like yeasts of *A. subnitidus* could probably have been colonized freely on wood and have been inhaled by stag beetle larvae (Suzuki et al. 2013). Such an association is facultative but is considered to be significant for the lifecycles of both beetles and yeasts.

Beetles of the family Passalidae, which are referred as bess beetles, bessbugs, or betsy beetles, are phylogenetically close to stag beetles (Boucher 2005, Hosoya and Araya 2005). Bess beetles are known as sub-social insects living in groups in rotting woods (Schuster and Schuster 1985). A pair of adult beetles constructs a multigeneration colony with their offspring larvae, which are fed and cared for by parental beetles. Both adults and larvae feed on adult fecal pellets plastered onto tunnel walls. Such coprophagy is possibly a means of providing microflora time and the proper conditions for further digestion of refractory wood tissues including lignocellulose (Ganter 2006), the process of which is sometimes described as an external rumen. *Scheffersomyces* (*Pichia*) *stipitis*-like yeasts, which have a xylose-fermenting ability, are consistently found in two bess beetles, *Odontotaenius disjunctus* and *Verres sternbergianus*, over a wide area including the eastern and mid-western USA and Panama (Suh et al. 2003). The consistent association of xylose-fermenting yeasts with almost identical genotypes with these bess beetles across a broad geographical distribution suggests a significant symbiotic association (Suh et al. 2003). Moreover, 771 yeast cultures were isolated from the gut of 16 species of bess beetles collected in nine localities in Guatemala (Urbina et al. 2013).

5. Yeast-like Symbionts Harbored by Insects

A typical example of an association with obligate intracellular yeast-like symbionts was found in rice plant hoppers, including *Nilaparvata lugens*, *Laodelphax striatellus*, and *Sogatella furcifera* (Hemiptera: Delphacidae). These insects are serious pests in paddy fields in East and Southeast Asian countries and are considered to have been one of the causative agents of famines in the history of these regions (Wada 2015). These insects possess YLSs in mycetocytes derived from fat body cells of the abdomen (Buchner 1965). The symbionts are transmitted maternally to the offspring by transovarial infection (Noda 1974, 1977, Mitsuhashi 1975, Cheng and Hou 2001) and are involved in sterol metabolism (Noda and Saito 1979a, b, Noda et al. 1979, Eya et al. 1989, Noda and Koizumi 2003), nitrogen recycling (Sasaki et al. 1996), or amino acid biosynthesis (Xue et al. 2014) in host insects. Heat treatment can reduce YLS populations harbored by plant hoppers and can generate aposymbiotic insects; heat-treated *L. striatellus* with decreased YLS populations were deficient in producing sufficient 24-methylenecholesterol and often failed to molt into adults (Noda et al. 1979). Eya et al. (1989) demonstrated that YLS isolated from *L. striatellus* eggs produce lanosterol, 24-methylenelanosterol, dihydroergosterol, and ergosterol in the culture medium. Obligatory symbioses with intracellular microbes including YLSs are found in many phytophagous hemipteran insects feeding on plant phloem or epidermal sap, which are deficient in some nutrition such as essential amino acids, vitamins, and sterols, and symbionts are generally considered to play roles of supplying essential nutrients (Vega and Dowd 2005).

A recent study that generated a draft genome of the *N. lugens* YLS demonstrated that this symbiont had adapted to an endosymbiotic lifecycle within the host insect (Fan et al. 2015). Some genes of the YLS, for example sexual reproduction-related genes, had been lost during the evolution of symbiosis with the plant hopper. Additionally, its genome size was smaller than free-living ascomycete fungi, although the reduction in size was not as pronounced in bacterial endosymbionts, which possess a fraction of the genome of free-living bacterial relatives (McCutcheon and Moran 2012). The YLS-specific genes were mainly involved in metabolism, degradation, and biosynthesis, reflecting the special role that YLSs play in providing the host with metabolic capabilities (Fan et al. 2015). In addition, phylogenetic analysis suggested that the *N. lugens* YLS was closely related to entomopathogenic ascomycetes (Fan et al. 2015), as hypothesized previously (Fukatsu and Ishikawa 1996). Divergence time estimation revealed that YLS diverged from the ancestors of *Cordyceps*, *Beauveria*, and *Metarhizium* fungi approximately 99–203 million years ago (Fan et al. 2015). Although the phyla from which the YLS was derived are still unclear, the closest relative is *Tolypocladium inflatum*, which is widely distributed in soils and is also a pathogen of beetles (Fan et al. 2015).

Another well-known association of YLSs in insects is found in two anobiid beetles, the cigarette beetle (*Lasioderma serricorne*) and the drugstore beetle (*Stegobium paniceum*) (Coleoptera: Anobiidae). YLSs (*Symbiotaphrina*) are harbored intracellularly in enlarged cells in grape-bunch-like proliferated tissues around the beetle midgut (Buchner 1965). Adult female beetles smear YLSs on their eggshells, which is consumed by the newborn larvae, resulting in vertical and obligatory transmission of YLSs (Buchner 1965). *Symbiotaphrina* symbionts have been demonstrated to supply host beetles with essential amino acids (Pant et al. 1960), sterols (Pant and Fraenkel 1954), and vitamins (Fraenkel and Blewett 1943). In addition, *S. kochii* possessed by the cigarette beetle was shown to produce digestive enzymes including lipases, α- and β-glucosidases, phosphatase, and trypsin to degrade complex polymers including polysaccharides in plant tissues and to support digestion in the host beetle (Shen and Dowd 1991). Moreover, these YLSs were likely to contribute to the detoxifications of plant toxins fed to host beetles. Cultures of *S. kochii* were demonstrated to be able to assimilate a variety of plant flavonoids, plant aromatic acids, and other plant toxins, as well as mycotoxins, insecticides, and herbicides as sole carbon sources (Shen and Dowd 1991). Esterase activity attributed to detoxification was shown to be induced by flavone, griseofulvin, *cis*-β-pinene, and malathion in *S. kochii* cultures (Shen and Dowd 1989).

6. Conclusion

In addition to the above, various taxa of insects are intimately associated with yeasts in the environment. Recent advances in molecular biological tools make it possible to comprehensively analyze the microbial fauna in insects and their habitat sphere, which will drive further research on the relationships among insects and microorganisms, including yeasts.

However, we still have few studies that have completely characterized the functions and features of individual symbionts in the lifecycle of its related insect. This may be attributed to the problem that only a proportion of symbionts can be cultured, and that most of the associated insects, except some model species including *Drosophila*, are difficult to manipulate.

Once we clarify the function and features of microorganisms that play a significant role in insect ecosystems, it cannot only lead to understand of the ecology and evolution of insects but it may also open the way to apply these microbes or their properties for practical uses. For example, the *Cryptococcus magnus*-like yeast, which is found in a stag beetle gut and is probably associated with digesting lignocellulosic wood tissues, was demonstrated to be used for decomposing biodegradable plastics more reliably and effectively; the yeasts produced substantial amounts of cutinase-like enzymes that successfully degrade poly(butylene succinate) (PBS) and poly(butylene succinate-*co*-adipate) (PBSA), which cannot be degraded by the type strain of *C. magnus* at all (Suzuki et al. 2013). Insects and their habitats are undoubtedly a promising source of biodiversity of microorganisms including yeasts (e.g. Suh and Blackwell 2005). It is highly likely that novel yeasts with more useful properties will be discovered in further investigations on previously less well-studied insects and their associated environments.

Acknowledgments

The authors thank Drs Yukinobu Nakatani and Daisuke Watabiki (NARO) for providing valuable comments and good photographs.

REFERENCES

Araya, K. 1993. Relationship between the decay types of dead wood and occurrence of lucanid beetles (Coleoptera: Lucanidae). Appl. Entomol. Zool. 28: 27–33.

Batra, L.R. 1966. Ambrosia fungi—extent of specificity to ambrosia beetles. Science 153: 193–195.

Becher, P.G., G. Flick, E. Rozpędowska, A. Schmidt, A. Hagman, S. Lebreton et al. 2012. Yeast, not fruit volatiles mediate *Drosophila melanogaster* attraction, oviposition and development. Funct. Ecol. 26: 822–828.

Berlec, A. 2012. Novel techniques and findings in the study of plant microbiota: search for plant probiotics. Plant Sci. 193–194: 96–102.

Blackwell, M. and C.P. Kurtzman. 2016. Social wasps promote social behavior in *Saccharomyces* spp. Proc. Natl. Acad. Sci. USA 113: 1971–1973.

Boucher, S. 2005. Évolution et phylogénie des Coléoptères Passalidae (Scarabaeoidea). Ann. Soc. Entomol. Fr. 41: 239–604.

Breed, M.D., E. Guzmán-Novoa and G.J. Hunt. 2004. Defensive behavior of honey bees: organization, genetics, and comparisons with other bees. Annu. Rev. Entomol. 49: 271–298.

Buchner, P. 1965. Endosymbiosis of Animals with Plant Microorganisms. InterScience Publishers, New York, NY, USA.

Buser, C.C., R.D. Newcomb, A.C. Gaskett and M.R. Goddard. 2014. Niche construction initiates the evolution of mutualistic interactions. Ecol. Lett. 17: 1257–1264.

Cavalieri, D., P.E. McGovern, D.L. Hartl, R. Mortimer and M. Polsinelli. 2003. Evidence for *S. cerevisiae* fermentation in ancient wine. J. Mol. Evol. 57(Suppl 1): S226–S232.

Cheng, D.J. and R.F. Hou. 2001. Histological observations on transovarial transmission of a yeast-like symbiote in *Nilaparvata lugens* Stål (Homoptera, Delphacidae). Tissue Cell. 33: 273–279.

Cini, A., C. Ioriatti and G. Anfora. 2012. A review of the invasion of *Drosophila suzukii* in Europe and a draft research agenda for integrated pest management. Bull. Insectol. 65: 149–160.

Cooke, R. 1977. The Biology of Symbiotic Fungi. John Wiley & Son, New York, NY, USA.

Cooper, D.M. 1960. Food preferences of larval and adult *Drosophila*. Evolution 14: 41–55.

Date, P., H.K.M. Dweck, M.C. Stensmyr, J. Shann, B.S. Hansson and S.M. Rollmann. 2013. Divergence in olfactory host plant preference in *D. mojavensis* in response to cactus host use. PLoS ONE 8: e70027

Day, S.E. and R.L. Jeanne. 2001. Food volatiles as attractants for yellowjackets (Hymenoptera: Vespidae). Environ. Entomol. 30: 157–165.

Devineni, A.V. and U. Heberlein. 2009. Preferential ethanol consumption in *Drosophila* models features of addiction. Curr. Biol. 19: 2126–2132.

Dobzhansky, Th., D.M. Cooper, H.J. Phaff, E.P. Knapp and H.L. Carson. 1956. Studies on the ecology of *Drosophila* in the Yosemite region of California. IV. Differential attraction of species of *Drosophila* to different species of yeasts. Ecology 37: 544–550.

Eya, B.K., P.T.M. Kenny, S.Y. Tamura, M. Ohnishi, Y. Naya, K. Nakanishi et al. 1989. Chemical association in symbiosis: sterol donor in planthoppers. J. Chem. Ecol. 15: 373–380.

Fan, H.-W., H. Noda, H.-Q. Xie, Y. Suetsugu, Q.-H. Zhu and C.-X. Zhang. 2015. Genomic analysis of an ascomycete fungus from the rice planthopper reveals how it adapts to an endosymbiotic lifestyle. Genome Biol. Evol. 7: 2623–2634.

Fogleman, J.C. and W.T. Starmer. 1985. Analysis of the community structure of yeasts associated with the decaying stems of cactus. III. *Stenocereus thurberi*. Microb. Ecol. 11: 165–173.

Fonseca, Á., T. Boekhout and J.W. Fell. 2011. *Cryptococcus* Vuillemin (1901). pp. 1661–1737. *In*: C.P. Kurtzman, J.W. Fell and T. Boekhout [eds.]. The Yeasts: A Taxonomic Study, 5th Edition. Elsevier, New York, NY, USA.

Fraenkel, G. and M. Blewett. 1943. Intracellular symbionts of insects as sources of vitamins. Nature 152: 506–507.

Fukatsu, T. and H. Ishikawa. 1996. Phylogenetic position of yeast-like symbiont of *Hamiltonaphis styraci* (Homoptera, Aphididae) based on 18S rDNA sequence. Insect Biochem. Mol. Biol. 26: 383–388.

Ganter, P.F. 2006. Yeast and invertebrate associations. pp. 303–370. *In*: C. Rosa and G. Péter [eds.]. Biodiversity and Ecophysiology of Yeasts. Springer-Verlag, Berlin, Germany.

Gilbert, D.C. 1980. Dispersal of yeasts and bacteria by *Drosophila* in a temperate forest. Oecologia 46: 135–137.

Gilliam, M. 1997. Identification and roles of non-pathogenic microflora associated with honey bees. FEMS Microbiol. Lett. 155: 1–10.

Goddard, M.R. and D. Greig. 2015. *Saccharomyces cerevisiae*: a nomadic yeast with no niche? FEMS Yeast Res. 15: fov009.

Goodrich, K.R. and R.A. Raguso. 2009. The olfactory component of floral display in *Asimina* and *Deeringothamnus* (Annonaceae). New Phytol. 183: 457–469.

Goodrich, K.R., M.L. Zjhra, C.A. Ley and R.A. Raguso. 2006. When flowers smell fermented: The chemistry and ontogeny of yeasty floral scent in Pawpaw (*Asimina triloba*: Annonaceae). Int. J. Plant Sci. 167: 33–46.

Goto, H. 2009. Taxonomic history of Japanese bark and ambrosia beetles with a check list of them. J. Jpn. For. Soc. 91: 479–485.

Günther, C.S., M.R. Goddard, R.D. Newcomb and C.C. Buser. 2015. The context of chemical communication driving a mutualism. J. Chem. Ecol. 41: 929–936.

Hamby, K.A., A. Hernandez, K. Boundy-Mills and F.G. Zalom. 2012. Associations of yeasts with spotted-wing drosophila (*Drosophila suzukii*; Diptera: Drosophilidae) in cherries and raspberries. Appl. Environ. Microbiol. 78: 4869–4873.

Hayashi, K., T. Ichikawa and Y. Yasui. 2010. Life history of the newly discovered Japanese tree sap mite, *Hericia sanukiensis* (Acari, Astigmata, Algophagidae). Exp. Appl. Acarol. 50: 35–49.

Heed, W.B. and R.L. Mangan. 1986. Community ecology of the Sonoran desert *Drosophila*. pp. 311–345. *In*: M. Ashburner, H.L. Carson and J.N. Thompson [eds.]. The Genetics and Biology of Drosophila. Academic Press, New York, NY, USA.

Herrera, C.M., C. de Vega, A. Canto and M.I. Pozo. 2009. Yeasts in floral nectar: a quantitative survey. Ann. Bot. 103: 1415–1423.

Hosoya, T. and K. Araya. 2005. Phylogeny of Japanese stag beetles (Coleoptera: Lucanidae) inferred from 16S mtrRNA gene sequences, with reference to the evolution of sexual dimorphism of mandibles. Zool. Sci. 22: 1205–1318.

Houseknecht, J.L., E.L. Hart, S.-O. Suh and J.J. Zhou. 2011. Yeasts in the *Sugiyamaella* clade associated with wood-ingesting beetles and the proposal of *Candida bullrunensis* sp. nov. Int. J. Syst. Evol. Microbiol. 61: 1751–1756.

Hulcr, J. and A.I. Cognato. 2010. Repeated evolution of crop theft in fungus-farming ambrosia beetles. Evolution 64: 3205–3212.

Hulcr, J., R. Mann and L.L. Stelinski. 2011. The scent of a partner: ambrosia beetles are attracted to volatiles from their fungal symbionts. J. Chem. Ecol. 37: 1374–1377.

Hunt, D.W.A. and J.H. Borden. 1989. Terpene alcohol pheromone production by *Dendroctonus ponderosae* and *Ips paraconfusus* (Coleoptera: Scolytidae) in the absence of readily culturable microorganisms. J. Chem. Ecol. 15: 1433–1463.

Hunt, D.W.A. and J.H. Borden. 1990. Conversion of verbenols to verbenone by yeasts isolated from *Dendroctonus ponderosae* (Coleoptera: Scolytidae). J. Chem. Ecol. 16: 1385–1397.

Ichikawa, T. and K. Ueda. 2010. Predation on exuded sap-dependant arthropods by the larvae of oriental carpenter moth, *Cossus jezoensis* (Matsumura) (Lepidoptera: Cossidae): preliminary observations. Techn. Bull. Fac. Agr. Kagawa Univ. 62: 39–58.

Johnson, S.D. and F.P. Schiestl. 2016. Floral Mimicry. Oxford University Press, Oxford, UK.

Jürgens, A. 2009. The hidden language of flowering plants: floral odours as a key for understanding angiosperm evolution? New Phytol. 183: 240–243.

Kirk, T.K. and E.B. Cowling. 1984. Biological decomposition of solid wood. pp. 455–487. *In*: R. Rowell [ed.]. The Chemistry of Solid Wood. American Chemical Society, Washington D.C., USA.

Kitamoto, H.K., Y. Shinozaki, X.H. Cao, T. Morita, M. Konishi, K. Tago et al. 2011. Phyllosphere yeasts rapidly break down biodegradable plastics. AMB Express 1: 44.

Klemm, D., H.-P. Schmauder and T. Heinze. 2002. Cellulose. pp. 275–319. *In*: E.J. Vandamme, S. De Baets and A. Steinbuchel [eds.]. Biopolymers. Vol. 6. Polysaccharides II: Polysaccharides from Eukaryotes. John Wiley & Sons, New York, NY, USA.

Knížek, M. and R. Beaver. 2004. Taxonomy and systematics of bark and ambrosia beetles. pp. 41–54. *In*: F. Lieutier, K.R. Day, A. Battisti, J.-C. Grégoire and H.F. Evans [eds.]. Bark and Wood Boring Insects in Living Trees in Europe, a Synthesis. Springer, Dordrecht, Netherlands.

Knoblauch, M. and A.J.E. van Bel. 1998. Sieve tubes in action. Plant Cell 10: 35–50.

Kurtzman, C.P. 2012. *Komagataella populi* sp. nov. and *Komagataella ulmi* sp. nov., two new methanol assimilating yeasts from exudates of deciduous trees. Antonie Van Leeuwenhoek 101: 859–868.

Kurtzman, C.P. and J.W. Fell. 2006. Yeast systematics and phylogeny – implications of molecular identification methods for studies in ecology. pp. 11–30. *In*: C. Rosa and G. Péter [eds.]. Biodiversity and Ecophysiology of Yeasts. Springer-Verlag, Berlin, Germany.

Kuwahara, M. 1991. Genetic engineering in utilizaiton of wood biomass. Wood Res. Tech. Note. 27: 12–23.

Lachance, M.-A., W.T. Starmer, C.A. Rosa, J.M. Bowles, J.S. Barker and D.H. Janzen. 2001. Biogeography of the yeasts of ephemeral flowers and their insects. FEMS Yeast Res. 1: 1–8.

Leroy, P.D., A. Sabri, S. Heuskin, P. Thonart, G. Lognay, F.J. Verheggen et al. 2012. Microorganisms from aphid honeydew attract and enhance the efficacy of natural enemies. Nat. Commun. 2: 348.

Leufvén, A., G. Bergström and E. Falsen. 1984. Interconversion of verbenols and verbenone by identified yeasts isolated from the spruce bark beetle *Ips typographus*. J. Chem. Ecol. 10: 1349–1361.

Limtong, S., N. Srisuk, W. Yongmanitchai, H. Kawasaki, H. Yurimoto, T. Nakase et al. 2004. Three new thermotolerant methylotrophic yeasts, *Candida krabiensis* sp. nov., *Candida sithepensis* sp. nov., and *Pichia siamensis* sp. nov., isolated in Thailand. J. Gen. Appl. Microbiol. 50: 119–127.

Lindow, S.E. and J.H.J. Leveau. 2002. Phyllosphere microbiology. Curr. Opin. Biotechnol. 13: 238–243.

McCutcheon, J.P. and N.A. Moran. 2012. Extreme genome reduction in symbiotic bacteria. Nat. Rev. Microbiol. 10: 13–26.

McGovern, P.E., J. Zhang, J. Tang, Z. Zhang, G.R. Hall, R.A. Moreau et al. 2004. Fermented beverages of pre- and proto-historic China. Proc. Natl. Acad. Sci. USA 101: 17593–17598.

McEvey, S.F. 2008. Taxonomic review of the Australian *Drosophila setifemur* species group, a new name for the *D. dispar* species group (Diptera: Drosophilidae). Rec. Aust. Mus. 61: 31–38.

Mitsuhashi, J. 1975. Cultivation of intracellular yeast-like organisms in the smaller brown planthopper, *Laodelphax striatellus* Fallén (Hemiptera: Delphacidae). Appl. Entomol. Zool. 10: 243–245.

Morais, P.B., L.C. Teixeira, J.M. Bowles, M.A. Lachance and C.A. Rosa. 2004. *Ogataea falcaomoraisii* sp. nov., a sporogenous methylotrophic yeast from tree exudates. FEMS Yeast Res. 5: 81–85.

Mueller, U.G., N.M. Gerardo, D.K. Aanen, D.L. Six and T.R. Schultz. 2005. The evolution of agriculture in insects. Annu. Rev. Ecol. Evol. Syst. 36: 563–595.

Noda, H. 1974. Preliminary histological observation and population dynamics of intracellular yeast-like symbiotes in the smaller brown planthopper, *Laodelphax striatellus* (Homoptera: Delphacidae). Appl. Entomol. Zool. 9: 257–277.

Noda, H. 1977. Histological and histochemical observation of intracellular yeastlike symbiotes in the fat body of the smaller brown planthopper, *Laodelphax striatellus* (Homoptera: Delphacidae). Appl. Entomol. Zool. 12: 134–141.

Noda, H. and T. Saito. 1979a. The role of intracellular yeastlike symbiotes in the development of *Laodelphax striatellus* (Homoptera: Delphacidae). Appl. Entomol. Zool. 14: 453–458.

Noda, H. and T. Saito. 1979b. Effects of high temperature on the development of *Laodelphax striatellus* (Homoptera: Delphacidae) and on its intracellular yeastlike symbiotes. Appl. Entomol. Zool. 14: 64–75.

Noda, H. and Y. Koizumi. 2003. Sterol biosynthesis by symbiotes: cytochrome P450 sterol C-22 desaturase genes from yeastlike symbiotes of rice planthoppers and anobiid beetles. Insect Biochem. Mol. Biol. 33: 649–658.

Noda, H., K. Wada and T. Saito. 1979. Sterols in *Laodelphax striatellus* with special reference to the intracellular yeastlike symbiotes as a sterol source. J. Insect Physiol. 25: 443–447.

Okabe, K. 2010. Key stone species for microbe biodiversity. Annual Report of Forestry and Forest Products Research Institute H22: 204.

Ômura, H. and K. Honda. 2003. Feeding responses of adult butterflies, *Nymphalis xanthomelas, Kaniska canace* and *Vanessa indica,* to components in tree sap and rotting fruits: synergistic effects of ethanol and acetic acid on sugar responsiveness. J. Insect Physiol. 49: 1031–1038.

Ômura, H., K. Honda and N. Hayashi. 2000. Identification of feeding attractants in oak sap for adults of two nymphalid butterflies, *Kaniska canace* and *Vanessa indica*. Physiol. Entomol. 25: 281–287.

Pant, N.C. and G. Fraenkel. 1954. Studies on the symbiotic yeasts of two insect species, *Lasioderma serricorne* F. and *Stegobium paniceum* L. Biol. Bull. 107: 420–432.

Pant, N.C., P. Gupta and J.K. Nayar. 1960. Physiology of intracellular symbionts of *Stegobium paniceum* L., with special reference to amino acid requirements of the host. Experientia 16: 311–312.

Péter, G., D. Dlauchy and J. Tornai-Lehoczki. 2006. *Candida floccosa* sp. nov., a novel methanol-assimilating yeast species. Int. J. Syst. Evol. Microbiol. 56: 2015–2018.

Péter, G., J. Tornai-Lehoczki, L. Fülöp and D. Dlauchy. 2003. Six new methanol assimilating yeast species from wood material. Antonie Van Leeuwenhoek 84: 147–159.

Phaff, H.J. and W.T. Starmer. 1987. Yeasts associated with plants, insects and soil. pp. 123–180. *In*: A.H. Rose and J.S. Harrison [eds.]. The Yeasts. Academic Press, New York, NY, USA.

Phaff, H.J., M.W. Miller and E.M. Mrak. 1978. The Life of Yeasts. 2nd Ed. Harvard University Press, Cambridge, MA, USA.

Pierce, N.E. 1995. Predatory and parasitic Lepidoptera: carnivores living on plants. J. Lepid. Soc. 49: 412–453.

Pitman, G.B., J.P. Vité, G.W. Kinzer and A.F. Fentiman Jr. 1968. Bark beetle attractants: Trans-verbenol isolated from *Dendroctonus*. Nature 218: 168–169.

Pozo, M.I., C. de Vega, A. Canto and C.M. Herrera. 2009. Presence of yeasts in floral nectar is consistent with the hypothesis of microbial-mediated signaling in plant-pollinator interactions. Plant Signal Behav. 4: 1102–1104.

Procheş, Ş. and S.D. Johnson. 2009. Beetle pollination of the fruit-scented cones of the South African cycad *Stangeria eriopus*. Am. J. Bot. 96: 1722–1730.

Ranger, C.M., M.E. Reding, A.B. Persad and D.A. Herms. 2010. Ability of stress-related volatiles to attract and induce attacks by *Xylosandrus germanus* and other ambrosia beetles (Coleoptera: Curculionidae, Scolytinae). Agric. For. Entomol. 12: 177–185.

Ranger, C.M., M.E. Reding, P.B. Schultz and J.B. Oliver. 2012. Ambrosia beetle (Coleoptera: Curculionidae) responses to volatile emissions associated with ethanol-injected *Magnolia virginiana* L. Environ. Entomol. 41: 636–647.

Ranger, C.M., M.E. Reding, P.B. Schultz, J.B. Oliver, S.D. Frank, K.M. Addesso et al. 2016. Biology, ecology, and management of nonnative ambrosia beetles (Coleoptera, Curculionidae, Scolytinae) in ornamental plant nurseries. J. Integr. Pest Manage. 7: 9.

Ranger, C.M., P.B. Schultz, S.D. Frank, J.H. Chong and M.E. Reding. 2015. Non-native ambrosia beetles as opportunistic exploiters of living but weakened trees. PLoS ONE 10: e0131496.

Ranger, C.M., P.C. Tobin, M.E. Reding, A.M. Bray, J.B. Oliver, P.B. Schultz et al. 2013. Interruption of semiochemical-based attraction of ambrosia beetles to ethanol-baited traps and ethanol-injected trap trees by verbenone. Environ. Entomol. 42: 539–547.

Rosa, C.A., M.-A. Lachance, J.O.C. Silva, A.C.P. Teixeira, M.M. Marini, Y. Antonini et al. 2003. Yeast communities associated with stingless bees. FEMS Yeast Res. 4: 271–275.

Ryker, L.C. and K.L. Yandell. 1983. Effect of verbenone on aggregation of *Dendroctonus ponderosae* Hopkins (Coleoptera, Scolytidae) to synthetic attractant. Z. Angew. Entomol. 96: 452–459.

Sasaki, T., M. Kawamura and H. Ishikawa. 1996. Nitrogen recycling in the brown planthopper, *Nilaparvata lugens*: involvement of yeast-like endosymbionts in uric acid metabolism. J. Insect Physiol. 42: 125–129.

Scheidler, N.H., C. Liu, K.A. Hamby, F.G. Zalom and Z. Syed. 2015. Volatile codes: correlation of olfactory signals and reception in *Drosophila*-yeast chemical communication. Sci. Rep. 5: 14059.

Schiestl, F.P. and S. Dötterl. 2012. The evolution of floral scent and olfactory preferences in pollinators: coevolution or pre-existing bias? Evolution 66: 2042–2055.

Schoonhoven, L.M., J.J.A. van Loon and M. Dicke. 2010. Insect-Plant Biology. 2nd Ed. Oxford University Press, Oxford, UK.

Schuster, J.C. and L.B. Schuster. 1985. Social behavior in passalid beetles (Coleoptera: Passalidae): cooperative brood care. Fla. Entomol. 68: 266–273.

Shen, S.K. and P.F. Dowd. 1989. Xenobiotic induction of esterases in cultures of the yeast-like symbiont from the cigarette beetle. Entomol. Exp. Appl. 52: 179–184.

Shen, S.K. and P.F. Dowd. 1991. Detoxification spectrum of the cigarette beetle symbiont *Symbiotaphrina kochii* in culture. Entomol. Exp. Appl. 60: 51–59.

Stamps, J.A., L.H. Yang, V.M. Morales and K.L. Boundy-Mills. 2012. Drosophila regulate yeast density and increase yeast community similarity in a natural substrate. PLoS ONE 7: e42238.

Starmer, W.T. 1982. Analysis of community structure of yeasts associated with the decaying stems of cactus. I. *Stenocereus gummosus*. Microb. Ecol. 8: 71–81.

Starmer, W.T. and J.C. Fogleman 1986. Coadaptation of *Drosophila* and yeasts in their natural habitat. J. Chem. Ecol. 12: 1037–1055.

Stefanini, I., L. Dapporto, J.-L. Legras, A. Calabretta, M. Di Paola, C. De Filippo et al. 2012. Role of social wasps in *Saccharomyces cerevisiae* ecology and evolution. Proc. Natl. Acad. Sci. USA 109: 13398–13403.

Stefanini, I., L. Dapporto, L. Berná, M. Polsinelli, S. Turillazzi and D. Cavalieri. 2016. Social wasps are a *Saccharomyces* mating nest. Proc. Natl. Acad. Sci. USA 113: 2247–2251.

Stratford, M., C. Bond, S. James, N. Roberts and H. Steels. 2002. *Candida davenportii* sp. nov., a potential soft-drinks spoilage yeast isolated from a wasp. Int. J. Syst. Evol. Microbiol. 52: 1369–1375.

Suh, S.-O. and M. Blackwell. 2005. The beetle gut as a habitat for new species of yeasts. pp. 244–256. *In*: F.E. Vega and M. Blackwell [eds.]. Insect–Fugal Associations Ecology and Evolution. Oxford University Press, Oxford, UK.

Suh, S.-O., C.J. Marshall, J.V. McHugh and M. Blackwell. 2003. Wood ingestion by passalid beetles in the presence of xylose-fermenting gut yeasts. Mol. Ecol. 12: 3137–3145.

Suzuki, K., H. Sakamoto, Y. Shinozaki, J. Tabata, T. Watanabe, A. Mochizuki et al. 2013. Affinity purification and characterization of a biodegradable plastic-degrading eyzyme from a yeast isolated from the larval midgut of a stag beetle, *Aegus laevicollis*. Appl. Microbiol. Biotechnol. 97: 7679–7688.

Suzuki, N., A. Sakamiya, H. Kanazawa, O. Kurita, T. Yano and S. Karita. 2016. The characterization and utility evaluation of flavor-producing wild yeast isolated from tree sap. Japan J. Food Eng. 17: 59–69.

Takahashi, M. 1986. Fungal decay types, their significance in wood preservation. Wood Res. Tech. Note 22: 19–36.

Tanahashi, M. 1983. Conversion and total utilization of forest-biomass by explosion process. Wood Res. Tech. Note. 18: 34–65.

Tanahashi, M., K. Kubota, N. Matsushita and K. Togashi. 2010. Discovery of mycangia and associated xylose-fermenting yeasts in stag beetles (Coleoptera: Lucanidae). Naturwissenschaften 97: 311–317.

Tanahashi, M. and C.J. Hawes. 2016. The presence of a mycangium in European *Sinodendron cylindricum* (Coleoptera: Lucanidae) and the associated yeast symbionts. J. Insect Sci. 16: 76.

Tanahashi, M., N. Matsushita and T. Togashi. 2009. Are stag beetles fungivorous? J. Insect Physiol. 55: 983–988.

Tanaka, Y. 2016. Do all of Lucanid species eat sap? Insect Nat. 51(6): 9–12.

Taylor, M.W., P. Tsai, N. Anfang, H.A. Ross and M.R. Goddard. 2014. Pyrosequencing reveals regional differences in fruit-associated fungal communities. Environ. Microbiol. 16: 2848-2858.

Torto, B., A. Suazo, P.E.A. Teal and J.H. Tumlinson. 2005. Response of the small hive beetle (*Aethina tumida*) to a blend of chemicals identified from honeybee (*Apis mellifera*) volatiles. Apidologie 36: 523–532.

Torto, B., D.G. Boucias, R.T. Arbogast, J.H. Tumlinson and P.E.A. Teal. 2007a. Multitrophic interaction facilitates parasite-host relationship between an invasive beetle and the honey bee. Proc. Natl. Acad. Sci. USA 104: 8374–8378.

Torto, B., R.T. Arbogast, H. Alborn, A. Suazo, D. Van Engelsdorp, D. Boucias et al. 2007b. Composition of volatiles from fermenting pollen dough and attractiveness to the small hive beetle *Aethina tumida*, a parasite of the honeybee *Apis mellifera*. Apidologie 38: 380–389.

Ueda, H., I. Mitsuhara, J. Tabata, S. Kugimiya, T. Watanabe, K. Suzuki et al. 2015. Extracellular esterases of phylloplane yeast *Pseudozyma antarctica* induce defect on cuticle layer structure and water-holding ability of plant leaves. Appl. Microbiol. Biotechnol. 99: 6405–6415.

Urbina, H., J. Schuster and M. Blackwell. 2013. The gut of Guatemalan passalid beetles: a habitat colonized by cellobiose- and xylose-fermenting yeasts. Fung. Ecol. 6: 339–355.

van Bel, A.J.E. 2003. The phloem, a miracle of ingenuity. Plant Cell Environ. 26: 125–149.

Vega, F.E. and P.F. Dowd. 2005. The role of yeasts as insect endosymbionts. pp. 211–243. *In*: F.E. Vega and M. Blackwell [eds.]. Insect–Fugal Associations Ecology and Evolution. Oxford University Press, Oxford, UK.

Wada, T. 2015. Rice planthoppers in tropics and temperate East Asia: difference in their biology. pp. 77–90. *In*: K.L. Heong, J. Cheng and M.M. Escalada [eds.]. Rice Planthoppers: Ecology, Management, Socio Economics and Policy. Zhejiang University Press, Hangzhou, China; Springer, Dordrecht, Netherlands.

Whitney, H.S. 1971. Association if *Dendroctonus ponderosae* (Coleoptera: Scolytidae) with blue stain fungi and yeasts during brood development in lodgepole pine. Can. Entomol. 103: 1495–1503.

Whitney, H.S. and S.H. Harris. 1970. Maxillary mycangium in the mountain pine beetle. Science 167: 54–55.

Xue, J., X. Zhou, C.-X. Zhang, L.-L. Yu, H.-W. Fan, Z. Wang et al. 2014. Genomes of the rice pest brown planthopper and its endosymbionts reveal complex complementary contributions for host adaptation. Genome Biol. 15: 521.

Yeats, T.H. and J.K.C. Rose. 2013. The formation and function of plant cuticles. Plant Physiol. 163: 5–20.

Yoshimoto, J. and T. Nishida. 2007. Boring effect of carpenterworms (Lepidoptera: Cossidae) on sap exudation of the oak, *Quercus acutissima*. Appl. Entomol. Zool. 42: 403–410.

Yoshimoto, J. and T. Nishida. 2008. Plant-mediated indirect effects of carpenterworms on the insect communities attracted to fermented tree sap. Popul. Ecol. 50: 25–34.

Yoshimoto, J., T. Kakutani and T. Nishida. 2005. Influence of resource abundance on the structure of the insect community attracted to fermented tree sap. Ecol. Res. 20: 405–414.

PART–B

Applications of Insect Chemical Ecology to Agriculture, Environment Conservation, and Public Health

CHAPTER

7

Application of Trail Pheromones to Management of Pest Ants

Eiriki Sunamura

Society of Scientific Photography, Goban-cho 5-6,
Chiyoda-ku, Tokyo 102-0076, Japan
E-mail: eirikisunamura@yahoo.com

Abstract

Ants exhibit complicated social behaviors via chemical communication. Use of trail pheromones enables ants to attract their colony members and form trails. The presumed properties of trail pheromones such as species specificity, low eco-toxicity, and low effective dosage, could be effective for eco-friendly ant management. Recent studies have explored the possibility of applying trail pheromones to ant management by two approaches. One is by exploiting the effect of trail pheromones to attract target ants. Inclusion of subtle amounts of trail pheromones into current insecticides may result in efficient attracting and killing of the ants. Another approach is to disrupt trail formation of the target ants by dispersing large amounts of synthetic trail pheromones on the ground. Trail disruption results in reduced foraging success of target ants, less competitive ability against other animals, as well as an inability to spread. To put trail pheromones into practical use in the future, development of suitable formulations, in-depth evaluation of eco-toxicity, and commercialization of trail pheromone components are needed.

1. Introduction

Insects use various forms of chemical communication, and these have been applied for pest management. Sex pheromones have been successfully applied for occurrence prediction and mating disruption of target pests (see Chapter 9). Other than sex pheromones, however, insect chemical communication has only been exploited in a limited manner in practical pest management.

Ants are social insects and they rely heavily on chemical communication to exhibit complicated group behavior (Vander Meer et al. 1988, Hölldobler and Wilson 1990). For example, caste deviation is regulated by queen pheromones; and recognition of colony members and castes are done by the scents of cuticular hydrocarbons; the colony is alerted to the presence of enemies by alarm pheromones; and directions toward food sources or destinations are informed by trail pheromones. Because ants live in dark nests, they have developed chemical communication rather than relying on vision.

This chapter reviews recent studies that have explored the possibility of applying trail pheromones for ant management.

2. Pest Status of Ants and Current Control Measures

Some ant species have become serious pests. In urban environments, various species of ants intrude into structures and cause disturbance (Santos 2016). Some species, such as pharaoh ant *Monomorium pharaonis* and carpenter ant *Camponotus* spp., even nest indoors or in building materials and are recognized as structural pests. Indeed, ants are ranked as the number one household pest in the USA (Field et al. 2007). In the agricultural environment, leaf-cutting ants of the genus *Atta* and *Acromyrmex* cause significant economic damage by cutting off the leaves, flowers, buds, and twigs of crops (Zanetti et al. 2014). Additionally, many other ant species build symbiotic relationship with aphids, scale insects, and mealy bugs, and sometimes cause outbreaks of these pests in crops (Flanders 1945). In the natural environment, invasive alien ant species disrupt local ecosystems. Five species of ants are listed among world's worst invasive species by IUCN, and their distribution is becoming global, including Argentine ant *Linepithema humile*; red imported fire ant *Solenopsis invicta*; yellow crazy ant *Anoplolepis gracilipes*; big-headed ant *Pheidole megacephala*; and little fire ant *Wasmannia auropunctata* (Lowe et al. 2000).

To manage such pest ants, chemical control measures have been developed (Soeprono and Rust 2004), with the major use being insecticidal bait. Ants are attracted to bait that includes slow-acting insecticide; they then carry it to their nest and share it with their nestmates. After that, the insecticide can kill the entire nest. Another measure is aerosols, with insecticide dissolved in a liquid and bottled in small cans with gas, and sprayed directly onto ants. Residual spray is also a major control measure. In this application, liquid concentrate including insecticide or repellent is diluted in water and sprayed over areas where ants are nesting or where ants should be kept away from. Ant infestation in treated areas is prevented for a certain period, but in some cases these measures do not provide satisfactory control of ants. For example, ants have different food preferences among species, and bait matrix in an insecticidal bait product may not be attractive to a target ant species. A typical aerosol kills only ants that are hit directly by the spray, and these individuals are only a small portion of the entire colony. Finally, for residual spray, the impact of insecticide on the environment, as well as the loss of efficacy by rainfall, are of concern.

3. Ant Trail Pheromone and Its Application

3.1 Ant Trail Pheromone

Ants form trails when they need to recruit many individuals, such as when collecting large food items, moving nests, and fighting against enemy ants (Czaczkes et al. 2015). Trail formation is induced by trail pheromone. For example, in the case of foraging, a worker scout that finds food outside of the nest goes back to the nest secreting trail pheromone (typically from the abdominal tip) and dropping it on the ground. Then, nestmates go out to collect the food by following the pheromone trail. The food is then carried to the nest quickly by many workers. Trail pheromone may be composed of a single compound, or a blend of several compounds (Morgan 2009). In general, pheromones are used for intraspecific communication, and so the components and effects are species specific. However, some trail pheromones of ants are known to be cross-active among related species.

The generally assumed profiles of trail pheromone, namely species specificity, low eco-toxicity, and low effective dose, may be used for eco-friendly ant management. Attempts at such applications have been made by two different approaches. One approach exploits trail-inducing effects of trail pheromone when applied at an adequate dose; and the other approach exploits the trail-disrupting effect of trail pheromone when applied at an excessive dose. Previous studies involving each approach are reviewed in the following sections.

It should be noted that most of the studies have been done with two invasive species, *L. humile* (Fig. 1) and *S. invicta*. Both are native to South America but have expanded their distributions over continents. In *L. humile*, chemical analyses of its gaster extracts and bioassays suggested that (Z)-9-hexadecenal is the sole trail pheromone component (Cavill et al. 1979, 1980). However, a recent study reported that dolichodial and iridomyrmecin also function as trail pheromones, and these two compounds but not (Z)-9-hexadecenal were detected from natural trails of *L. humile* (Choe et al. 2012). In *S. invicta*, (*Z*,*E*)-α-farnesene is one of its trail pheromone components and is responsible for orientation of workers along trails (Vander Meer et al. 1988). (Z)-9-hexadecenal has already been industrialized and can be purchased in a high-purity form, whereas (*Z*,*E*)-α-farnesene is not available.

Fig. 1. A worker of the Argentine ant, *Linepithema humile*.

3.2 Exploiting the Trail-inducing Effect of Trail Pheromone

In previous studies, the trail-inducing effect of trail pheromone was incorporated in "attract and kill" tactics, in combination with two killing agents, insecticidal bait and residual spray. Synthetic trail pheromone incorporated in bait may act as an attractant and increase bait consumption by target species, leading to increased mortality. Greenberg and Klotz (2000) demonstrated in laboratory studies that sucrose solution mixed with (Z)-9-hexadecenal in an adequate concentration attracted >150% *L. humile* compared with just the sucrose solution. In further field trials, 10% sucrose water bottled in 50 mL vials treated with 1 μg (Z)-9-hexadecenal was consumed by 29% more *L. humile* than without (Z)-9-hexadecenal treatment. Welzel and Choe (2016) incorporated (Z)-9-hexadecenal in insecticidal bait and conducted a field test. Residential houses were treated with 20 gel baits (20 mL each; including 0.01% thiamethoxam as the active ingredient; and 0.2 μg (Z)-9-hexadecenal as the attractant) and the efficacy was monitored for four weeks. The pheromone-treated gel bait achieved a 74% reduction in *L. humile* activity, while untreated gel bait resulted in a 42% reduction. A similar line of research has been done in leaf-cutting ants (Robinson et al. 1982).

Synthetic trail pheromone was also mixed with liquid concentrate for residual spray. Choe et al. (2014) put sand treated with an ethanol solution of (Z)-9-hexadecenal in close proximity to a trail or nest entrance of *L. humile* in the field, and showed that the treated sand attracted more than two-fold or five-fold more workers within several minutes compared with untreated sand. The dose of (Z)-9-hexadecenal was 12 ng/cm^2, which was high enough to attract ants, but low enough that humans did not sense the characteristic smell of the compound. The authors further treated a partial surface of a 86 cm × 42 cm container inhabited by *L. humile* with insecticidal spray mixed with (Z)-9-hexadecenal, and the pheromone-mixed treatment achieved 57–142% more ant mortality compared with spray treatment without pheromone. The required dose of insecticide spray was much less than that instructed on the product label, and the dose of (Z)-9-hexadecenal was 12 ng/cm^2. The authors suggested several merits of attract and kill tactics: reduced amounts of insecticide required but increased chance of ant contact with the insecticide; no need to treat the whole ground evenly; prolonged exposure of ants to the insecticide by keeping them attracted to the treated surface; and promotion of ants coming to the treated surface before the loss of insecticide residue. In their experiment, the attractiveness of trail pheromone lasted for less than 60 min, so the formulation should be improved for practical use.

The above studies suggest that synthetic trail pheromone can be incorporated with killing agents economically, based on the cost of (Z)-9-hexadecenal (≈37 USD/g: Choe et al. 2014).

3.3 Exploiting the Trail-disrupting Effects of Excess Trail Pheromone

The basic idea of the trail disruption approach is that ants may not be able to detect or orient to natural trail pheromone when excess amounts of synthetic trail pheromone is dispersed over the ground (Fig. 2). This was inspired by the mating disruption technique in which an excess amount of synthetic sex pheromone is dispersed in the air to disrupt mate location by a target species (Tatsuki et al. 2005, Suckling et al. 2008).

There is an adequate range of concentrations for trail pheromone. If the concentration is too low, ants cannot detect it, but if the concentration is too high, the ants seem unable to grasp the exact location of the pheromone source. Using *L. humile,* Tanaka et al. (2008) investigated the number of trail following ants and the accuracy of orientation with various concentrations of (Z)-9-hexadecenal linearly treated on drawing paper. They found that the optimum concentration for the surface was 5 ng per 20 cm. In experiments with higher concentrations, 50 ng or more (Z)-9-hexadecenal per 20 cm, *L. humile* workers walked near the line in a zig-zag manner and a normal trail was not formed. Van Vorhis Key et al.

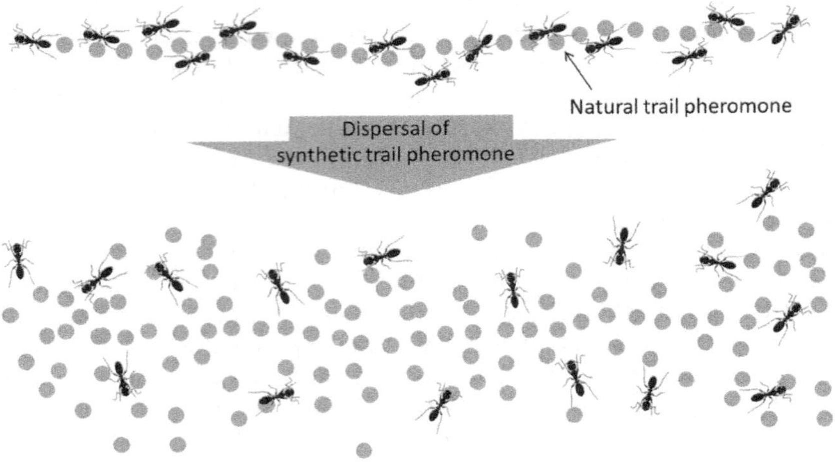

Fig. 2. Trail disruption by synthetic trail pheromone. Circles denote trail pheromone.

(1981) reported similar results using gaster extracts of *L. humile*, which included (*Z*)-9-hexadecenal as well as other components.

Trail following behavior of ants can actually be disrupted by treating with an excess amount of synthetic trail pheromone near natural trails. Tanaka et al. (2009) treated a 20 cm-long tube dispenser that released (*Z*)-9-hexadecenal at a rate of 59–83 µg/h at a point 10 cm away from a natural trail of *L. humile*. Immediately, nearby ants (46–465 cm range of the trail) scattered in all directions, and number of ants that passed a predetermined point on the natural trail decreased to 13–34%. Similarly, Suckling et al. (2008) reported trail following disruption of *L. humile* using a vial containing (*Z*)-9-hexadecenal (release rate not described). Suckling et al. (2010, 2011) demonstrated trail disruption of *S. invicta* using an ethanol solution of (*Z,E*)-α-farnesene or intermittent aerosol. In their experiment, the optimum concentration of (*Z,E*)-α-farnesene for trail following was 110 ng/cm, and the concentration required for trail disruption was 1600 ng/cm^2. Both were much higher than those required for *L. humile*. The authors discussed that the relatively low purity of their (*Z,E*)-α-farnesene (91%) was responsible.

Trail disruption by synthetic trail pheromone led to reduced foraging success of the target species via trail formation. Suckling et al. (2008) dispersed sand containing (*Z*)-9-hexadecenal over a 4 m^2 area. (*Z*)-9-Hexadecenal that evaporated from the sand reduced *L. humile* recruitment to food sources within the treated area for two hours. The release rate of (*Z*)-9-hexadecenal required for inhibiting ant foraging was estimated to be more than 200 µg/h/m^2 but less than 260 µg/h/m^2. Tanaka et al. (2009) treated with long-life (*Z*)-9-hexadecenal dispensers every 1 m over a 100

m^2 area: each dispenser was a 20 cm tube; releasing (Z)-9-hexadecenal at a rate of 59–83 μg/h; and for more than one month. Recruitment of *L. humile* to sugar water was reduced by 60–89% during the pheromone-treated period (three days). Differences in the required amount of (Z)-9-hexadecenal between Suckling et al. (2008) and Tanaka et al. (2009) may stem from differences in the environment and size of the treated area.

If foraging success of ants could be reduced for a long period, the ants would fall into food shortage and the nest could decline. This "starve out" tactic was tested in the field by Nishisue et al. (2010), but the result was not good. They treated with long-life (Z)-9-hexadecenal dispensers as described above over 100 m^2 areas (total pheromone release rate 71–100 μg/h/m^2) of house gardens from spring to autumn. Reduced foraging of *L. humile* in the treated areas was confirmed throughout the experimental period, but density of *L. humile* in the treated area remained at the same level as that in a control area. The reason why foraging disruption failed to decrease ant density is unclear, but it may be difficult to have strong impact on ant nest performance by treatment with trail pheromone alone.

Greater success may come from combination treatments with a trail disruption technique and some other control agents. Sunamura et al. (2011) treated with trail pheromone, insecticidal bait, or both of them, over 100 m^2 areas of house gardens from spring to autumn. The formulations used were long-life (Z)-9-hexadecenal dispensers, as described above (total pheromone release rate 71–100 μg/h/m^2), and commercial bait product containing lithium perfluorooctane sulfonate as the active ingredient (25 bait boxes; 2.1 g each). As a result, the density of *L. humile* was suppressed only by the combination treatment throughout the experimental period. Sole treatment with trail pheromone or insecticidal bait could not prevent outbreaks of *L. humile* after summer. The mechanism of synergy with trail pheromone and insecticidal bait was discussed as killing by the bait plus inhibiting re-infestation from nearby area by the pheromone.

Suzuki and colleagues applied the results of Sunamura et al. (2011) in small area to a much larger field test (Suzuki 2015). In their *L. humile* control project a 6.7 ha invaded area in Yokohama Port, Japan, was subjected to combined treatment with (Z)-9-hexadecenal dispensers and insecticidal bait over a 420 m invasion front. *L. humile* never spread beyond the combination-treatment area, and the density declined to such a low level that eradication was possible. Although their trial did not have an experimental control, the trail disruption technique was considered to have contributed to the prevention of *L. humile* range expansion.

Westermann et al. (2014) showed that, in a zone with *L. humile* co-occurring with three other ant species, excess trail pheromone of *L. humile* (100 μg or more (Z)-9-hexadecenal on bait cards with a 10 cm radius) resulted in improved foraging success of the other ants. As the concentration of trail pheromone increased, recruitment of *L. humile* to the bait became slower. In addition, the frequency of aggressive behavior by *L. humile* to other species also decreased. If a high dose of synthetic trail pheromone was applied over a co-occurring area of the target invasive ant and other native ants for a long period, the combination of trail disruption technique and native ants as bio-control agents could deter invasive ants from the treated area.

4. Future Directions

In this chapter, we reviewed studies that attempted to apply trail-inducing or trail-disrupting effects with synthetic trail pheromone for ant control. Regarding the exploitation of trail-inducing effects, trail pheromone could be easily incorporated in attract and kill tactics. Because trail-inducing effect can be obtained with a small amount of trail pheromone, synthetic trail pheromone in such usage would cause little impact on the environment. In addition, economical feasibility can be anticipated through the usage of industrialized compounds such as the trail pheromone component of *L. humile*, (Z)-9-hexadecenal. However, species-specific effects of the pheromone may cause difficulty in putting synthetic trail pheromones into practical use as attractants. A variety of ant species invade gardens and houses, and their trail pheromones are different from species to species, so the target species to be controlled must be identified by home owners or pest management professionals, who may not be familiar with ant taxonomy. Of course, the dominant and characteristic invasive species such as *S. invicta* could be identified with a relatively high probability, and for such species synthetic trail pheromones could be used as attractants in combination with insecticides.

Trail disruption is a relatively novel and unique approach compared with the "attract and kill" strategy. In the future, population suppression by foraging suppression could be achieved by sole treatment with trail pheromone. Besides, trail pheromone in combination with other control agents could enable a highly sophisticated IPM. However, to put trail disruption into practical use, there seem to be more hurdles than the attract and kill strategy. First, there are technical problems to be solved. For example, regarding the possibility of achieving population decline by sole treatment with trail pheromone, recent studies have elucidated that many factors other than trail pheromone are involved in trail formation, such as memory and behavioral interactions (Czaczkes et al. 2015), thus complete trail disruption may not be achieved by simply applying synthetic trail pheromone over the ground. In addition, a fundamental question of "how is foraging in a group by trail formation relatively important for an ant nest compared with foraging individually?" should be addressed. Development of suitable formulations is also necessary. Second, trail disruption techniques require much larger amounts of synthetic trail pheromone than the attract and kill technique, both in the short and long term. This could become an obstacle in terms of securing economic feasibility. Furthermore, trail pheromones are thought to be eco-friendly, but because a high amount will need to be applied, toxicity as well as the effect on behaviors of non-target organisms would merit investigation.

Although there are problems to be solved, both the attract and kill and trail disruption techniques, in combination with insecticides, can control target ant species efficiently with reduced amounts of insecticide. Against persistent species such as invasive ants that we need to control in the long term, it is desirable to put the techniques into practical use as a component of IPM.

Acknowledgments

I thank Sadahiro Tatsuki, Mamoru Terayama, Hironori Sakamoto, Koji Nishisue, Shun Suzuki, Toshio Kishimoto, Hideaki Mori, Takehiko Fukumoto, and Yosaburo Utsumi for collaboration and deepening my knowledge on the topic of this chapter.

REFERENCES

Cavill, G.W.K., P.L. Robertson and N.W. Davies. 1979. An Argentine ant aggregation factor. Experientia 35: 989–990.

Cavill, G.W.K., N.W. Davies and F.J. McDonald. 1980. Characterization of aggregation factors and associated compounds from the Argentine ant, *Iridomyrmex humilis*. J. Chem. Ecol. 6: 371–384.

Choe, D.-H., D.B. Villafuert and N.D. Tsutsui. 2012. Trail pheromone of the Argentine ant, *Linepithema humile* (Mayr) (Hymenoptera: Formicidae). PLoS ONE 7: e45016.

Choe, D.-H., K. Tsai, C.M. Lopez and K. Campbell. 2014. Pheromone-assisted techniques to improve the efficacy of insecticide sprays against *Linepithema humile* (Hymenoptera: Formicidae). J. Econ. Entomol. 107: 319–325.

Czaczkes, T.J., C. Grüter and F.L.W. Ratnieks. 2015. Trail pheromones: an integrative view of their role in social insect colony organization. Annu. Rev. Entomol. 60: 581–599.

Field, H.C., W.E. Evans Sr., R. Hartley, L.D. Hansen and J.H. Klotz. 2007. A survey of structural ant pest in the Southwestern U.S.A. (Hymenoptera: Formicidae). Sociobiology 49: 151–164.

Flanders, S.E. 1945. Coincident infestations of *Aonidiella citrina* and *Coccus hesperidum*, a result of ant activity. J. Econ. Entomol. 38: 711–712.

Greenberg, L. and J.H. Klotz. 2000. Argentine ant (Hymenoptera: Formicidae) trail pheromone enhances consumption of liquid sucrose solution. J. Econ. Entomol. 93: 119–122.

Hölldobler, B. and E.O. Wilson. 1990. The Ants. Harvard University Press, Cambridge, MA, USA.

Lowe, S., M. Browne and S. Boudlejas. 2000. 100 of the world's worst invasive alien species. Aliens 12: 1–12.

Morgan, E.D. 2009. Trail pheromones of ants. Physiol. Entomol. 34: 1–17.

Nishisue, K., E. Sunamura, Y. Tanaka, H. Sakamoto, S. Suzuki, T. Fukumoto, et al. 2010. A long term field trial to control the invasive Argentine ant (Hymenoptera: Formicidae) with synthetic trail pheromone. J. Econ. Entomol. 103: 1784–1789.

Robinson, S.W., A.R. Jutsum, J.M. Cherrett and R.J. Quinlan. 1982. Field evaluation of methyl 4-methylpyrrole-2-carboxylate, an ant trail pheromone, as a component of baits for leaf-cutting ant (Hymenoptera: Formicidae) control. Bull. Entomol. Res. 72: 345–356.

Santos, M.N. 2016. Research on urban ants: approaches and gaps. Insect. Soc. 63: 359–371.

Soeprono, A.M. and M.K. Rust. 2004. Strategies for controlling Argentine ants (Hymenoptera: Formicidae). Sociobiology 44: 669–682.

Suckling, D.M., R.W. Peck, L.M. Manning, L.D. Stringer, J. Cappadonna and A.M. El-Sayed. 2008. Pheromone disruption of Argentine ant trail integrity. J. Chem. Ecol. 34: 1602–1609.

Suckling, D.M., R.W. Peck, J.D. Stringer, K. Snook and P.C. Banko. 2010. Trail pheromone disruption of Argentine ant trail formation and foraging. J. Chem. Ecol. 36: 122–128.

Suckling, D.M., L.D. Stringer and J.E. Corn. 2011. Argentine ant trail pheromone disruption is mediated by trail concentration. J. Chem. Ecol. 37: 1143–1149.

Sunamura, E., S. Suzuki, K. Nishisue, H. Sakamoto, M. Otsuka, Y. Utsumi et al. 2011. Combined use of synthetic trail pheromone and insecticidal bait provides effective control of an invasive ant. Pest Manag. Sci. 67: 1230–1236.

Suzuki, S. 2015. Management aiming at eradication—the case of Yokohama Port. pp. 287–306. *In*: S. Tatsuki [ed.]. Argentine Ant: The Strongest Invasive Alien Species. University of Tokyo Press, Tokyo, Japan.

Tanaka, Y., E. Sunamura, K. Nishisue, M. Terayama, H. Sakamoto, S. Suzuki et al. 2008. Response of the invasive Argentine ant to high concentration of a synthetic trail pheromone component suggests a potential control strategy of pest ants. Ari 31: 43–50.

Tanaka, Y., K. Nishisue, E. Sunamura, S. Suzuki, H. Sakamoto, T. Fukumoto et al. 2009. Trail-following disruption in the invasive Argentine ant with a synthetic trail pheromone component (Z)-9-hexadecenal. Sociobiology, 54: 139–152.

Tatsuki, S., M. Terayama, Y. Tanaka and T. Fukumoto. 2005. Behavior-disrupting agent and behavior disrupting method of Argentine ant. US Patent No. US2005/0209344A1.

Vander Meer, R.K., F. Alvarez and C.S. Lofgren. 1988. Isolation of the trail recruitment pheromone of *Solenopsis invicta*. J. Chem. Ecol. 14: 825–838.

Van Vorhis Key, S.E., L.K. Gaston and T.C. Baker. 1981. Effects of gaster extract concentration on the trail following behaviour of the Argentine ant, *Iridomyrmex humilis* (Mayr). J. Insect Physiol. 27: 363–370.

Welzel, K.F. and D.-H. Choe. 2016. Development of a pheromone-assisted baiting technique for Argentine ants (Hymenoptera: Formicidae). J. Econ. Entomol. 109: 1303–1309.

Westermann, F.L., D.M. Suckling and P.J. Lester. 2014. Disruption of foraging by a dominant invasive species to decrease its competitive ability. PLoS ONE 9: e90173.

Zanetti, R., J.C. Zanuncio, J.C. Santos, W.L. Paiva da Silva, G.T. Ribeiro and P.G. Lemes. 2014. An overview of integrated management of leaf-cutting ants (Hymenoptera: Formicidae) in Brazilian forest plantations. Forests 5: 439–454.

CHAPTER

8

Female Sex Pheromones and Mating Behavior in Diurnal Moths: Implications for Conservation Biology

Hideshi Naka
Laboratory of Applied Entomology, Tottori University
Koyama-Minami 4-101, Tottori City, Tottori 680-8553, Japan
E-mail: chun@muses.tottori-u.ac.jp

Abstract

The Lepidoptera are one of the largest groups of insects, comprising approximately 150,000 described species, about 94% of which are moths. Although moths are highly diverged in morphology and biology, most species use a common reproduction system that involves long-distance attraction of males with volatile chemicals, i.e. sex pheromones, produced by females. Because most moths are nocturnal, sex pheromone-mediated olfactory cues are a reliable tool for recognitions of conspecific mates. Conversely, diurnal insects may use visual signals instead of volatile pheromones; diurnal insects generally have more conspicuous, with sexual dimorphisms in their wing or body patterns compared to nocturnal species. Our studies on several diurnal moths, including clearwing moths, hawk moths, and Japanese nine-spotted moth, substantiated these moths use chemical signals which are common in other moths as well as visual cues with species-specific morphological features. Moreover, we demonstrated that their pheromone chemicals identified in these studies are very useful and essential to understand their lifecycle, behavior, ecology, or evolution. We could discover and monitor a rare or even a novel species by using these synthetic pheromones. In this chapter, I review diurnal moth pheromone studies and show some examples of their potential applied uses for biodiversity assessment, conservation of the species in population declines, and environmental protection.

1. Introduction: Sex Pheromones in Diurnal Lepidopteran Species

Most lepidopteran species use volatile female sex pheromones for mate recognition. Volatile pheromones can deliver information farther away than visual or acoustic signals in smaller insects. Conversely, diurnal insects such as butterflies may use visual signals instead of volatile pheromones (Hidaka and Yamashita 1975, Wago et al. 1976, Kato and Yoshioka 2003, Robertson and Monteiro 2005). Generally, diurnal insects have more conspicuous, with sexual dimorphisms in their wing or body patterns compared to nocturnal species. Moreover, males have larger compound eyes than females in some diurnal insects in order to search for females during mate location (e.g., Zeil 1983, Rutowski 2000, Moser et al. 2004). These morphological features suggest that many diurnal insects use wing or body patterns for mate recognition.

In Lepidoptera, diurnal species retain the ancestral use of volatile sex pheromones in their mating communication systems. Diurnal or nocturnal mating may partly determine the cue or sensory strategy that a species uses (visual cues, tactual signs, calling songs, etc.) as a supplement of volatile sex pheromones for mate location and recognition. For example, at least four species belonging to an ancestral lepidopteran family Eriocranidae use female sex pheromones to attract males (Zhu et al. 1995, Kozlov et al. 1996), but there are no reports of female sex pheromones in Agathiphagidae or Micropterigidae, the most ancestral lepidopteran family (Ando 2016, El-Sayed 2016). It is known that most of these ancestral moths fly in the daytime (Hashimoto 2011).

On the other hand, few studies have reported female sex pheromones of butterflies, i.e., Papilionoidea and Hesperioidea species. Lundgren and Bergstrom (1975) found that both males and females of a lycaenid butterfly *Lycaeides argyrognomon* need scents from butterflies of the opposite sex to mate successfully. Takanashi et al. (2001) reported that female adults of the sulfur butterfly *Eurema mandarina* (formally *E. hecabe*) have a sex pheromone; male courtship behavior was elicited by exposure to crude extract from wings of females. However, the "female sex pheromone" in *L. argyrognomon* and *E. mandarina* is different from the female sex pheromone in moths; females of these species are not volatile and have never been found to attract males from a distance.

In moths closely related to butterflies, about half of the taxa in Obtectomera ("Butterfly", Pterophoroidea, Copromorphoidea, Epermenioidea, Alucitoidea, Calliduloidea, Hyblaeoidea, and

Thyridoidea) clade, sensu Regier et al. (2013), members of Pterophoridae and Carposinidae use female sex pheromones. Pterophoridae species typically use what are called Type I compounds, but Carposinidae species use unsaturated ketones (e.g., Honma et al. 1978, Klun et al. 1981, Haynes 1987, Foster and Thomas 2000, Fujii et al. 2010, Gibb et al. 2006). Compounds for moth female sex pheromones are classified into groups of primary fatty alcohols and their derivatives (mainly acetates and aldehydes) with a long straight chain (Type I), polyunsaturated hydrocarbons and the epoxy derivatives with a longer straight chain (Type II), and others (ketones, secondary alcohols, esters of saturated fatty acids, etc.). Type I compounds are prevalently found in moths, Type II in Erebidae and Geometridae species, and others in some groups, e.g., Eriocranidae, Psycidae and Carposinidae species (Ando et al. 2004, Ando 2016). However, the evolution and use of female sex pheromones in the taxa closely related to butterflies is poorly understood at this time. In contrast, males in these taxa use male sex pheromones for various situations of their mating behavior. For example, many Pieridae butterflies have male sex pheromones in their androconia, specialized wing scales with scent compounds used during courtship mainly as an aphrodisiac pheromone (e.g., Rutowski 1983, Vane-Wright and Boppré 1993, Andersson et al. 2003, 2007, Yildizhan et al. 2009, Ômura and Yotsuzuka 2015). Danaid butterflies (Nymphalidae: Danainae) also use male sex pheromones. In danaid butterflies, adults of most genera or larvae in only the genus *Idea* ingest pyrroridizine alkaloids from floral nectar, dried or rotten plants, or their food plant, and then use the alkaloids as the precursors of their male sex pheromones. The male sex pheromones are secreted from two types of scent-producing organs, i.e. a pair of abdominal brushes (hairpencils) and alar patch- or pouch-like glands (sex gland). This type of male sex pheromone also functions as an aphrodisiac pheromone (e.g., Pliske and Eisner 1969, Schneider et al. 1975, Boppré et al. 1978, Honda et al. 1995, Nishida et al. 1996).

Both Callidulidae and Hyblaeidae are small taxa closely related to butterflies. Many species in these families fly in daytime, and the resting posture of Callidulinae species is butterfly-like, with the wings held closely over the back (Nasu 2011). Nishio (1983, 1999) reported that males of *Pterodecta felderi* (Callidulidae) showed a pair of hairpencils from their abdominal tips while in a calling posture, and then females were attracted to them. Similarly, some species of Callidulidae may have male sex pheromones to attract females. Males of *Hyblaea* (Hyblaeidae) species have a pair of tibial hairpencils on their hind legs, and it was suggested that the hairpencils are organs that secrete male sex pheromone (e.g., Dugdale et al. 1998, Yoshiyasu 2011).

The clearwing moths, in the family Sesiidae, is a large family that includes about 1,040 described species worldwide (Heppner and

Duckworth 1982, Heppner 1991), and several thousand undescribed species which will be added from Australasia, Afrotrophic and Neotropical region (Arita and Nasu 2011). To the best of my knowledge, all clearwing moths are diurnal and mimic wasps or bees; they have species-specific wasp-mimic body patterns (Fig. 1). Since the original study on the lesser peachtree borer *Synanthedon pictipes* and the peachtree borer *S. exitiosa* by Tumlinson et al. (1974), the female sex pheromone of 27 species in this family have subsequently been identified, and sex attractants of about 100 species were reported (Ando 2016, El-Sayed 2016). In addition to using female sex pheromones, clearwing moths may use visual cues such as wasp-mimicking body patterns for mate recognition. Detailed studies on female sex pheromones of clearwing moths are described below.

Castniidae is a small family that includes about 150 species, and is distributed mainly in Neotropical regions, with some in Australasia and a

Fig. 1. Asian Sesiidae species: (a) a male adult of *Synantedon multitarsus* (Aomori Pref., Japan); (b) a mating pair of *Melittia inouei* (Aomori Pref., Japan); (c) an adult of *Macrotarsipus* sp. (Langkawi Is., Malaysia); (d) a male of *Glossosphecia romanovi*, a serious pest in viticulture (Aomori Pref., Japan); (e) a female *Nokona pernix* showing calling posture (Tottori Pref., Japan). Photos (a, b, c, and d) courtesy of Seiya Kudo and Tadashi Kudo.

few in Southeast Asia (Edwards et al. 1998). Most of castniid species have wide, beautiful wings and all species have clubbed antennae. Because of these forms and their diurnal nature, they are often mistaken for butterflies. In fact, Neotropical species are called "Giant butterfly-moths". The family Castniidae have been placed in the superfamily Sesioidea (Minet 1991), but Regier et al. (2013) re-included them in the superfamly Cossoidea. Sesioidea, Cossoidea and Zyganoidea are grouped as a large and near-monophyletic assemblage, and many species in this group are diurnal. Sarto i Monteys et al. (2012) reported detailed, butterfly-like mating behavior of a castniid *Paysandisia archon*, a serious pest of palm trees. Males show perching or patrolling behaviors to locate conspecific females which enter his territory, and they pursue the females with a courtship flight. These females never attract males using female sex pheromones. Scanning electron microscopy (SEM) and chemical analysis of crude extract of female abdomens indicated that females may not have any pheromone glands. Moreover, three compounds (*Z*,*E*)-farnesal, (*E*,*E*)-farnesal, and (2*E*,13*Z*)-2,13-octadecadien-1-ol (E2,Z13-18:OH) were isolated from male forewings and hindwings. As the three compounds elicited female antennal responses, these compounds may act as male sex pheromones for courtship. Surprisingly, E2,Z13-18:OH is used as a female sex pheromone in some Sesiidae species (Mozûraitis and Karalius 2007, Naka et al. 2007). Compounds with the 2,13-octadecadienyl skeleton have been reported as female sex pheromones in several species in the lepidopteran families of Sesiidae, Cossidae, and Tineidae (Ando 2016, El-Sayed 2016). It is interesting that the similarity of the biosynthesis pathway in males and olfactory receptors in female antennae for detecting E2,Z13-18:OH in *P. archon* and female Sesiidae and/or Cossidae species.

A similar case also was observed in species in a genus *Cystidia* (Geometridae: Ennominae). All *Cystidia* species fly and mate in the daytime. Though *Cystidia couaggaria* females attract males using a sex pheromone in the evening, the closely related species *C. truncangulata* may never use any sex pheromone as we were unable to find pheromone compounds in the crude extracts of female abdomens (Yamakawa et al 2012). Moreover, *C. truncangulata* males fly around their host plant, *Celastrus orbiculatus* (Celastraceae), as though patrolling, and occasionally flying males chase each other. In addition, females never showed calling behaviors when I observed females during the day or night. This suggests that *C. truncangulata* males mainly use visual cues to search for females, not a female sex pheromone (Naka, unpublished data). Because almost all moths in the family Geometridae use female sex pheromones, ancestral *Cystidia* species also may have a female sex pheromone. It is interesting why *C. truncangulata* lost the sex pheromone in their mating communication.

The burnet moths, family Zygaenidae, comprises about 1,000 described species worldwide (Tarmann 2004, Owada et al. 2011). Many zygaeniid species have conspicuous and beautiful wing patterns like butterflies, and therefore many lepidopterists collect specimens of zygaeniids, especially species in the subfamily Chalcosiinae or genus *Zygaena*. Most species of this family are known as day-flying; at least 26 of 29 species in Japan are diurnal (Nasu 2013). Subchev (2014) reviewed female sex pheromones in Zygaenidae. According to the review, most zygaeniid species use female sex pheromones to attract conspecific males; Zygaeninae typically use Type I compounds, whereas Procridinae use 2-butyl esters such as (2*R*)-butyl (9*Z*)-tetradecenoate. Interestingly, Procridinae females have a pheromone gland in a peculiar position. Generally pheromone glands of female moths are located on the intersegmental membrane between the 8th and 9th segments, but in Procridinae species the pheromone glands are located on the anterior part of the 3th-5th abdominal tergites (Hallberg and Subchev 1997).

Finding females using visual cues has been reported in various zygaeniid species. In European and Japanese species of genus *Zygaena* (Zygaeninae), males first use long range attraction to female sex pheromones, and then search for females using visual cues and hovering in the short range (Zagatti and Renou 1984, Koshio 2003, Hofmann and Kia-Hofmann 2010). Adding to that, some Procridinae species (Tanaka and Koshio 2002, Toshova et al. 2007) and the white-tailed zygaenid moth *Elcysma westwoodii* (Chalcosiinae) (Koshio and Hidaka 1995) show similar mating behaviors; they use both chemical and visual cues in mate recognition. In *Zygaena trifolii* and *Z. filipendulae*, two different mate-locating strategies for copulation have been reported. In the morning males search for females using visual cues, and in the late afternoon they use female sex pheromones as a chemical cue. In the morning, females do not release sex pheromones and males can find them using exclusively visual cues (Naumann 1988, Naumann et al. 1999). These are the only reports of moth species that use separate mating strategies depending on time periods.

The family Noctuoidea is a very large taxon that includes over 70,000 species (7,200 genera) distributed all over the world (Kitching and Rawlins 1998). Noctuoidea, especially the family Noctuidae, have many agricultural pests, hence identification of female sex pheromones have been vigorously conducted since the dawn of pheromone study. As far as we know, almost all species of Noctuoidea have female sex pheromones. However, some species use various other modes of communication such as visual and/or acoustic cues in addition to sex pheromones. For example, males of Japanese nine-spotted moth *Amata fortunei* use both female sex pheromone and black-and-orange body stripes (KonDo et al. 2012; details are described below). Some Noctuoidea species rely on acoustic signals for mating. The oleander moth (the polka-dot wasp moth) *Syntomeida epilais* (Erebidae: Arctiinae) uses an acoustic courtship song. Sanderford and Conner (1990, 1995) reported that males and females of *S. epilais* produced sexually dimorphic songs during courtship in antiphonal calling when males responded to female sex pheromone. Similarly, antiphonal calling in mating behaviors were reported from some Nearctic/Neotrophic Arctiinae species (e.g., Sanderford and Conner 1990, Sanderford et al. 1998). Recently, Rowland et al. (2011) found that some species in the genus *Lymantria* (Erebidae: Lymantriinae) use similar antiphonal calling for short range mate-finding and possibly recognition. These suggest that antiphonal calling may be common in Erebidae species.

Species in the genus *Heliocheilus* (Noctuidae: Heliothinae) lost the use of female sex pheromones, while other Heliothinae species typically use female sex pheromones (Ando 2016, El-Sayed 2016). Males of the millet head miner moth *Heliocheilus albipunctella* aggregate (leks or multi-male aggregations) and produce buzzing sound from their forewings, and then females are attracted to males by the sound (Matthews 1987a, b, Green et al. 2000a). Males may use only a calling song without any long-range male sex pheromone (Green et al. 2000b). The closely related species *H. fervens* in Japan also uses a male calling song (Sato, personal communication).

2. Diurnal Moths Using Chemical and Visual Cues for Mate Recognition

From 2002, Drs. Tetsu Ando, Le Van Vang, Yutaka Arita, and I started a comprehensive identification of the female sex pheromones of clearwing moths in Japan to reveal their mating behaviors, especially usage of chemical and/or visual cues for mate recognition. First, to identify female moth sex pheromones, in most cases we must prepare virgin females because only virgin females release sex pheromones in most moth species. In Sesiidae species, obtaining virgin females (= rearing larvae) is difficult because larvae bore into trunks, branches or roots of trees or herbaceous plants. A collaborator, Dr. Arita is a taxonomist who studies microlepidoptera, especially Sesioidea. Fortunately, he had already determined how to obtain virgin females of clearwing moths: collect the last instar larvae from host plants and rear them for a short period of time under laboratory conditions (25±1°C, 16L:8D) (Arita and Ikeda 2000). For example, we collected some muddy cocoons of the clearwing moth *Glossosphecia* (formally *Toleria*) *romanovi* from the ground in vineyards during the winter using small rakes (Naka et al. 2010). We also collected galls of hibernating larvae of the clearwing moth *Nokona pernix* on vines of their host plant *Paederia scandens* (Rubiaceae: Rubioideae) in the winter, and vernalized under winter conditions (8±1°C, dark condition) for one month (Naka et al. 2006). In the case of the Japanese cherry treeborer *Synanthedon hector*, we collected the last instar larvae with the tree bark and resin of *Amygdalus* (formally *Prunus*) *persica* (Rosaceae: Amygdaloideae), and then placed them in a plastic cage under laboratory conditions until adult emergence. Moistened sphagnum was spread on the bottom of the plastic cage, and the larvae were kept with the tree bark and resin (Naka et al. 2008).

After females successfully emerged, we observed them every hour in photophase. Generally, clearwing moths are diurnal; therefore females release sex pheromone only in photophase. Females show a characteristic calling posture when they release sex pheromone; exposing their pheromone gland (between the 8th and 9th abdominal segments) while raising the tip of their abdomen (Fig. 1e). For example, *N. pernix* showed this calling posture only on sunny mornings, and *S. hector* did this during twilight (Naka et al. 2006, 2008). We excised the abdominal ends (8~10th abdominal segments includes pheromone gland) of one- or two-day-old virgin females maintaining the calling posture and separately immersed them in hexane (250 ml/female) for 5~10 min to extract the pheromone. The crude extract was kept at −20°C until chemical analysis.

Only seven compounds, (3*E*,13*Z*)-3,13-octadecadien-1-ol (E3,Z13-18:OH), (3*Z*,13*Z*)-3,13-octadecadien-1-ol (Z3,Z13-18:OH), E2,Z13-18:OH, and the acetate esters of these three compounds (E3,Z13-18:OAc, Z3,Z13-18:OAc and E2,Z13-18:OAc, respectively), and (2*E*,13*Z*)-2,13-octadecadienal (E2,Z13-18:Ald) have been found as female sex pheromones of clearwing moths so far (Ando 2016). Therefore, we prepared these seven compounds and related compounds, e.g. (3*Z*,13*E*)-3,13-octadecadien-1-ol and its acetate before chemical analysis. E3,Z13-18:OAc, Z3,Z13-18:OAc, and E2,Z13-18:OAc, which were supplied by Shin-etsu Chemical Co., Ltd. with purity levels greater than 98%, were utilized as analytical standards and lures for field tests. E3,Z13-18:OH, Z3,Z13-18:OH and E2,Z13-18:OH were prepared by saponification of the corresponding acetates. E2,Z13-18:Ald was obtained by oxidation of E2,Z13-18:OH with pyridinium chlorochromate in CH_2Cl_2 (Islam et al. 2007). (3*E*,13*E*)- and (3*Z*,13*E*)-isomers, which had been synthesized starting from 1,10-decanediol (Naka et al. 2006, 2007; Islam et al. 2007), were also used for these studies.

For identification of female sex pheromones, we performed mainly two analyses using gas chromatography (GC). One is GC equipped with an electroantennographic detector (GC-EAD, Struble and Arn 1984; Fig. 2a), and the other is GC combined with mass spectrometry (GC/MS). The GC-EAD system can discriminate specific chemicals which stimulate the olfactory sensillae of an insect antennae. GC-EAD makes it possible to identify volatile pheromones rapidly. To explain more specifically, one example, the identification of the female sex pheromone of *N. pernix* from crude gland extract, is shown as follows. Figure 2b illustrates the GC-EAD analysis of the crude gland extract (0.5 female equivalents; FE) on a DB-23 capillary column. Male antennae responded to two components (I and II). It indicated that EAG-active components I and II may be candidate female sex pheromone components of *N. pernix*. Following GC-EAD analysis, we conducted GC/MS analysis of the crude extract (1 FE) in an electron impact ionization (EI) mode (70 eV) under the same conditions as those for the GC-EAD measurement. The mass spectra of components I (t_R 11.55 min) and II (t_R 11.84 min) were successfully recorded, and the mass spectra and t_R of both I and II were quite similar to those of synthetic E3,Z13-

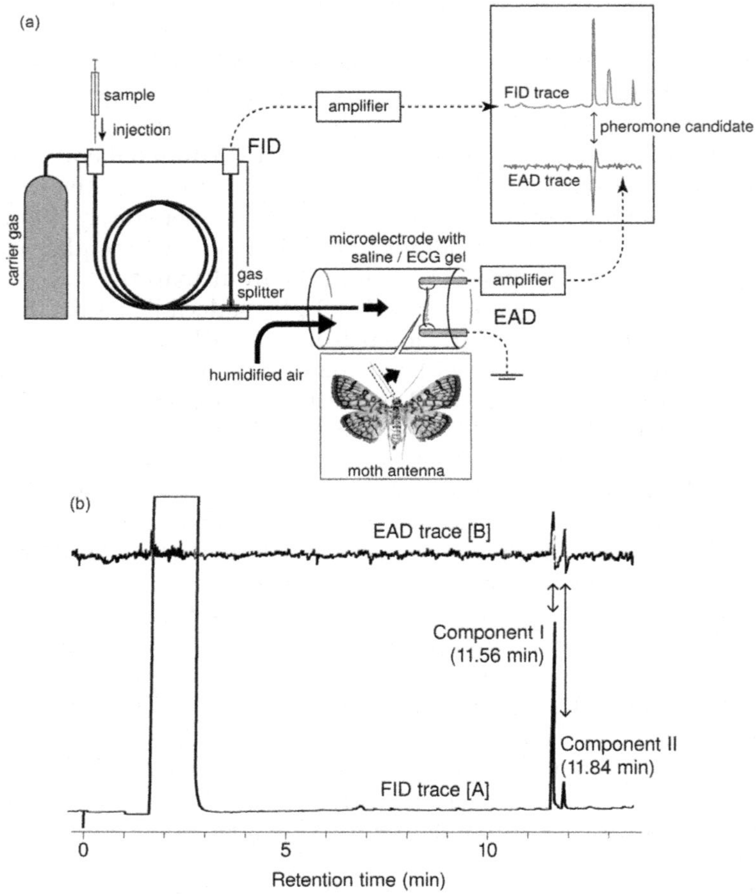

Fig. 2. (a) GC-EAD system and (b) chromatogram of a crude extract of *Nokona pernix* (0.5 FE) in GC-FID (DB-23 column) (A) and GC-EAD (B). The male antennae responded to two components, I (t_R 11.55 min) and and II (t_R 11.84 min; Naka et al. 2006).

18:OH and Z3,Z13-18:OH. Therefore we estimated that component I may be E3,Z13-18:OH and II may be Z3,Z13-18:OH. The ratio of components I and II in crude extract was uniformly found to be approximately 9:1 (Fig. 3).

Once the structures of candidate pheromone components are determined, their attractiveness for males should be tested in the field (and/or wind tunnel). We prepared some pheromone lures, a rubber septum (white rubber, 8 mm O.D., Sigma-Aldrich Co., St. Louis, MO, USA) containing various ratios of synthetic E3,Z13-18:OH and Z3,Z13-18:OH (totally 1.0 mg each), and performed field attraction tests using pheromone baited sticky-type trap (SE-trap®, 30 cm × 27 cm bottom plate with a roof, Sankei Chemical Co., Ltd., Kagoshima, Japan). In the test, many males were attracted to the 9:1 mixture of E3,Z13-18:OH and Z3,Z13-18:OH (Naka et al 2006; Fig. 4). From the results of these chemical analyses and field attraction tests, we concluded that the female sex pheromone of *N. pernix* is a 9:1 mixture of E3,Z13-18:OH and Z3,Z13-18:OH. Thus, we have

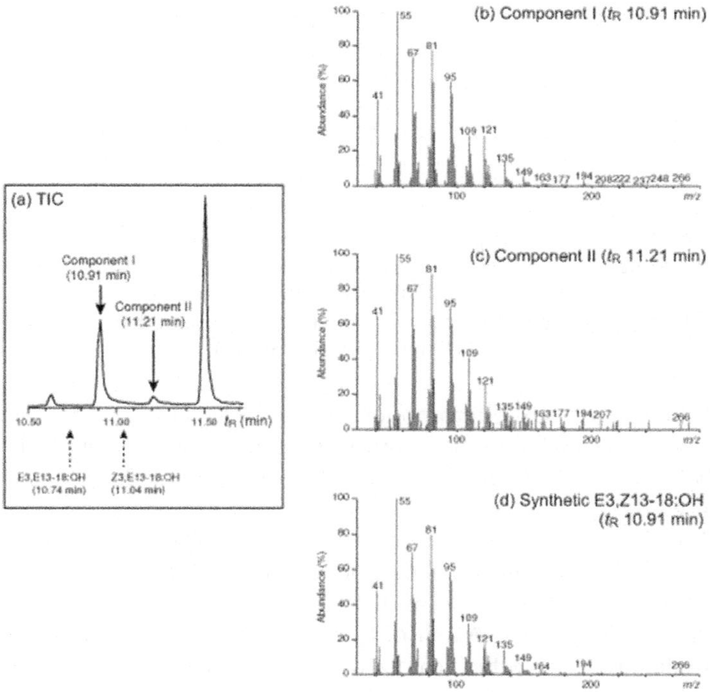

Fig. 3. GC-MS analysis of the sex pheromone components of *Nokona pernix* females; (a) TIC; (b, c, d) mass spectra of compound I (t_R 10.91 min), compound II (t_R 11.21 min), and synthetic E3,Z13-18:OH (t_R 10.91 min), respectively.

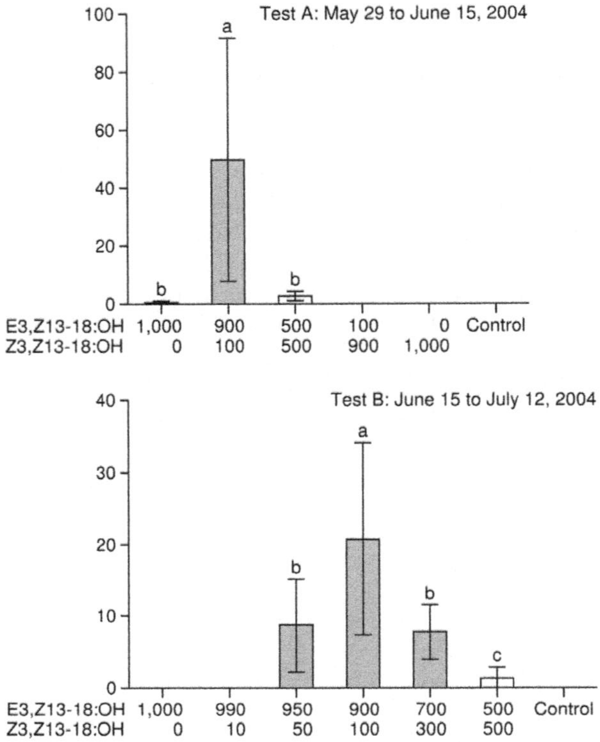

Fig. 4. Attraction of *Nokona pernix* males by traps baited with two synthetic pheromone components, (3*E*,13*Z*)- and (3*Z*,13*Z*)-3,13-octadecadien-1-ol. Tests A and B were performed at the Experimental Farm of Meijo University (Kasugai City, Aichi Prefecture, Japan). Scores followed by a different letter are significantly different ($P<0.05$; Naka et al. 2006).

revealed the female sex pheromones or sex attractants of 25 clearwing moths distributed in Japan and Southeast Asia so far (Naka et al. 2006, 2007, 2008, 2009, 2010, 2013, Shoji et al. 2009, Vang et al. 2012, Naka et al. unpublished data).

Mr. Yoshiteru Horie, a student of Dr. Arita, and I performed some additional field tests using visual models and a pheromone lure containing a 9:1 mixture of E3,Z13-18:OH and Z3,Z13-18:OH (totally 1.0 mg) after the identification process. When only a pheromone lure was hung to a branch, males were attracted from a distance but did not touch the lure. In contrast, when a dried specimen of a female of *N. pernix* (with cuticular hydrocarbons removed beforehand by rinsing with *n*-hexane or acetone) was put near the lure, attracted males touched the specimen and tried to copulate (Horie and Naka in preparation). This indicates that at short-range males recognize females using visual cues, not chemical cues such as cuticular hydrocarbons.

In another experiment, we made female models using stone-powder clay and painted them with acryl gouache (Turner Acryl Gouache, Turner Colour Works Ltd., Osaka, Japan). Models were painted as follows: the "black model" was painted only black, the "yellow model" was painted only yellow, and the "female-like model" had a black body with two yellow bands like as *N. pernix* female (Fig. 5). These models were stone-powder clay (Fando®, Artclay Co. Ltd., Tokyo, Japan) shaped like a female's body without any wings. First, we put a female-like model near a pheromone lure as with the aforementioned dried specimen. Though the model lacked any wings, males tried to copulate to the model with eagerness at the same frequency as males attempted copulation with a dried specimen. When we showed a black or yellow model to males, they attempted to copulate with the black model, but no males were attracted to the yellow model. When we offered both a black model and a female-like model 10 cm apart from each other, most males tried to copulate with the female-like model. This indicates that *N. pernix* males used females' black-and-yellow body pattern in mate recognition, regardless of female wings, after males were attracted to the female sex pheromone (Horie and Naka in preparation; Fig. 6).

Nokona pernix males required the presence of a female-like body to attempt copulation. Barry and Nielsen (1984) also reported that *S. exitiosa* males recognize females using visual cues (red band on the abdomen) much like *N. pernix*. However, the raspberry clearwing *Pennisetia hylaeiformis* (Priesner et al. 1986) and *S. hector* (Horie and Naka in preparation) attempt to copulate with pheromone lures directly. *N. pernix* and *S. exitiosa* mate

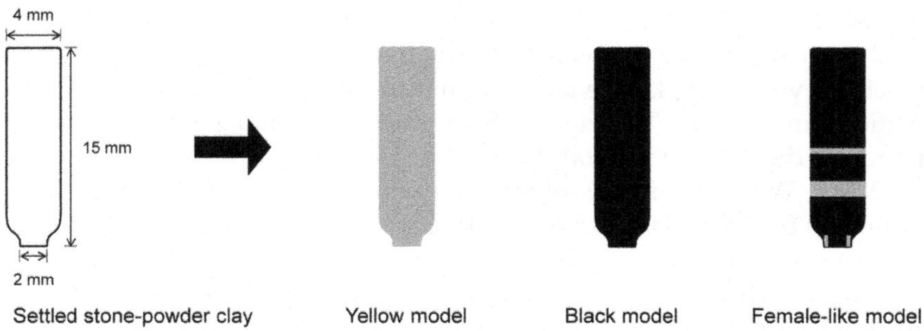

Fig. 5. Visual models used for behavioral assays. The models were figured to 4 mm width × 15 mm height by stone-powder clay (Fando®, Artclay Co. Ltd., Tokyo, Japan) and then painted with acryl gouache (Turner Acryl Gouache, Turner Colour Works Ltd., Osaka, Japan). Models were painted as follows: the "black model" was painted only black, the "yellow model" was painted only yellow, and the "female-like model" was painted to black body with two yellow bands like as a body pattern of female *Nokona pernix*.

Fig. 6. Male *Nokona pernix* attracted to a "female-like model" rather than a "black model".

in the daytime, but *P. hylaeiformis* and *S. hector* mate later in the afternoon or at twilight. The mating times of the respective species are different; and the latter species mates under low illuminance conditions such as during overcast weather conditions or at sunset. The different responses to the pheromone lure suggests the use of visual cues in mate recognition, and may be related to light intensity in the environment which equates to mating time.

Drs. Yûsuke KonDo, Koji Tsuchida, and I performed some laboratory bioassays using males of *A. fortunei*. Fortunately, we did not encounter trouble in supplying these insects for bioassays because breeding this species is very easy; larvae feed on mixtures of a commercially produced artificial moth diet (Insecta® LFS, Nosan Co., Kanagawa, Japan) and a mouse diet (MF, Oriental Yeast Co., Ltd., Tokyo, Japan) (KonDo et al. 2011). We successfully obtained many adults and some breeding strains. While females showed calling posture within four hours after photophase, extractions of female sex pheromone and all bioassays were conducted in the early morning. First, we tested whether males approach the pheromone source (1 female equivalent crude extract to a 2.0 cm width, 5.0 cm length piece of filter paper) in a clear acrylic wind tunnel (0.5 m diameter, 2.0 m length) under laboratory conditions (25°C, 50–60% humidity, wind speed 0.8–1.2 m/s, 100–150 lux). Though over 80% of males approached and about 60% of males contacted the pheromone source, no males tried copulating with the pheromone source. Next, Dr. KonDo made several female models using an epoxy resin (Woodworking Pate

A; Cemedine Co., Ltd., Tokyo, Japan), and we presented them to males with a pheromone source in the wind tunnel. The models had different black-and-orange body patterns. Males were successfully attracted to the models with the pheromone source, and they bent their abdomen toward the models and attempted copulation. Males responded more to models with the same number and similar area of yellow bands as the natural conspecific females than to those with more bands and greater total band area, suggesting that dissimilarity in band number and area to conspecific females could interrupt male mating behavior. (KonDo et al. 2012). We concluded that *A. fortunei* males rely on both female sex pheromone and a specific black-and-orange striped body pattern as a visual cue to locate and identify conspecific females for mating; the sex pheromone plays a major role in mate finding and recognition, whereas visual cues play a supplementary role.

3. Discovery of New Species Using Synthetic Pheromone Compounds

Female sex pheromones can contribute to the discovery of a new species. Historically, collectors have used virgin females to attract conspecific males for some species. Recently, lures baited with synthetic female sex pheromones have been manufactured, particularly for clearwing moths. For example, in United Kingdom, pheromone lures to observe and/or collect clearwing moth species are commercially available (see http://www.angleps.com/pheromones.php). European moth taxonomists and enthusiasts have been collecting clearwing moths using pheromone lures since about 1977, and subsequently, many novel species were described by them. In Japan, although the sex attractant of *S. hector* was reported (Yaginuma et al. 1986) and screening tests using four isomers of 3,13-octadecadienyl acetate were conducted (Tamaki et al. 1977), the pheromone lure was not available to collect clearwing moths for collection. In 1987, some pheromone lures for North American clearwing species became available in Japan. Dr. Arita and his colleague tried to collect Japanese clearwing moth species using these lures, but none were caught. The following year, Dr. Oleg Gorbunov, from Rossiiskaya Akademiya Nauk, provided Dr. Arita with pheromone lures for European clearwing moth species. When Drs. Gorbunov and Arita tested these lures in Japanese fields, males of various clearwing species, especially *Nokona feralis* and *Sesia yezoensis*, were attracted to the lures (Arita and Ikeda 2000; Arita personal communication).

Observing or collecting adults of clearwing moths without pheromone lures is difficult. Since clearwing moths are diurnal, they are not attracted to light traps. Moreover, populations might be very low. If we want to collect adult clearwing moths by only visually searching for and catching them, we will suffer from encounters with numerous wasps and bees. In addition, rearing them is very difficult since their larvae bore into the trunks of trees or roots of herbaceous plants, as described above.

Why are pheromone lures so useful to observe or collect clearwing moths? The low diversity of pheromone compounds in Sesiidae is one of the answers for this question. Only seven compounds described above were found as female sex pheromones of clearwing moths. Although sex pheromones were identified for 27 species and sex attractants were reported for about 100 species (Ando 2016, El-Sayed 2016), the structural diversity of female sex pheromones of clearwing moths is considerably lower than in other taxa. In the other taxa, e.g., Tortricidae or Crambidae, a species frequently use three or more essential components for the female sex pheromone (Ando 2016, El-Sayed 2016). However, the clearwing moth species for which pheromones have been described use either two of seven or single compounds for their female sex pheromones.

Due to the low structural diversity of female sex pheromones, sometimes several Sesiidae species are attracted to the same pheromone lure. Therefore, we can collect various species of clearwing moths much more easily than visual manual collection techniques when using pheromone lures containing two of the seven aforementioned compounds in the field. For example, three lures containing 5:5, 1:9 and 0:10 mixtures of E3,Z13-18:OAc and Z3,Z13-18:OAc attract males of at least six Sesiidae species in Japan: *Synanthedon hector*, *S. tenuis*, *S. quercus*, *Melittia formosana nagaii*, *Pennisetia fixseni fixseni* and *Sesia yezoensis* (Naka et al. 2008, 2013; Fukuda unpublished data; Naka unpublished data). Thus, pheromone lure-based collecting have uncovered many novel species of clearwing moths. Drs. Arita, Gorbunov and Axel Kallies tried collecting clearwing moths using pheromone lures from 2000 to 2006 in North Vietnam. During seven years, they discovered over 80 undescribed species (e.g., Arita and Gorbunov 2000, Kallies and Arita 2001, 2004). Now I distribute 35 different pheromone lures, each of which containing two of the seven compounds, for taxonomists and enthusiasts to collect clearwing moths. Similar trials as in Vietnam were also performed in Japan using the lures, and consequently several undescribed species were found (e.g., Arita et al. 2009, 2016, Kishida et al. 2014). Moreover, two mysterious species were (re)discovered using the lures. *Scalarignathia montis* (Fig. 7), a clearwing moth species which was recorded by Leech in 1897, and for which the holotype specimen was the only one in the world in over a century, was rediscovered by Mr. Kazuya Fukuzumi at a plateau in Nagano Prefecture, Central region of Japan, in 2008 (Yano 2011, Fukuzumi 2012). Moreover, some collaborators discovered a new habitat of this species using these lures (Yano, Aoki, and Kudo personal communications). Mr. Shingo Onodera collected a male of an unknown *Synanthedon* species in Hokkaido

Fig. 7. Photo of male *Scalarignathia montis*, by courtesy of Takahiro Yano.

Prefecture, Japan, in 2002, and he collected some males of the same species again in summer of 2016. At first, no one could identify them because no *Synanthedon* species identical to this species were present in Japan, but, finally, Mr. Walter Garrevoet in Belgium identified them as *Synanthedon martjanovi*. These males also were collected using these lures (Onodera et al. 2016). This species seemed once recorded from Japan (De Freina 1997), however details were unclear and the specimen was not figured. However, collecting females of clearwing moth species is still difficult for us.

Why is the structural diversity of the female sex pheromone very low in clearwing moths? Compared to nocturnal moths, clearwing moths may coexist, and therefore compete, with fewer moth species that use a female sex pheromone in conflict with their own. In nocturnal moths, there are more opportunities for conflicts between ratios of the same pheromone components and/or time of calling or mating between sympatric species because a great majority of moths are nocturnal. Therefore, species of nocturnal moths must utilize different pheromone components and ratios, calling times, and habitats in order to avoid reproductive interference from other sympatric species. Indeed, in genera *Yponomeuta* (Yponomeutidae: Yponomeutinae) and *Ostrinia* (Crambidae: Pyraustinae), pheromone components, ratios, and calling time are generally differed among sympatric species (e.g., Löfstedt et al. 1991, Ishikawa et al. 1999). These phenomena may be explained as ecological character displacements in mating strategies. When one species has more competitors, the selection pressure to the species might be greater. On the other hand, moth species, which have unusual structures for female sex pheromones, are not bound by component ratio of pheromone components or mating time in their mating strategies. For example, the potato tuber moth *Phthorimaea operculella* uses two C13 compounds, (4*E*,7Z)-4,7-tridecadienyl acetate and (4*E*,7Z,10Z)-4,7,10-tridecatrienyl acetate, for their female sex pheromone. As far as known, *P. operculella* is the only species which uses these two compounds as a female sex pheromone, which means that *P. operculella* may have no competitors which utilize this unusual structure in their female sex pheromone. Ono et al. (1990) reported that the component ratio and titer of these two compounds in crude extract from females varied, and males were attracted various ratios of the two compounds. Moreover, females release the sex pheromone during the whole periods of scotophase. This suggests that *P. operculella* may not have stringent requirements for its ratio of pheromone components and time of mating. Because clearwing moths are diurnal, they may have compete with fewer species for reproductive constraints such as these, and the selection pressure for ecological character displacements for mating strategies might be low.

Alternatively, clearwing moths may assure premating isolation enough to avoid these conflicts by using a combination of chemical cues (female sex pheromone) and supplementary visual cues (females' body pattern). For example, two *Synanthedon* species, *S. tenuis* and *S. quercus*, use a single compound (Z3,Z13-18:OAc) for their female sex pheromone, and both species mate during the same window of time (in the evening) (Naka et al. 2013). However, there have been no reports so far of attraction to the wrong species, interspecific mating, or hybridization between the two species. *Synanthedon tenuis* has a black body with three yellow stripes, and *S. quercus* has dark-brown body with orange stripes on every abdominal segment. Moreover, body sizes of these two species are quite different; *S. tenuis* is approximately 18 mm, and *S. quercus* is approximately 24–27 mm. If the two species use female body coloration as a visual cue for mating as *N. pernix* does, the difference between body patterns and sizes might contribute to premating isolation between them.

I am interested in whether such a decrease of structural diversity of female sex pheromone components occurs in other diurnal moth groups.

4. Elucidation of Life Histories of Diurnal Moths

Female sex pheromones also contribute to revealing the life histories of moth species. Males of the arctiid moth *Epatolmis caesarea japonica* are common and often are attracted to light traps. But females have never been attracted to light traps, therefore few dried specimens are present in Japan. Similarly, females of a hawk-moth, *Kentrochrysalis consimilis*, are also very rare. Of course, I have no dried specimen of females of these two species. Finding and rearing the larvae of these species may be almost the only way to obtain the females. Although these are extreme cases, generally we can collect more males than females when collecting moths or butterflies. It is known that males are more active when searching for females, whereas females are usually motionless. In addition, the difference in daily flight activity between females and males may determine our opportunity to encounter them for collection. If females are diurnal and males fly throughout the day, only males will be attracted to light traps.

In contrast, in some moth species, collecting males is very difficult even though females are common. The saturniid silkworm *Rhodinia fugax* is distributed in most regions of Hokkaido, Honshu, Shikoku and Kyushu in Japan, and flies in late autumn. We can easily collect females of this species using light traps, but males are rarely attracted to the traps. I hypothesized that the daily flight activity may be different between females and males, and only males may be diurnal. First, I collected females and tried rearing their offspring. After the field caught females oviposited, I reared larvae on fresh leaves of cherry trees (*Cerasus* spp.; Rosaceae: Amygdaloideae) and finally succeeded to obtain adult females and males. Adults emerged in the evening. Males flew a little at the night, but females rested on their cocoons until dawn. When the day was beginning to break, females adopted the calling posture while still on the cocoons. The flight activity of the males was also high at daybreak and synchronized to the calling activities of the females, and copulation was observed in the same time frame. Next, I extracted female sex pheromone from females in the calling posture at dawn. Drs. Qi Yan, Tetsu Ando and I conducted GC-EAD and GC/MS analysis, and, consequently, we identified (6*E*,11*Z*)-6,11-hexadecadienal (E6,Z11-16:Ald) as a female sex pheromone of *R. fugax*. The pheromone lure containing E6,Z11-16:Ald successfully attracted many males in the field (Yan et al. 2015).

We offered the lure to some enthusiasts of moths. Mr. Yoshichika Aoki took an amazing picture of *R. fugax* males which were attracted to the lure. The photograph received the honor of publication on the cover of the January 2015 issue of *Journal of Chemical Ecology*. Dr. Seiya Kudo and his father Mr. Tadashi Kudo observed *R. fugax* males at coppices in Aomori and Iwate Prefecture, northernmost region of Honshu, Japan. They found two peaks of male flight activity, one at dawn and one between 2 pm~3 pm. Although males only fly in the daytime, the time frame is not always synchronized to the calling time of females. It is interesting why males were attracted to lures at those times. Millar et al. (2016) identified the sex pheromone of another saturniid species, the luna moth (*Actias luna*), as (6*E*,11*Z*)-6,11-octadecadienal, (*E*)-6-octadecenal, and (*Z*)-11-octadecenal, and states that these chemicals are powerful sampling tools that can be used in operational conservation and monitoring programs for insect in population declines.

The diurnal hummingbird hawk-moth *Hemaris affinis* (Bombycoidea: Sphingidae: Macroglossinae) inhabits domestic woodlands, and this species mimics bumblebees (*Bombus* spp.). We can occasionally encounter an individual while nectar feeding or laying eggs. Drs. Takuya Uehara, Hiroshi Honda and I planned to identify the female sex pheromones of all Sphingidae species in Japan. We aimed to identify the female sex pheromone of *H. affinis* as the first study of our Sphingidae pheromone project. The most difficult problem for us was how to obtain the eggs, larvae, or pupae of this species. Though several moth enthusiasts successfully breed *Hemaris* species in Europe, no one has bred *H. affinis* in Japan. Therefore I searched for the eggs or larvae of *H. affinis* on their host plant *Weigela hortensis* (Caprifoliaceae: Diervilloideae) which inhabits ravines and woodland paths. At the beginning, I could not find any eggs or larvae, but finally obtained about 20 eggs and larvae. Finding adults was difficult, but fortunately finding eggs and larvae was not difficult. I sent the resulting pupae to Dr. Uehara (University of Tsukuba), who attempted to extract the female sex pheromone. Troublingly, emerged adults appeared unable to feed under artificial conditions (25±1°C, 16L:8D). This problem was rectified by directly placing 5% sugar solution onto the proboscis of each individual adult. Calling behavior was observed by placing newly emerged females into a mesh cage (23.5 × 30 × 33.5 cm). Female calling behavior was observed 13–16 hours after the beginning of photophase. Analysis of the crude extract was conducted by GC-EAD, and male antennae responded to three compounds. From the t_Rs of the three components, one of them was thought to be (*Z*)-11-hexadecenal (Z11-16:Ald), and the other two may be hexadecadienyl aldehydes such as bombykal, (10*E*,12*Z*)-10,12-hexadecadienal. Following GC-EAD analysis, we conducted GC/MS analysis of crude extract and 4-methyl-1,2,4-triazoline-3,5-dione (MTAD) adducts of candidate pheromone components looked like bombykal. MTAD is a useful tool to determine the position of conjugated double-bonds (Young et al. 1990, Reaney et al. 2001). Finally, we identified these components based on their mass spectra and retention indices on two GC columns, as Z11-16:Ald, bombykal and (10*E*,12*E*)-10,12-hexadecadienal with a ratio of 45:20:35. We made some pheromone lures, and I performed some field trapping tests. Before the tests, I had two hypotheses. The first was that males would be attracted to a mixture containing either one or two of the three identified compounds. Similar to clearwing moths, I thought the diversity of female sex pheromones of diurnal hawk-wing moths may be low. The second was that only a few males will be caught in the traps because adults of this species are rare. However, both hypotheses were found to be false. In field experiments, males were attracted only to the lure which contained all three compounds. Moreover, in total 142 males, a much higher number than I expected, were caught in the traps in a month. Surprisingly, males were attracted to pheromone traps that were set in places far distant from moth habitats, e.g., central area of Tokyo, although we rarely see adults in such places and we can find larvae only in domestic woodlands. This demonstrated that the pheromone lure is a powerful tool to monitor/ discover diurnal moths. In addition, attracted males never touched the pheromone lure, but males tried to copulate with a dried specimen of a female when placed near the lure (Yano personal communication). *Hemaris affinis* males may rely on female sex pheromone for orientation from a distance and visual cues at short range for mate recognition, as do other diurnal lepidopteran species, e.g., clearwing moths and burnet moths.

Prismosticta hyalinata (Bombycoidea: Prismostictidae) was thought to be a rare species because female and male adults of this species are not phototactic. By a happy coincidence, Mr. Keiichiro Shikata found a pair of newly mated adults, and the female successfully oviposited fertilized eggs onto cuttings of *Symplocos sawafutagi* (Ericales: Symplocaceae, formally *S. chinensis*). I received some eggs from him and started breeding them using fresh leaves of *S. coreana*. However, *S. coreana* does not occur in the vicinity of our university, so in order to collect fresh *S. coreana* leaves, I had to drive over an hour away to the mountains every other day. My efforts to collect fresh leaves were rewarded and I finally reared enough pupae to observe adult mating behaviors and analyze the female sex pheromone. I found that adults flew only during the daytime and showed mating behavior mainly in the afternoon; that is, *P. hyalinata* is a diurnal moth. Analysis of the crude extract of female glands using GC-EAD and GC/MS revealed that females produced a 10,12-diene (compound X; Naka et al. in preparation) and this was thought to be a female sex pheromone for this species. Drs. Tetsu Ando and Qi Yan synthesized compound X and related compounds, and made some pheromone lures. I placed sticky traps baited with the lures at coppices in Tottori Prefecture, Western region of Honshu, Japan. Surprisingly, all traps baited with the lures containing compound X caught large numbers of males, although this species was thought to be a rare species. Moreover, many males flew around a lure in the afternoon as I hung it on a twig. These field tests revealed three novel findings. First, *P. hyalinata* is not a rare species; this species may be a common species distributed in various environments where *Symplocos* trees are distributed. Second, males were attracted to the female sex pheromone in the afternoon, and catching adults using a butterfly net was very difficult because the males fly so fast. Third, this species is multivoltine; the males were caught in traps in late May, June and early September in Tottori Prefecture (Naka et al. in preparation).

A rare lappet moth *Pyrosis idiota* appears in early spring in Japan. On rare occasions we can collect females using light traps at plateaus in western Japan, but only one specimen suspected as the male of this species has been recorded from Okayama Prefecture (Miyake 2009). Miyake

(2010) successfully reared this species using fresh leaves of *Quercus* spp. (Fagaceae: Quercoideae), and confirmed the specimen is a true *P. idiota* male. I think males of this species may be diurnal, and females may release sex pheromone in the day time. Once the female sex pheromone of this species is identified, we can reveal various aspects of the biology of this species, e.g., distribution, habitat, rarity/risk of extinction, daily rhythms and seasonal prevalence.

Acknowledgments

I thank Jun Tabata for providing an opportunity to write this review, and Miriam F. Cooperband for correcting my English with critical readings. I also thank Toshimasa Mitamura, Atsushi Mochizuki, Fumiaki Mochizuki, Takeshi Fujii, Yume Imada, Seiji Miyake, Takashi Sato, Teruhiko Fukuda, Kazuya Fukuzumi, Naoki Yata for supplying relevant information. Thanks are owed to Yutaka Arita, Yoshichika Aoki, Seiya Kudo, Tadashi Kudo, and Takahiro Yano for their permission to use beautiful photographs and unpublished data. We gratefully acknowledge the work of all great collaborators for our studies mentioned in the body.

REFERENCES

Andersson, J., A.-K. Borg-Karlson and C. Wiklund. 2003. Antiaphrodisiacs in pierid butterflies: a theme with variation! J. Chem. Ecol. 29: 1489–1499.

Andersson, J., A.-K. Borg-Karlson, N. Vongvanich and C. Wiklund. 2007. Male sex pheromone release and female mate choice in a butterfly. J. Exp. Biol. 210: 964–970.

Ando, T. 2016. Sex pheromones of moths. (http://lepipheromone.sakura.ne.jp/PheromoneList/List_of_Sex_Pheromones.pdf)

Ando, T., S. Inomata and M. Yamamoto. 2004. Lepidopteran sex pheromones. Top. Curr. Chem. 239: 51–96.

Arita, Y. and O. Gorbunov. 2000. On the tribe Melittini (Lepidoptera: Sesiidae) of Vietnam. Tinea 16: 252–291.

Arita, Y. and M. Ikeda. 2000. Sesiidae of Japan. Mushisha, Tokyo, Japan.

Arita, Y., M. Kimura and M. Owada. 2009. Two new species of the clearwing moth (Sesiidae) from Okinawa-jima, the Ryukyus. Trans. Lepid. Soc. Japan 60: 189–192.

Arita, Y. and Y. Nasu. 2011. Sesioidea. pp. 255–264. *In*: F. Komai, Y. Yoshiyasu, Y. Nasu and T. Saito [eds.]. A Guide to the Lepidoptera of Japan. Tokai University Press, Hiratsuka, Japan.

Arita, Y., M. Kimura, N. Yata and M. Nagase. 2016. Vicariance in the *Macroscelesia japona* species-group (Lepidoptera: Sesiidae) in the Ryukyus, Japan. Tinea 23: 184–198.

Barry, M.W. and D.G. Nielsen. 1984. Behavior of adult peachtree borer (Lepidoptera: Sesiidae). Ann. Entomol. Soc. Am. 77: 246–250.

Boppré, M., R.L. Petty, D. Schneider and J. Meinwald. 1978. Behaviorally mediated contacts between scent organs: another prerequisite for pheromone production in *Danaus chrysippus* males (Lepidoptera). J. Comp. Physiol. 126: 97–103.

De Freina, J.J. 1997. Die Bombyces und Sphinges der Westpalaearktis (Insecta, Lepidoptera). Band IV. Sesiidae. Edition Forschung & Wissenschaft Verlag Gmb H., München, Germany.

Dugdale, J.S., N.P. Kristensen, G.S. Robinson and M.J. Scoble. 1998. The smaller microlepidoptera-grade superfamilies. pp. 217–232. *In*: N.P. Kristensen [ed.]. Lepidoptera, Moths and Butterflies 1: Evolution, Systematics, and Biogeography. Handbook of Zoology/Handbuch der Zoologie 35. De Gruyter, Berlin, Germany.

Edwards, E.D., P. Gentili, M. Horak, N.P. Kristensen and E.S. Nielsen. 1998. The cossoid/sesioid assemblage. pp. 181–195. *In*: N.P. Kristensen [ed.]. Lepidoptera, Moths and Butterflies 1: Evolution, Systematics, and Biogeography. Handbook of Zoology/Handbuch der Zoologie 35. De Gruyter, Berlin, Germany.

El-Sayed, A.M. 2016. Pherobase. (http://www.pherobase.com/)

Foster, S.P. and W.P. Thomas. 2000. Identification of a sex pheromone component of the raspberry budmoth, *Heterocrossa rubophaga*. J. Chem. Ecol. 26: 2549–2555.

Fujii, T., D. Iwai, F. Mochizuki and T. Fukumoto. 2010. Sex pheromones and sex attractants from *Platyptillia ignifera*. Japan Patent. JP2010018543.

Fukuzumi, K. 2012. *Scalarignathia montis* (Leech [1889]) (Sesiidae) from Nagano Pref. for the first time in one hundred twenty years. Japan Heterocerist. J. 264: 355.

Gibb, A.R., D.M. Suckling, B.D. Morris, T.E. Dawson, N. Bunn, D. Comeskey et al. 2006. (Z)-7-tricosene and monounsaturated ketones as sex pheromone components of the Australian guava moth *Coscinoptycha improbana*: identification, field trapping, and phenology. J. Chem. Ecol. 32: 221–237.

Green, S.V., E.O. Owusu, O. Youm and D.R. Hall. 2000a. Mating behaviour of the millet headminer moth, *Heliocheilus albipunctella* (de Joannis) (Lepidoptera: Noctuidae: Heliothinae). (https://assets.publishing.service.gov.uk/media/57a08d7e40f0b649740018bc/R6693_FTRappend3.pdf)

Green, S.V., E.O. Owusu, O. Youm, T.J. Bruce and D.R. Hall. 2000b. Investigation of a long range male sex pheromone in the millet headminer moth, *Heliocheilus albipunctella* de Joannis (Lepidoptera: Noctuidae: Heliothinae). (https://assets.publishing.service.gov.uk/media/57a08d7f40f0b652dd0019d4/R6693_FTRappend5.pdf)

Hallberg, E. and M. Subchev. 1997. Unusual location and structure of female pheromone glands in *Ino ampelophaga* Bayle (Lepidoptera, Zygaenidae). Int. J. Insect Morphol. Embryol. 25: 381–389.

Hashimoto, S. 2011. Micropterigoidea, Agathiphagoidea, Eriocranioidea. pp. 65–78. *In*: F. Komai, Y. Yoshiyasu, Y. Nasu and T. Saito [eds.]. A Guide to the Lepidoptera of Japan. Tokai University Press, Hiratsuka, Japan.

Haynes, K.F. 1987. Identification of sex pheromone of calendula plume moth, *Platyptilia williamsii*. J. Chem. Ecol. 13: 907–916.

Heppner, J.B. 1991. Faunal regions and the diversity of Lepidoptera. Trop. Lepid. 2 (Suppl. 1): 1–85.

Heppner, J.B. and W.D. Duckworth. 1982. Addendum and corrigenda to "Classification of the superfamily Sesioidea". J. Lepid. Soc. 36: 119–120.

Hidaka, T. and K. Yamashita. 1975. Wing color pattern as the releaser of mating behavior in the swallowtail butterfly, *Papilio xuthus* L. (Lepidoptera: Papilionidae). Appl. Entomol. Zool. 10: 263–267.

Hofmann, A. and T. Kia-Hofmann. 2010. Experiments and observations on pheromone attraction and mating in burnet moths (*Zygaena* Fabricius, 1777) (Lepidoptera: Zygaenidae). Entomol. Gazette 61: 83–93.

Honda, K., A. Tada and N. Hayashi. 1995. Dihydropyrrolizines from the male scent-producing organs of a danaid butterfly, *Ideopsis similis* (Lepidoptera: Danaidae) and the morphology of alar scent organs. Appl. Entomol. Zool. 30: 471–477.

Honma, K., K. Kawasaki and Y. Tamaki. 1978. Sex-pheromone activities of synthetic 7-alken-11-ones for male peach fruit moths, *Carposina niponensis* Walsingham (Lepidoptera: Carposinidae). Jpn. J. Appl. Entomol. Zool. 22: 87–91.

Ishikawa, Y., T. Takanashi, C. Kim, S. Hoshizaki, S. Tatsuki and Y. Huang. 1999. *Ostrinia* spp. in Japan: their host plants and sex pheromones. Entomol. Exp. Appl. 91: 237–244.

Kallies, A. and Y. Arita. 2001. The Tintiinae of North Vietnam (Lepidoptera, Sesiidae). Trans. Lepid. Soc. Japan 52: 187–235.

Kallies, A. and Y. Arita. 2004. A survey of the clearwing moths of the tribe Sesiini (Lepidoptera, Sesiidae) from Vietnam and adjacent countries with a synopsis of the Oriental Sesiini fauna. Tinea 18: 65–95.

Kato, Y. and Y. Yoshioka. 2003. Visual stimuli affecting male mating behavior of *Graphium sarpedon* (Lepidoptera, Papilionidae). Trans. Lepid. Soc. Japan 54: 209–219.

Kishida, Y., T. Kudo and S. Kudo. 2014. A new species of *Nokona* Matsumura from Japan (Lepidoptera: Sesiidae). Tinea 23: 4–9.

Kitching, I.J. and J.E. Rawlins. 1998. The smaller microlepidoptera grade superfamilies. pp. 217–232. *In*: N.P. Kristensen [ed.]. Lepidoptera, Moths and Butterflies 1: Evolution, Systematics, and Biogeography. Handbook of Zoology/Handbuch der Zoologie 35. De Gruyter, Berlin, Germany.

Klun, J.A., K.F. Haynes, B.A. Bierl-Leonhardt, M.C. Birch and J.R. Plimmer. 1981. Sex pheromone of the female artichoke plume moth, *Platyptilia carduidactyla*. Environ. Entomol. 10: 763–765.

KonDo, Y., H. Naka and K. Tsuchida. 2011. Artificial diet for the Japanese nine-spotted moth *Amata fortunei fortunei* (Arctiidae: Syntominae). Entomol. Sci. 14: 387–391.

KonDo, Y., H. Naka and K. Tsuchida. 2012. Pheromones and body coloration affect mate recognition in the Japanese nine-spotted moth *Amata fortunei* (Lepidoptera: Arctiidae). J. Ethol. 30: 301–308.

Koshio, C. 2003. Mating behaviour and activity patterns of the Japanese burnet moth *Zygaena niphona* Butler, 1877 (Lepidoptera: Zygaenidae: Zygaeninae). Proceedings of the 7th International Symposium on Zygaenidae (Lepidoptera), pp. 85–98.

Koshio, C. and T. Hidaka. 2005. Reproductive behavior of the white zygaenid moth, *Elcysma westwoodi* (Lepidoptera, Zygaenidae) I. Mating sequences. J. Ethol. 13: 159–163.

Kozlov, M.V., J. Zhu, P. Philipp, W. Francke, E.L. Zvereva, B.S. Hansson et al. 1996. Pheromone specificity in *Eriocrania semipurpurella* (Stephens) and *E. sangii* (Wood) (Lepidoptera: Eriocraniidae) based on chirality of semiochemicals. J. Chem. Ecol. 22: 431–454.

Löfstedt, C., W.M. Herrebout and S.B.J. Menken. 1991. Sex pheromones and their potential role in the evolution of reproductive isolation in small ermine moths (Yponomeutidae). Chemoecology 2: 20–28.

Lundgren, L. and G. Bergstrom. 1975. Wing scents and scent-released phases in the courtship behavior of *Lycaeides argyrognomon* (Lepidoptera: Lycaenidae). J. Chem. Ecol. 1: 399–412.

Matthews, M. 1987a. Moths make inroads into Mali's crop of millet. New Sci. 114: 50.

Matthews, M. 1987b. The African species of *Heliocheilus* Grote (Lepidoptera: Noctuidae). Syst. Entomol. 12: 459–473.

Millar, J.G., K.F. Haynes, A.T. Dossey, J.S. McElfresh and J.D. Allison. 2016. Sex attractant pheromone of the luna moth, *Actias luna* (Linnaeus). J. Chem. Ecol. 42: 869–876.

Minet, J. 1991. Tentative reconstruction of the ditrysian phylogeny (Lepidoptera: Glossata). Entomol. Scand. 22: 69–95.

Miyake, S. 2009. A male specimen putative *Pyrosis idiota* Graese. Yadoriga 222: 30.

Miyake, S. 2010. Early stages of *Pyrosis idiota* Graese. Yadoriga 225: 19–22.

Moser, J.C., J.D. Reeve, J.M.S. Bento, T.M.C. Della Lucia, R.S. Cameron and N.M. Heck. 2004. Eye size and behaviour of day- and night-flying leafcutting ant alates. J. Zool. 264: 69–75.

Mozûraitis, R. and V. Karalius. 2007. Identification of minor sex pheromone components of the poplar clearwing moth *Paranthrene tabaniformis* (Lepidoptera, Sesiidae). Zeit. Naturforsch. C 62: 138–142.

Naka, H., T. Nakazawa, M. Sugie, M. Yamamoto, Y. Horie, R. Wakasugi et al. 2006. Synthesis and characterization of 3,13- and 2,13-octadecadienyl compounds for identification of the sex pheromone secreted by a clearwing moth, *Nokona pernix*. Biosci. Biotech. Biochem. 70: 508–516.

Naka, H., S. Inomata, K. Matsuoka, M. Yamamoto, H. Sugie, K. Tsuchida et al. 2007. Sex pheromones of two Melittini species, *Macroscelesia japona* and *M. longipes* (Lepidoptera: Sesiidae): identification and field attraction. J. Chem. Ecol. 33: 591–601.

Naka, H., Y. Horie, F. Mochizuki, L.V. Vang, Y. Yamamoto, T. Saito et al. 2008. Identification of the sex pheromone secreted by *Synanthedon hector* (Lepidoptera: Sesiidae). Appl. Entomol. Zool. 43: 467–474.

Naka, H., Y. Horie, T. Ando and Y. Arita. 2009. Sex attractant of *Melittia sangaica* (Lepidoptera: Sesiidae). Bull. Meijo Univ. Farm 10: 63–65.

Naka, H., M. Mochizuki, K. Nakada, N.D. Do, T. Yamauchi, Y. Arita et al. 2010. Female sex pheromone of *Glossosphecia romanovi* (Lepidoptera: Sesiidae): identification and field attraction. Biosci. Biotech. Biochem. 74: 1943–1946.

Naka, H., T. Suzuki, T. Watarai, Y. Horie, F. Mochizuki, A. Mochizuki et al. 2013. Identification of the sex pheromone secreted by *Synanthedon tenuis* (Lepidoptera: Sesiidae). Appl. Entomol. Zool. 48: 27–33.

Nasu, Y. 2011. Calliduloidea. pp. 414–417. *In*: F. Komai, Y. Yoshiyasu, Y. Nasu and T. Saito [eds.]. A Guide to the Lepidoptera of Japan. Tokai University Press, Hiratsuka, Japan.

Nasu, Y. 2013. Zygaenoidea. pp. 317–331. *In*: T. Hirowatari, Y. Nasu, Y. Sakamaki and Y. Kishida [eds.]. The Standard of Moths in Japan III. Gakken Education Publishing, Tokyo, Japan.

Naumann, C.M. 1988. Zur Evolution und adaptiven Bedeutung zweier unterschiedlicher Partnerfindungsstrategien bei *Zygaena trifolii* (Esper, 1783) (Lepidoptera, Zygaenidae). Verh. Dtsch. Zool. Ges. 81: 257–258.

Naumann, C.M., G.M. Tarmann and W.G. Tremewan. 1999. The Western Palaearctic Zygaenidae (Lepidoptera). Apollo Books, Stenstrup, Denmark.

Nishida, R., S. Schulz, C.S. Kim, H. Fukami, Y. Kuwahara, K. Honda et al. 1996. Male sex pheromone of a giant danaine butterfly, *Idea leuconoe*. J. Chem. Ecol. 22: 949–972.

Nishio, N. 1983. Mating behavior of *Pterodecta felderi*. New Entomol. 32: 21–22.

Nishio, N. 1999. Life history of *Pterodecta felderi*. Insect Nat. 34(13): 14–17.

Ômura, H. and S. Yotsuzuka. 2015. Male-specific epicuticular compounds of the sulfur butterfly *Colias erate poliographus* (Lepidoptera: Pieridae). Appl. Entomol. Zool. 50: 191–199.

Ono, T., R.E. Charlton and R.T. Cardé. 1990. Variability in pheromone composition and periodicity of pheromone titer in potato tuberworm moth, *Phthorimaea operculella* (Lepidoptera: Gelechiidae). J. Chem. Ecol. 16: 531–542.

Onodera, S., H. Onodera, N. Yata and Y. Arita. 2016. *Synanthedon martjanovi* Sheljuzhko, 1918 (Lepidoptera, Sesiidae) from Hokkaido in Japan. Japan Heterocerist. J. 279: 97–98.

Owada, M., Y. Nasu and N. Teramoto. 2011. Zygaenoidea. pp. 230–254. *In*: F. Komai, Y. Yoshiyasu, Y. Nasu and T. Saito [eds.]. A Guide to the Lepidoptera of Japan. Tokai University Press, Hiratsuka, Japan.

Priesner, E., P. Witzgall and S. Voerman. 1986. Field attraction response of raspberry clearwing moths, *Pennisetia hylaeiformis* Lasp. (Lepidoptera: Sesiidae), to candidate pheromone chemicals. J. Appl. Entomol. 102: 195–210.

Pliske, T.E. and T. Eisner. 1969. Sex pheromone of the queen butterfly: biology. Science 164: 1170–1172.

Reaney, M.J.T., Y.-D. Liu and W.G. Taylor. 2001. Gas chromatographic analysis of diels-alder adducts of geometrical and positional isomers of conjugated linoleic acid. J. Am. Oil Chem. Soc. 78: 1083–1086.

Regier, J.C., C. Mitter, A. Zwick, A.L. Bazinet, M.P. Cummings, A.Y. Kawahara et al. 2013. A large-scale, higher-level, molecular phylogenetic study of the insect order Lepidoptera (moths and butterflies). PLoS ONE 8: e58568.

Robertson, K.A. and A. Monteiro. 2005. Female *Bicyclus anynana* butterflies choose males on the basis of their dorsal UV reflective eyespot pupils. Proc. R. Soc. B 272: 1541–1546.

Rowland, E., P.W. Schaefer, P. Belton and G. Gries. 2011. Evidence for short-range sonic communication in lymantriine moths. J. Insect Physiol. 57: 292–299.

Rutowski, R.L. 1983. The wing-waving display of *Eurema daira* males (Lepidoptera: Pieridae): its structure and role in successful courtship. Anim. Behav. 31: 985–989.

Rutowski, R.L. 2000. Variation in eye size in butterflies: inter- and intraspecific patterns. J. Zool. 252: 187–195.

Sanderford, M.V. and W.E. Conner. 1990. Courtship sounds of the polka-dot wasp moth, *Syntomeida epilais*. Naturwissenschaften 77: 345–347.

Sanderford, M.V. and W.E. Conner. 1995. Acoustic courtship communication in *Syntomeida epilais* Wlk. (Lepidoptera: Arctiidae, Ctenuchinae). J. Insect Behav. 8: 19–31.

Sanderford, M.V., F. Coro and W.E. Conner. 1998. Courtship behavior in *Empyreuma affinis* Roths. (Lepidoptera, Arctiidae, Ctenuchinae): acoustic signals and tympanic organ response. Naturwissenschaften 85: 82–87.

Sarto i Monteys, V., P. Acín, G. Rosell, C. Quero, M.A. Jiménez and A. Guerrero. 2012. Moths behaving like butterflies. Evolutionary loss of long range attractant pheromones in castniid moths: a *Paysandisia archon* model. PLoS ONE 7: e29282.

Schneider, D., M. Boppré, H. Schneider, W.R. Thompson, C.J. Boriak, R.L. Petty et al. 1975. A pheromone precursor and its uptake in male *Danaus* butterflies. J. Comp. Physiol. A 97: 245–256.

Shoji, A., Y. Adachi, H. Naka, T. Ando and Y. Arita. 2009. *Nokona regalis* (Lepidoptera: Sesiidae): a late collection record in autumn and a sex attractant. Trans. Lepid. Soc. Japan 60: 193–195.

Struble, D.L. and H. Arn. 1984. Combined gas chromatography and electroantennogram recording of insect olfactory responses. pp. 161–178 *In*: H.E. Hummel and T.A. Miller [eds.]. Techniques in Pheromone Research. Springer-Verlag, New York, USA.

Subchev, M. 2014. Sex pheromone communication in the family Zygaenidae (Insecta: Lepidoptera): a review. Acta Zool. Bulgar. 66: 147–157.

Takanashi, T., M. Hiroki and Y. Obara. 2001. Evidence for male and female sex pheromones in the sulfur butterfly, *Eurema hecabe*. Entomol. Exp. Appl. 101: 89–92.

Tamaki, Y., T. Yushima, M. Oda, K. Kida, K. Kitamura, S. Yabuki et al. 1977. Attractiveness of 3,13-octadecadienyl acetates for males of clearwing moths. Jpn. J. Appl. Entomol. Zool. 21: 106–107.

Tanaka, Y. and C. Koshio. 2002. Activity patterns and reproductive behaviour of the plum moths *Illiberis rotundata* Jordan (Lepidoptera: Zygaenidae: Procridinae). Jpn. J. Entomol. 5: 70–80.

Tarmann, G.M. 2004. Zygaenid Moths of Australia. A Revision of the Australian Zygaenidae (Procridinae: Artonini). CSIRO Publishing, Collingwood, Australia.

Toshova, T.B., M.A. Subchev and M. Tóth. 2007. Role of olfactory and visual stimuli in the mating behaviour of male vine bud moths, *Theresimima ampellophaga* (Lepidoptera: Zygaenidae). Eur. J. Entomol. 104: 57–65.

Tumlinson, J.H., C.E. Yonce, R.E. Doolittle, R.R. Heath, C.R. Gentry and E.R. Mitchell. 1974. Sex pheromones and reproductive isolation of the lesser peachtree borer and the peachtree borer. Science 185: 614–616.

Vane-Wright, R.I. and M. Boppré. 1993. Visual and chemical signaling in butterflies: functional and phylogenetic perspectives. Phil. Trans. R. Soc. B 340: 197–205.

Vang, L.V., C.N.Q. Khanh, H. Shibasaki and T. Ando. 2012. Female sex pheromone secreted by *Carmenta mimosa* (Lepidoptera: Sesiidae), a biological control agent for an invasive weed in Vietnam. Biosci. Biotech. Biochem. 76: 2153–2155.

Wago, H., K. Unno and Y. Suzuki. 1976. Studies on the mating behavior of the pale grass blue, *Zizeeria maha argia* (Lepidoptera: Lycaenidae) I. Recognition of conspecific individuals by flying males. Appl. Entomol. Zool. 11: 302–311.

Yaginuma, K., M. Kumakura, Y. Tamaki, T. Yushima and J.H. Tumlinson. 1976. Sex attractant for the cherry tree borer, *Synanthedon hector* Butler (Lepidoptera: Sesiidae). Appl. Entomol. Zool. 11: 266–268.

Yamakawa, R., Y. Takubo, K. Ohbayashi, H. Naka and T. Ando. 2012. Female sex pheromone of *Cystidia couaggaria couaggaria* (Lepidoptera: Geometridae): identification and field attraction. Biosci. Biotech. Biochem. 76: 1303–1307.

Yan, Q., A. Kanegae, T. Miyachi, H. Naka, H. Tatsuta and T. Ando. 2015. Female sex pheromones of two Japanese saturniid species, *Rhodinia fugax* and *Loepa sakaei*: identification, synthesis, and field evaluation. J. Chem. Ecol. 41: 1–8.

Yano, T. 2011. *Scalarignathia montis* in Takabotchi. Yadoriga 230: 6–7.

Yildizhan, S., J. van Loon, A. Sramkova, M. Ayasse, C. Arsene, C. tenBroeke et al. 2009. Aphrodisiac pheromones from the wings of the small cabbage white and large cabbage white butterflies, *Pieris rapae* and *Pieris brassicae*. Chem BioChem 10: 1666–1677.

Yoshiyasu, Y. 2011. Hyblaeoidea. pp. 725–726. *In*: F. Komai, Y. Yoshiyasu, Y. Nasu and T. Saito [eds.]. A Guide to the Lepidoptera of Japan. Tokai University Press, Hiratsuka, Japan.

Young, D.C., P. Vouros and M.F. Holick. 1990. Gas chromatographic-mass spectrometry of conjugated dienes by derivatization with 4-methyl-1,2,4-triazoline-3,5-dione. J. Chromatogr. A 522: 295–302.

Zagatti, P. and M. Renou. 1984. Les pheromones sexuelles des zygènes. Le comportement de *Zygaena filipendulae* L. (Lepidoptera, Zygaenidae). Ann. Soc. Entomol. Fr. 20: 439–454.

Zeil, J. 1983. Sexual dimorphism in the visual system of flies: the free flight behaviour of male Bibionidae (Diptera). J. Comp. Physiol. 150: 395–412.

Zhu, J., M.V. Kozlov, P. Philipp, W. Francke and C. Löfstedt. 1995. Identification of a novel moth sex pheromone in *Eriocrania cicatricella* (Zett.) (Lepidoptera: Eriocraniidae) and its phylogenetic implications. J. Chem. Ecol. 21: 29–43.

CHAPTER

9

Mating Disruption: Concepts and Keys for Effective Application

Jun Tabata

National Agriculture and Food Research Organization,
3-1-3 Kannondai, Tsukuba City, Ibaraki 305-8604, Japan
E-mail: jtabata@affrc.go.jp

Abstract

Mating disruption is a strategy to suppress the population growth of pests by mimicking their sex pheromones and interfering with mate-finding and copulation. Studies on mating disruption have been carried out since the 1960s, but many mysteries and obstacles still remain to be solved. Particularly in non-lepidopteran pests, commercially available products for mating disruption are currently very scarce. In this chapter, the concepts of mating disruption are summarized in brief, and subsequently some case studies including moths as well as beetles and mealybugs are reviewed. Based on these previous studies, we seek to deduce some keys for the further development, application, and promotion of mating disruption in pest management programs.

1. Introduction

Pheromones are ubiquitously found in intra-specific communication in animals, including vertebrates and invertebrates. Although pheromones have many functions, for example aggregation, trail generation, regulation of caste differentiation, etc., one of the most prevalent forms are probably sex pheromones, which control mate-finding and copulation behaviors. Particularly in some taxa of insects, females produce and emit volatile pheromones to attract males from a long distance. These sex-attractant pheromones generally elicit drastic behavioral and/or physiological responses in conspecific mates even in minute amounts. The pheromone-mediated mate-finding systems utilized by pest insects have long been recognized to be of potential value in their management, even before the identification of any pest pheromones (Wright 1964a, b, Carde 1990). The chemical constituents of pheromonal messages could have potential applications as environmentally safe substitutes for conventional insecticides in the protection of crops or forests (Silverstein 1990, Baker et al. 2016, Millar et al. 2016). Thus, many agronomists and chemical ecologists have examined the use of synthetic "mimics" in interference of communications via natural pheromones, which lead to the suppression of population growth in pests. Such tactics are called "communication disruption" or "communication interference". In particular, sex pheromone-based communication disruption (mating disruption) is currently being used successfully against many pest species. In this chapter, the concept of mating disruption is reviewed along with case studies, and we deduce some key points for applying this eco-friendly strategy in pest management programs.

2. Concept and Mode of Action

2.1 Concept

Sex attractant pheromones in insects are generally volatile compounds with low to medium molecular weights. Unlike general insecticides that work when they are touched or digested by pests, volatile pheromones do not generally remain in the environment and are unlikely to cause environmental pollution. Moreover, because pheromones induce species-specific responses only in the target, there is very little concern about toxicity in humans, livestock, fish, or most other organisms, including beneficial insects such as pollinators or natural enemies. Thus, mating disruption based on pheromones is an environmentally safe substitute for insecticides. It is particularly important to rescue and preserve beneficial insects in agricultural fields. Successive use of mating disruption and reduction in insecticide application can enhance ecological functions promoted by biodiversity, which lead to more ecosystem services and is indispensable for sustainable agriculture. However, some species of natural enemies use pheromones emitted by host pests in order to locate them. Hence mating disruption has the potential to affect the behavior and activities of natural enemies. This will be discussed later.

2.2 Mode of Action

The possible mechanisms of mating disruption have been reviewed by e.g.

Bartell (1982), Cardé (1990), Sanders (1997), and Evenden (2016), and can be largely divided into "competitive" and "non-competitive" mechanisms (Miller et al. 2006a, b). A competitive mechanism means that the frequency with which male moths find calling females under conditions of disruption is reduced because males are preoccupied by nearby attractive or arrestive plumes from dispensers of synthetic pheromone (Miller et al. 2006a). Non-competitive disruption mechanisms are effects without attraction to synthetic pheromone sources and include (i) camouflage, (ii) desensitization (sensory fatigue), and (iii) sensory imbalance. However, the evidence for which mechanism is mainly responsible for communicational disruption for particular pests with particular disruption regimens remains largely circumstantial, and we have very limited examples where a particular mechanism has been definitively proven to be the leading cause of mating disruption for any insect in field conditions (Miller et al. 2006a). Some of these mechanisms are not exclusive and can operate together in the field.

2.2.1 Competitive Attraction

Competitive attraction is also referred to as confusion (Bartell 1982) or false-plume-following (Cardé 1990, Stelinski et al. 2004). This mechanism involves the males locking-on to and following a "false" pheromone instead of the natural pheromone emitted by calling females is the most straightforward disruption process (Sanders 1997). For the success of competitive attraction, synthetic pheromone sources should be as attractive as or more attractive than natural pheromones. Moreover, there must be a sufficient number of source points of synthetic pheromone to lower the probability of males successfully encountering females. Efficacy will depend on the size of pest populations, and more points would be required with larger populations. However, it may not be necessary to completely deprive males of opportunities for mating, because loss of time from visiting "false" pheromone sources before eventually finding a "true" female pheromone can lead to the aging of males and delayed copulation, which generally reduces fecundity in mated females. Many studies have emphasized competitive attraction as a primary contributor to mating disruption. For example, Stelinski et al. (2004, 2005a, b) revealed using direct behavioral observations that male moths approached sources of their respective synthetic pheromones in the oriental fruit moth *Grapholita molesta*, oblique-banded leafroller *Choristoneura rosaceana*, redbanded leafroller *Argyrotaenia velutinana*, and codling moth *Cydia pomonella*. Competitive attraction can theoretically enhance some non-competitive actions, for example sensory fatigue, because more effects can be expected when males are attracted by a synthetic pheromone source and come closer to it.

2.2.2 Noncompetitive Disruption

Camouflage: If we set many points of synthetic pheromone sources with sufficient doses, it is assumed a uniform pheromone fog will be created and will mask or camouflage a true pheromone plume emitted by a calling female. Males will fly around in a disoriented manner and cannot pick out natural pheromone plumes against the background fog. Thus the effectiveness of camouflage depends on the concentration of the synthetic pheromone fog. It is, however, questioned whether it is possible to achieve "uniform permeation" in field conditions (Sanders 1997).

Desensitization (sensory fatigue): This mode of action can be caused at two different levels, namely adaptation and habituation. Adaptation is a reduction in sensitivity of chemoreceptors at a peripheral level, and habituation is a reduction in responses at the level of the central nervous system. These are neurophysiologically different processes, but their consequence at a behavior level is the same reduction in responses. Sensory fatigue in olfactory systems caused by continuous stimuli can be found in many organisms. Sex pheromones are generally emitted by females only for a limited period of a day, and the neurosystems for pheromone receptions of males are considered to be activated during the corresponding period. On the other hand, synthetic pheromones for mating disruption are always in wafting through the air and stimulate pest olfaction systems continuously. Thus, sensory fatigue could often play a role in mating disruption.

Sensory Imbalance: In moths, most pheromones are generally composed of multi-component blends of fatty alcohols or their derivatives with a straight, even-numbered carbon chain, typically with one or two positions of unsaturation along the chain. Although the compounds in a pheromone blend are often shared with other species, the particular compositions and their blend ratios provide a unique and unambiguous signal. The pheromone blend ratio produced by a female moth is generally species specific and highly critical in eliciting responses in conspecific males. By releasing part of the pheromone blends from synthetic pheromone sources, a natural pheromone plume can be deviated, and/or the input of a pheromone blend to chemoreceptors of males can be imbalanced. Consequently, males fail to respond to natural pheromone from a calling female. Even if the blend ratio of pheromone chemicals in the atmosphere is similar to that of the natural pheromone, dose imbalance in pheromonal input can occur in pheromone perception and response systems. Most insects have a range of a pheromone dose to which they can respond normally, and excess amounts of pheromones often elicit only weak attractiveness. This may partly be attributed to sensory fatigue effects, but, even when sensors are adequately active, abnormal behavior such as dispersing from a disrupted field can be observed in some leafrollers.

2.3 Practical Applications

In practice, synthetic pheromones formulated by industries are used for mating disruption. Major formulations are (i) dispersible type that can be applied using conventional spraying equipment and (ii) reservoirs that are generally set by hand.

2.3.1 Dispersible Type

In dispersible types, microparticles, such as microcapsules or beads, containing minute quantities of synthetic pheromones are used. The individual particles release pheromone far less than natural insects calling mates. Since the early 1970s, a gelatin-based microcapsule and a nylon microcapsule have been employed for mating disruption of gypsy moth. The merit of the microdispersible type is that microparticles with pheromones can be applied by conventional spraying equipment, which is used regularly for spraying agro-chemicals. Spraying of microparticles is also advantageous for application over large areas with relatively low labor. Similarly, aqueous emulsions comprising a biodegradable carrier for pheromones has been developed for spraying techniques. The most popular carrier is probably paraffin, which can be applied at an ambient temperature as an aqueous emulsion, adheres to tree bark or foliage, releases pheromone for an extended period of time, and will slowly erode from bark and biodegrade in soil (Atterholt et al. 1999). Emulsions often include volatility suppressants such as soy oil or vitamin E. However, with microdispersible formulations it is difficult to maintain constant release rates over long periods. Thus repeated treatments could be required to control pest populations for a whole season.

2.3.2 Reservoirs

Reservoir-type formulations use plastic flakes, tubes, or chopped fibers impregnated with synthetic pheromones. The amounts of pheromones are generally many times larger than those produced by natural females. Therefore, reservoirs provide long-lasting pheromone release at a controlled rate. Even from a small number of reservoirs, sufficient concentrations of pheromone in the air can be achieved. However, they cannot be effectively scattered from aircraft or other machines. Thus this formulation type is more labor intensive if it has to be placed uniformly across agricultural fields. Recently, a novel emulsified wax dispenser of pheromone (Specialized Pheromone & Lure Application Technology; SPLAT®) was developed by ISCA Tech. (Riverside, CA), which can potentially save time and labor. SPLAT® emulsion is a unique, controlled-release technology that can be adapted to dispense and protect a wide variety of compounds from degradation, including semiochemicals, pesticides, and phagostimulants, in diverse environments (Mafra-Neto et al. 2013). SPLAT® can be applied using a custom-built, tractor-mounted applicator (Stelinski et al. 2007). The mode of action of reservoir types with excessively high amount of synthetic pheromones is suggested as follows (Sanders 1997). At some distance from a synthetic pheromone source, the concentration of pheromone in the air is low enough to allow males to "lock-on" and to orientate toward the pheromone source, but, in many cases, it is probably that the increasingly high pheromone concentration around the synthetic pheromone source causes sensory fatigue. Thus combined effects of false-female-following and sensory fatigue could be expected in the reservoir types.

2.3.3 Multiplex Mating Disruptant

In some regions, several different species of pests attack the same crop. Because mating disruption is based on pheromones that are fundamentally species specific, it is necessary to apply multiple kinds of disruptants in such cases. This is a disadvantage of pheromone-based techniques compared with insecticides, which are generally applicable to a broad range of pests. To save the cost and labor of applying different kinds of disruptants, "multiplex mating disruptants" that include pheromone compounds of several different species in the same reservoirs are currently being produced by Shin-Etsu Chemical Co. (Tokyo, Japan) and are available in Japan (Table 1; Japan Plant Protection Association 2014). All of these disruptants are formulated in a reservoir made from a fine polyethylene plastic tube. These multiplex mating disruptants, named Confuser®, show efficacy equivalent to that of a single mating disruptant targeting each pest. However, the modes of operation may be different among multiplex and single disruptants. Multiplex mating disruptants contains pheromone(s) of other species, which generally act as an antagonist for the attraction of conspecific mates. Thus attractiveness, i.e. the effect of false-female-following, of multiplex disruptants is potentially inferior to that of single disruptants. Nevertheless, many multiplex mating disruptants have been proven to suppress population growth of several pest species simultaneously, which implies noncompetitive actions including sensory fatigue have an impact on the total output of mating disruption. In New Zealand, a four-species mating disruption product, ISOMATE® 4-Play™, which consists of a range of pheromone components used by the codling moth, lightbrown apple moth *Epiphyas postvittana*, green-headed leafroller *Planotortrix octo*, and brown-headed leafroller *Ctenopseustis obliquana*, was commercialized in 2012 after a relaxation of regulatory requirements (Suckling et al. 2016).

Table 1. Multiplex mating disruptants registered in Japan*

Trade name	Crop	Target pest	Active ingredient (%)	Total amount per dispenser	Recommended number of dispensers
Confuser® AA	Deciduous fruit trees including apple	*Phyllonorycter ringoniella* *Grapholita molesta* *Adoxophyes orana* *Archips breviplicana* *Archips fuscocupreanus* *Carposina sasakii*	(Z)-10-Tetradecenyl acetate (28.0%) (*E*,*Z*)-4,10-Tetradecadienyl acetate (12.2%) (Z)-8-Dodecenyl acetate (5.4%) (Z)-11-Tetradecenyl acetate (19.9%) (Z)-9-Tetradecenyl acetate (3.9%) 10-Methyldodecyl acetate (0.5%) (Z)-9-Dodecenyl acetate (1.0%) 11-Dodecenyl acetate (0.6%) (Z)-11-Tetradecenol (0.24%) (Z)-13-Eicosen-10-one (18.1%) Other stabilizers (10.16%)	0.52 g	120–150 dispensers per 1,000 m^2
Confuser® V	Vegetables Potatoes Beans Ornamental plants	*Plutella xylostella* *Helicoverpa armigera* *Spodoptera litura* *Spodoptera exigua* *Autographa nigrisigna* *Trichoplusia ni* *Mamestra brassicae*	(Z)-9-Hexadecenal (1.0%) (Z)-11-Hexadecenol (0.5%) (Z)-7-Dodecenyl acetate (5.9%) (Z)-7-Dodecenol (3.8%) (Z)-11-Hexadecenyl acetate (19.2%) (Z)-11-Hexadecenal (24.0%) (*Z*,*E*)-9,12-Tetradecadienyl acetate (8.1%) (Z)-9-Tetradecenol (4.2%) (*Z*,*E*)-9,11-Tetradecadienyl acetate (19.5%) Other stabilizers (13.8%)	0.41 g	100–200 dispensers per 1,000 m^2

(Contd.)

Table 1 (*Contd.*)

Trade name	Crop	Target pest	Active ingredient (%)	Total amount per dispenser	Recommended number of dispensers
Confuser® MM	Deciduous fruit trees including peach	*Grapholita molesta* *Adoxophyes orana* *Carposina sasakii* *Lyonetia clerkella*	(Z)-8-Dodecenyl acetate (21.1%) (Z)-11-Tetradecenyl acetate (13.5%) (Z)-9-Tetradecenyl acetate (2.7%) 10-Methyldodecyl acetate (0.36%) (Z)-9-Dodecenyl acetate (0.71%) 11-Dodecenyl acetate (0.36%) (Z)-11-Tetradecenol (0.19%) (Z)-13-Eicosen-10-one (17.9%) 14-Methyl-1-octadecene (31.3%) Other stabilizers (11.88%)	0.55 g	100–120 dispensers per 1,000 m^2
Confuser® R	Deciduous fruit trees including apple, pear, etc.	*Grapholita molesta* *Carposina sasakii* *Adoxophyes orana* *Archips breviplicana* *Archips fuscocupreanus*	(Z)-8-Dodecenyl acetate (13.2%) (Z)-11-Tetradecenyl acetate (34.1%) (Z)-9-Tetradecenyl acetate (6.9%) 10-Methyldodecyl acetate (0.91%) (Z)-9-Dodecenyl acetate (1.8%) 11-Dodecenyl acetate (0.93%) (Z)-11-Tetradecenol (0.47%) (Z)-13-Eicosen-10-one (30.6%) Other stabilizers (11.09%)	0.36 g	100–200 dispensers per 1,000 m^2
Confuser® N	Deciduous fruit trees including pear, plum, etc.	*Grapholita molesta* *Carposina sasakii* *Homona magnanima* *Adoxophyes honmai* *Adoxophyes orana*	(Z)-8-Dodecenyl acetate (36.2%) (Z)-11-Tetradecenyl acetate (23.9%) (Z)-9-Tetradecenyl acetate (4.8%) 10-Methyldodecyl acetate (0.62%) (Z)-9-Dodecenyl acetate (1.2%)	0.26 g	50–200 dispensers per 1,000 m^2

(*Contd.*)

		Archips breviplicana *Grapholita delineana*	11-Dodecenyl acetate (0.65%) (Z)-11-Tetradecenol (0.28%) (Z)-13-Eicosen-10-one (21.3%) Other stabilizers (11.03%)		
Confuser® G	Turf grass	*Parapediasia teterella* *Spodoptera depravata*	(Z)-11-Hexadecenal (32.0%) (Z)-9-Hexadecenal (1.5%) (Z)-11-Hexadecenol (1.5%) (Z)-9-Tetradecenyl acetate (30.0%) (Z,*E*)-9,12-Tetradecadienyl acetate (7.0%) Other stabilizers (28.0%)	0.148 g	100–200 dispensers per 1,000 m^2

* Details of these disruptants are shown from a guidebook for pheromone-based agrochemicals published by Japan Plant Protection Association (2014).

3. Case Study: Moths

The great majority of mating disruptants available for farmers are aimed at controlling moth pests (Table 2). Most moths are strongly dependent on pheromone systems for copulation and reproduction. Because most moths are generally nocturnal and small, mate-finding behavior mediated by pheromones without reliance on visual signals is a rational strategy. A number of studies have analyzed and identified pheromone chemicals in several hundred moths and revealed that most moth pheromones are composed of multi-component blends of relatively simple and similar aliphatic compounds. They are generally even-numbered 10- to 18-carbon long-chain acetates, alcohols, and aldehydes, typically with one or two positions of unsaturation along the chain. This structural feature of moth pheromones is favorable for the industrial synthesis of pheromone compounds; many constituents can be generated by the same or similar process with different combinations of starting materials. Moreover, the volatilities of these fatty compounds are moderate and easy to control. In addition, some moths are economically and environmentally serious pests for agricultural crops or forests. As a consequence, studies of mating disruption using synthetic pheromones started in the early 1960s, a few years after the first chemical characterization of an insect pheromone, bombykol, from the silkworm moth. Since then, many chemical ecologists and agronomists have sought to manage moth pests by mating disruption, and this strategy is currently adopted across more than 600,000 ha (Table 2).

3.1 Gypsy Moth

The disruptant against the gypsy moth, *Lymantria dispar* (Lepidoptera: Lymantriidae), is probably used over the largest area among all of the previous trials of mating disruption. This species is extensively polyphagous and attacks more than 300 species of trees and shrubs, and current taxonomy divides this species into several subspecies including *L. d. dispar* (the European gypsy moth), which is native to temperate forests in Western Europe, *L. d. asiatica* (the Asian gypsy moth) native to Eastern Asia, and *L. d. japonica* native to Japan (Pogue and Schaefer 2007). The European gypsy moth was originally brought to the United States by the French artist Étienne Léopold Trouvelot to be used in hybridization experiments with native silk producing moths, the goal being to develop a strain resistant to the protozoan disease, *Nosema bombycis*, which devastated the European silk industry, and several gypsy moths escaped from Trouvelot's home in 1868 or 1869 (Pogue and Schaefer 2007). More recently, the Asian gypsy moth had invaded the United States in late 1991 (Pogue and Schaefer 2007). This species defoliates both broad-leaf and

Table 2. Mating disruptants used around the world in 2006*

Target pest	Crop	Treated area (ha)	Region
Lymantria dispar	Forest	200,000	US
Cydia pomonella	Apple, pear	152,500	US, Europe, Australia, South Africa, Argentina, etc.
Lobesia botrana *Eupoecilia ambiguella*	Grape	115,000	Europe
Paralobesia viteana			US
Epiphyas postvittana			Australia
Grapholita molesta	Pear, peach, nectarine	40,000	US, Europe, Australia, etc.
Pectinophora gossypiella	Cotton	49,500	US, Mexico, India, Pakistan, etc.
Thaumatotibia leucotreta	Citrus	4,000	South Africa
Anarsia lineatella	Peach, nectarine	4,000	Europe
Tortricids	Apple	7,000	US
Zeuzera pyrina	Pear, olive	3,000	Europe
Keiferia lycopersicella	Tomato	10,000	Mexico
Chilo suppressalis	Rice	2,000	Spain
Homona magnanima *Adoxophyes honmai*	Tea	3,070	Japan
Synanthedon hector *Synanthedon tenuis*	Apricot, peach, persimmon, cherry	3,860	Japan
Tortricids	Deciduous fruit trees	1,360	Japan
Carposina sasakii		240	Japan
(Multiplex disruptants)		10,170	Japan
Plutella xylostella *Helicoverpa armigera*	Vegetables	560	Japan
Spodoptera litura		380	Japan
Spodoptera exigua		120	Japan
(Multiplex disruptants)		760	Japan
	Total	607,520	

* Data are cited from Fukumoto and Mochizuki (2007).

conifer trees in forests and sometimes causes outbreaks. Particularly in North America where this species had invaded, serious damage can occur continuously because of a lower abundance of natural enemies.

Females of the European gypsy moth are winged but flightless, whereas females of the Asian gypsy moth are able to fly. Because the mobility of adult females of the European gypsy moth is limited, there is little risk of immigration of pregnant females from outside of forests where mating disruption treatment is being employed. Moreover, adult moths emerge only during limited periods, typically a few weeks in July to August. Therefore, microdispersible formulations that release pheromones can be used for a short period but are still suitable for a broad-range application. These features are favorable for controlling by mating disruption.

Studies of the sex pheromone of the gypsy moth started in the early 1930s in the United States. Collins and Potts (1932) extracted female abdominal tips, which are displayed during the night when for calling males, of the gypsy moth using organic solvent and demonstrated the extract exhibited attractiveness to males. This work was prior to the study of bombykol by Butenandt et al. (1959). Subsequently, Jacobson et al. (1960) isolated and elucidated the sex pheromone compound of the gypsy moth as 10-acetoxy-(Z)-7-hexadecenol (gyptol). Then Jacobson et al. (1961, 1962) synthesized this compound and reported its activity. However, gyptol had no ability to attract male gypsy moths (Eiter et al. 1967). Finally, Bierl et al. (1970) isolated the natural pheromone from 78,000 females and determined its structure as (Z)-7,8-epoxy-2-methyloctadecane (disparlure). Disparlure has two enantiomers: (+)-disparlure and (−)-disparlure. The natural pheromone was confirmed to be (7*R*,8*S*)-(+)-disparlure by enantioselective synthesis (Klimetzek et al. 1976). The opposite enantiomer, (7*S*,8*R*)-(−)-disparlure, showed an antagonistic effect on male attraction, although racemic (±)-disparlure can attract a substantial number of males (Vité et al. 1976, Cardé et al. 1977).

Mating disruption against gypsy moth in the United States has been carried out since 1971 (Beroza and Knipling 1972). A pioneering trial used (±)-disparlure-treated hydrophobic filter paper, which was distributed by aircraft over 40-acre plots at the rate of 20 mg/acre (Stevens and Beroza 1972). This treatment successfully prevented released males from locating females for the first six days, but the effects decreased by three weeks after applying the lure-treated paper. Subsequently, numerous formulations containing disparlure were developed, for example, hollow plastic fibers, gelatin microcapsules, and plastic laminated flakes, but disparlure release was inefficient, and major problems were encountered in the aerial application of these formulations due to the spray systems available for aircraft in these early trials (Thorpe et al. 2006). Improvement of the dosage, release rate control, application equipment, etc. has been challenging. The current formulation of disparlure flakes, modified

Disrupt® II (Hercon Environmental Co., Emigsville, Pennsylvania), was registered in the United States in 1992 and used (e.g. Tcheslavskaia et al. 2005). More recently, a new liquid SPLAT® GM™ (ISCA Tech) organic formulation was approved for its ability to disrupt gypsy moth mating and was shown to have persistent effects (e.g. Onufrieva et al. 2015). Mating disruption is the primary control tactic used against the gypsy moth as the gypsy moth Slow the Spread (STS) program, which aims to reduce spread rates by targeting isolated low-density colonies most likely to contribute to range expansion (Sharov et al. 2002).

3.2 Tea Leafroller Moths in Japan

Tea tree, *Camellia sinensis* (Theaceae), is an evergreen shrub native to Asia and is planted across more than 2,500 ha worldwide (Japan Tea Central Public Interest Incorporated Association 2016). Many moths are known to infest tea leaves, and among them, two leafrollers are notorious in Japan. The oriental tea tortrix moth, *Homona magnanima* (Lepidoptera: Tortricidae), is a relatively large, with a wingspan of 19–28 mm for males and of 26–37 mm for females. This species is found in Japan, China, Taiwan, and Vietnam and is polyphagous with a broad range of host plants. The other is the smaller tea tortrix moth, *Adoxophyes honmai* (Lepidoptera: Tortricidae), with a forewing length of 7.3 mm for males and 8 mm for females (Yasuda 1998). This species is found only in Japan and is very closely related to the summer fruit tortrix, *A. orana*. As well as *A. orana*, *A. honmai* attacks many fruit trees, including apple, pear, and grape, but prefers tea above all. The smaller tea tortrix moth was reported to have acquired resistance against many kinds of organophosphate insecticides in 1960s (Noguchi 1991). More recently, this species has shown resistance even against insect growth regulators (IGRs), which are designed to target juvenile harmful pests while causing less damage to beneficial insects and are known as pesticides with "reduced risk" (Uchiyama and Ozawa 2014). Thus alternative tools rather than insecticides have been required for many years, and pioneering chemical ecologists have been challenged to elucidate the pheromones of these leafrollers and apply them for pest management programs.

Tamaki et al. (1971) first identified two sex pheromone components of *A. honmai* as (Z)-9-tetradecenyl acetate (Z9-14:OAc) and (Z)-11-tetradecenyl acetate (Z11-14:OAc). These two compounds did not show any activities in male moths when they were displayed singly. However, very strong attractiveness was observed when the two were mixed and presented together (Tamaki et al. 1971). This was an epoch-making discovery to show that moth pheromones are comprised of multiple components that elicit no behavioral responses individually. Subsequently, Z9-14:OAc and Z11-14:OAc have also been isolated and identified from *A. orana* (Meijer et al. 1972). The blend ratio of Z9-14:OAc and Z11-14:OAc was ≈7:3 in *A. honmai* whereas it was ≈9:1 in *A. orana*. In addition to these two major components that are indispensable for male-attraction, a further study discovered the two minor components, (*E*)-11-tetradecenyl acetate (E11-14:OAc) and 10-methyldodecyl acetate (10Me-12:OAc), which enhance the activity of *A. honmai* pheromone. The pheromone of *A. orana* does not include E11-14:OAc and 10Me-12:OAc. The natural blend ratio of the four components, Z9-14:OAc, Z11-14:OAc, E11-14:OAc, and 10Me-12:OAc was 63:31:4:2 in *A. honmai* pheromone (Tamaki et al. 1979). Later, Noguchi et al. (1979) determined the *H. magnanima* pheromone as a 30:3:1 blend of Z11-14:OAc, (*Z*)-9-dodecenyl acetate (Z9-12:OAc), and 11-dodecenyl acetate (11-12:OAc). Similar to the case of *A. honmai* pheromone, each component showed no attractant activity individually but elicited strong attractiveness in a mixture of the three.

Based on these results, Tamaki et al. (1983) conducted a screening experiment to evaluate the mating disruption efficacy of pheromone components and several mixtures of these chemicals. Three compounds, Z9-14:OAc, Z11-14:OAc, and Z9-12:OAc, were selected as potential disruptants that induced disorientation of both *A. honmai* and *H. magnanima* using conditions of evaporation within and outside pheromone traps. Moreover, Z11-14:OAc alone, as well as a 1:1:1 mixture of Z9-14:OAc, Z11-14:OAc, and Z9-12:OAc, successfully suppressed mating of tethered females of both species in 1-m^3 field cages. Surprisingly, the mating disruption efficacy of Z11-14:OAc, which cannot attract males singly, was significantly superior to that of the natural blends of *A. honmai* four-component or *H. magnanima* three-component pheromone (Tamaki et al. 1983). Thus, Z11-14:OAc was selected and registered in 1983 as the first mating disruptant in Japan. Because Z11-14:OAc is frequently found as a key component in pheromones of the subfamily Tortricinae (Ando 2016, El-Sayed 2016), disruptant with Z11-14:OAc as an active ingredient was considered to be potentially applicable for a broad range of tortrix species.

The Z11-14:OAc disruptant successfully kept populations of moths at low levels until the early 1990s; however, infestations of *A. honmai* increased again and the efficacy of control declined markedly after 10 years of application in one locality in Shizuoka Prefecture in Japan, where this disruptant was used repeatedly every year (Mochizuki et al. 2002). This was the first documented example of "resistance" to mating disruption (Cardé and Haynes 2004). Mating disruption based on pheromones was fundamentally considered as a pest control method in which the insect does not become resistant, because changes in female pheromone biosynthesis or male response were unlikely to lead to a new communication channel that was unaffected by synthetic pheromone treatments that do not precisely match the female-produced blend (Witzgall et al. 2010). Nevertheless for such an assumption, *A. honmai* evolved the resistance against Z11-14:OAc rapidly. Use of the full four-pheromone-component blend restored the disruptive effect on this "resistant" population (Mochizuki et al. 2002), and most of the disruptants now coming onto the market are comprised of multiple components. However, it is essential for improvements of mating disruption tactics to understand how and why the resistance evolved in the case of Z11-14:OAc in *A. honmai*.

To elucidate the mechanism(s) by which the Z11-14:OAc disruptant lost its effectiveness, Tabata et al. (2007a) imposed further selection by rearing resistant moths collected in the field with Z11-14:OAc in the laboratory. After more than 70 generations of selection, a strain with quite strong resistance was established, the males of which could find and copulate with their mates even in the presence of 1 mg/l Z11-14:OAc. Although the mating ability of this strain under mating disruption was greatly increased, the composition and blend ratio of the sex pheromone produced and emitted by females was not obviously changed in comparison with those of females sensitive to mating disruption. However, male responses to the pheromone blend were markedly broadened after selection; wind-tunnel experiments indicated that the male response range to pheromone compositions could be much broader in the resistant (R) strain than in the non-resistant (S) culture, although R males could be less responsive to a low concentration of sex pheromone than S males (Tabata et al. 2007b). Of note, R males could locate the source of a pheromone blend lacking a main component, Z9-14:OAc or Z11-14:OAc, both of which are normally indispensable for male-attraction in this species (Tamaki et al. 1971, 1979). Wild-type males in the original resistant colony before laboratory selection were also reported to have been attracted to a lure baited with an off-ratio pheromone in an atmosphere treated with the Z11-14:OAc disruptant in a field trapping experiment (Mochizuki et al. 2008). Considering that both S and R moths originated in the same tea field despite their clear difference in response to pheromones and that only the R strain has been exposed to the disruptant, the extremely broad response range of the R strain is likely to be caused by the selection imposed by Z11-14:OAc in the original tea field and in the subsequent laboratory trials.

Because Z11-14:OAc alone had no attractiveness, the mode of action of Z11-14:OAc disruptant for tea leafrollers should be non-competitive. One of the proposed mechanisms was a "camouflage" effect; i.e. inability of responding moths to distinguish individual odor trails from an odor background (Bartell 1982). Mochizuki et al. (2008) indicated it would be unnecessary for wild-type males that obtained resistance to the Z11-14:OAc disruptant in Shizuoka to distinguish between natural female odor and synthetic background odor with regard to Z11-14:OAc, because they could utilize Z11-14:OAc released from the disruptant as if it was one of the attractive compounds and could respond to a lure without Z11-14:OAc in an atmosphere treated with the Z11-14:OAc disruptant. This ability was speculated to be necessary to overcome the "camouflage" effect (Mochizuki et al. 2008). Using a partial blend of pheromone as a disruptant was also suggested to assign the targeted moths another effect, a "sensory imbalance" (Rothschild 1981, Sanders 1997). The peripheral input from the component in excess in the atmosphere lowered the sensitivity of antennal receptors attuned to that component with a consequence that the natural blend ratio from a female was interpreted by the male as being of an unnatural ratio, or modified the ability of the central nervous system to discriminate the natural ratio (Cardé 1990). Males who perceived an "imbalanced" pheromone signal cannot recognize the signal source to be their mates. Although these proposed effects of mating disruption, "camouflage" or "sensory imbalance", were categorized as different mechanisms, they may be caused via similar processes in an olfactory neurosystem rather than working independently. In addition to the initial resistance to "camouflage" observed in the field colony (Mochizuki et al. 2008), the males of the highly resistant strain selected in the presence of extremely high dosages of Z11-14:OAc in a confined space in the laboratory could obtain resistance to "sensory imbalance", because males with the ability to respond to an off-ratio pheromone blend might be able to locate their mates even if their sensors received an "imbalanced" signal created by disruptants with a partial pheromone blend (Cardé and Minks 1995).

The broadened response of males may further drive evolution regarding sex pheromone communication due to the promotion of "asymmetric tracking", a process hypothesized to explain how a new pheromone system could evolve in moths (Phelan 1992). According to this hypothesis, the female signal could experience only weak selection, which should reduce the cost of producing abnormal pheromone signals, because variation in the pheromone signal could be "tracked" by males that possessed a wide breadth of responses, and thus a change in the pheromone communication system could occur. In the case of *A. honmai*, a large variation in the ratio of two major components, Z9-14:OAc and Z11-14:OAc, was reported; for example, 15–30% (%Z11-14:OAc) in Kagoshima Prefecture, 25–80% in Kyoto Prefecture, or 45–75% in Mie Prefecture (Noguchi and Sugie 2004). These data suggested that a broad response to off-ratio pheromone blends could be potentially occur in *A. honmai* males and support the "asymmetric tracking hypothesis". If the application of mating disruption induced resistance by enhancing the breadth of the response to pheromones in male moths, the evolution of sex pheromone communication may be further driven due to the promotion of "asymmetric tracking" (Tabata et al. 2007a, b). Mochizuki et al. (2008) reported that the pheromone blend ratio of *A. honmai* developed resistance to the Z11-14:OAc disruptant was different from the blend ratio analyzed before the use of the disruptant in the Shizuoka population. Evenden and Haynes (2001) also documented that selection with mating disruption resulted in a divergence in the proportion of the pheromone phenotypes in another moth species, *Trichoplusia ni*; when two pheromone types of *T. ni*, a normal and a mutant type, were contained in a large field cage and treated with the normal pheromone as a mating disruptant, the frequency of the mutant type was significantly altered by the fourth generation compared with that in a control cage treated with clean-air (Evenden and Haynes 2001). Thus, studies on selection with mating disruption could provide further understanding of the evolutionary process of the pheromonal communication system.

4. Case Study: Beetles

Many coleopteran insects use pheromones for aggregation and reproduction, but studies and examples of mating disruption in these insects are limited. Several constraints for the applications of mating disruption are assumed: generally, coleopteran insects are long-lived in the adult stage and are able to reproduce for longer periods than moths, which would require the controlled release of pheromones for a longer period in order for them to be successful as mating disruptants. In addition, their intrinsic longer reproductive-periods can diminish effects of mating disruptants that impose loss of time and energy for copulations. Moreover, some beetles can find mates and copulate using visual cues or odors other than pheromones, such as host plant odors, without any reliance on pheromones. Furthermore, the structures of some coleopteran pheromones are more complicated than those of moths and are difficult to synthesize on a commercial scale and at an acceptable cost. Nevertheless despite these constraints, some studies succeeded in mating disruption in coleopteran pests, and a few of these have been developed into commercially available disruptants.

4.1 Sugarcane Wireworm

The first example of a commercially available and practical to use coleopteran mating disruption was probably against sugarcane wireworm, *Melanotus okinawensis* (Coleoptera: Elateridae), in Southwestern islands of Japan. Larvae of *M. okinawensis* and its allied, *M. sakishimensis*, live under the ground in sugarcane fields and preferentially attack underground shoots, resulting in failures of germination or sprouting ratoons. They generally spend 2 to 5 years as larvae under the ground and, therefore, damage sugarcanes for long periods. To eliminate these wireworms, a large amount of insecticides had been applied to soils of sugarcane fields, but insecticides are not desirable particularly in island environments because it has a greater risk of pollutions of groundwater. Thus pheromone-based management techniques are required as an alternative tactic.

Females of some click beetles (adults of wireworms) have a paired ball-like structure in the abdomen, which fills up during the first days of adult life and has the function of being a pheromone gland or reservoir (Tóth 2013). Adult females of *M. okinawensis* produce *n*-dodecyl acetate (12:OAc) as a sex pheromone to attract males (Tamaki et al. 1986). This compound is structurally simple and very inexpensive, which is preferable for wide-scale applications (Fig. 1a). Similarly, *M. sakishimensis* females emit (*E*)-9,11-dodecadienyl butyrate (E9,11-12:OBt) and (*E*)-9,11-dodecadienyl hexanoate (E9,11-12:OHx) as its sex pheromone (Tamaki et al. 1990) (Fig. 1b).

At first, mass trapping strategies rather than mating disruption were adopted and evaluated to control *M. okinawensis* populations on several islands. Nagamine and Kinjo (1990) reported the efficacy of mass trapping using 4.4 traps per hectare in ≈25 ha of sugarcane fields on Okinawa Island from 1985 to 1989. Since this study, many efforts to control *M.*

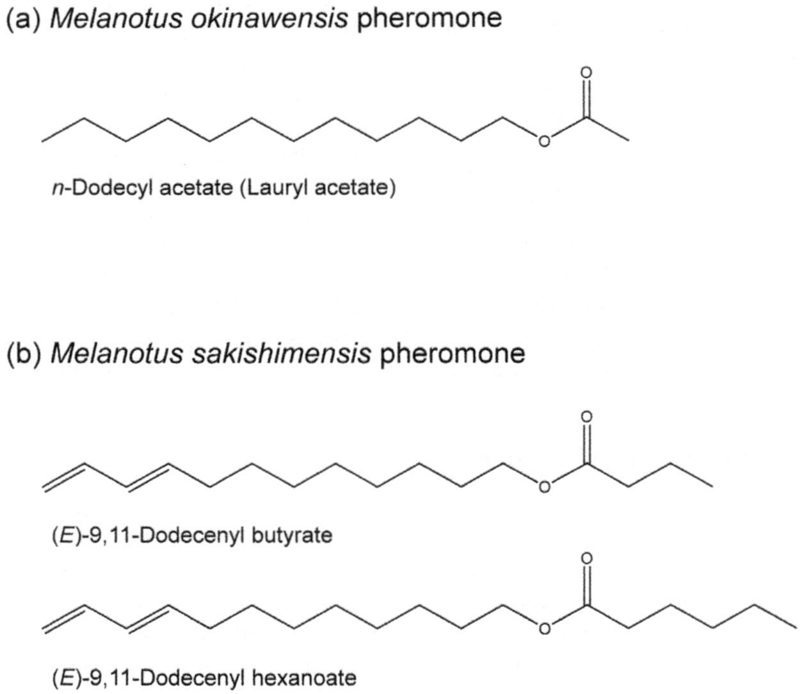

Fig. 1. Sex pheromones of two closely related click beetles, *Melanotus okinawesnsis* (a) and *M. sakishimensis* (b). The pheromone of *M. okinawensis* is structurally simple and suitable for the use of mating disruption. However, relatively high costs are required to produce the *M. sakishimensis* pheromone compounds, which are thus difficult to be used as a mating disruptant.

okinawensis using one trap per 1–1.5 ha have been conducted for more than 10 years, but damage to sugarcanes remains an unsolved problem (Nakamori and Kawamura 1997). Three possible constraints for mass trapping in this case have been suggested: (i) insufficient trap density, (ii) insufficient efficacy for male removal, and (iii) significant immigration of gravid females from outsides of the treated areas (Arakaki et al. 2008a). Arakaki et al. (2008a) evaluated mass trapping efficacy on a small isolated island with ≈1,000 funnel vane traps with 1 ml of 12:OAc (more than 10 traps per hectare) to cover an entire agricultural field on the island. After successive treatments for six years, mean trap captures were decreased to ≈27%. Moreover, release-recapture experiments and by-hand collection surveys indicated the population density in the treated area decreased to ≈10% after the project, although no significant decrease was observed in a population of *M. okinawensis* on an untreated neighboring island. On the other hand, the frequencies of mated females on the treatment island were 74–86%, which was only slightly reduced compared with that on the untreated island (93–97%). Estimated male removal rates, which represent the mass trapping efficacy, were 0.32–0.35 and also appeared too small to explain the reduction in population density on the treated area. Namely, the relatively low estimates of interference with copulation and male removal were inconsistent with the substantial reduction in the *M. okinawensis* population. Arakaki et al. (2008a) suggested that the omnipresence of synthetic pheromone sources may have led to confusion with male orientations to natural females and caused "delayed" mating, which potentially resulted in deleterious effects on reproduction. This study implied effects of mating disruption (behavioral interference) rather than mass trapping (male removal) could be critical for the suppression of *M. okinawensis* population growth.

Mating disruption is more convenient and cost-effective because of there is no need for trap apparatus. Thus, a mating disruptant composed of 12:OAc in a polyethylene tube was developed, and, using this disruptant, the number of *M. okinawensis* monitored by traps in the sugarcane fields decreased by 96.1% in 2001 from the previous year (Arakaki et al. 2008b). Moreover, the trap catches in the treated area decreased further by 74.0% from 2001 until 2007 as a cumulative effect, and the number of adults captured by hand decreased from 4.7 per field in 2001 to 0.5 in 2007 (Arakaki et al. 2008b). Currently, a product containing 17 g of 12:OAc per 100 polyethylene tubes is commercially available for *M. okinawensis* mating disruption. Subsequently, an attractant that contains 0.5 g of E9,11-12:OBt (77%) and E9,11-12:OHx (2.7%) in a 1.4 m polyethylene tube was registered in 2008 in Japan for *M. sakishimensis* management, although this attractant is not currently aimed at mating disruption. E9,11-12:OBt and E9,11-12:OHx are carboxylic esters of unsaturated fatty alcohols and can be synthesized using a similar process to that of moth pheromones, but the cost for their syntheses is much higher than that for 12:OAc. A mating disruptant for *M. sakishimensis*, which requires more amounts of E9,11-12:OBt and E9,11-12:OHx, is therefore not commercially available to date.

4.2 White Grub Beetle

More recently, another sugarcane pest, the white grub beetle *Dasylepida ishigakiensis* (Coleoptera: Scarabaeidae) has been targeted for control by mating disruption. This species is endemic to the southwestern islands of Japan and was previously known as a non-pest inhabiting natural forests in a few islands of the Yaeyama region. However, in the late 1990s, an outbreak occurred in sugarcane fields on Miyako Island, which is approximately 95 km northeast of the Yaeyama region (Muraji et al. 2010). Currently, *D. ishigakiensis* in the Yaeyama and Miyako populations is suggested to be categorized into different taxa based on findings of morphology, chemical features, and DNA sequences (Hirai et al. 2008, Muraji et al. 2008, 2010), although there is no apparent reproductive barrier, and hybridization and introgression are likely to occur (Muraji et al. 2010). The farmland type that causes severe damage to sugarcane on Miyako Island diverged from the forest type in other regions more than one million years ago according to substitutions in mitochondrial DNA sequences (Muraji et al. 2008). This species spends two years underground during one generation cycle, and the larvae infest the roots of sugarcane. Adults emerge in the evening during February to March and copulate in sugarcane fields, but they do not feed because of retrogressed mouth parts (Oyafuso et al. 2002). Once adult females have copulated, they burrow into the ground and never reappear aboveground, and they are considered to be monogamous (Arakaki et al. 2004).

Adult females emit sex pheromone to call for males only within 30 minutes of sunset, when the light intensity decreases from approximately 500 lux to 1 lux. Moreover, mating occurred only when the temperature at sunset was higher than 18°C (Arakaki et al. 2004). The sex pheromone of *D. ishigakiensis* was identified as (*R*)-2-butanol by means of solid-phase micro-extraction and gas chromatography-mass spectrometry (Wakamura et al. 2009a). In addition to (*R*)-2-butanol, (*S*)-2-butanol and 2-propanol were discovered in volatiles emitted by females, although pure (*S*)-2-butanol showed antagonistic activities against male attraction and 2-propanol elicited neither synergistic nor inhibitory effects (Wakamura et al. 2009a). A racemic mixture of 2-butanol attracted no males in a field trap experiment (Wakamura et al. 2009b).

Optically pure (*R*)-2-butanol is relatively expensive and is hard to use for mating disruption in sugarcane fields. However, Yasui et al. (2012) demonstrated that racemic 2-butanol, which is much cheaper and is of acceptable cost for large-scale application, can generate mating disruption effects both in laboratory and field observations. Yasui et al. (2012) used tubes made of high density polyethylene mixed with ethylene vinyl acetate copolymer. Approximately 0.09 mg/day of the ingredient was released from 1 cm of this tube with 12 mg of 2-butanol (Yasui et al. 2012). By using 1,200 to 2,500 dispensers, which contain ≈240 mg/dispenser (20 cm in length) of racemic 2-butanol, the frequencies of mated females could be reduced less than 40% in 800 m^2 of sugarcane fields (Yasui et al. 2012). Subsequently, Arakaki et al. (2013) reported that racemic 2-butanol released from 10,000 dispensers into 3,200 m^2 sugarcane fields significantly reduced the mating rate of feral females and catches of feral males with pheromone traps compared with those of untreated sugarcane fields. Moreover, the larval density for the treated fields was found to be nearly zero the following winter (Arakaki et al. 2013). More recently, Arakaki et al. (2017) used 2,000 to 8,000 short dispensers (2 cm in length), which were made of biodegradable polyethylene tubes and can be distributed more easily by hand, and showed that their efficacy in 800 m^2 of sugarcane fields was equivalent to those of the dispensers used in previous studies (Yasui et al. 2012, Arakaki et al. 2013).

Because racemic 2-butanol had no male attractiveness (Wakamura et al. 2009b), non-competitive modes of action were imposed on *D. ishigakiensis*. According to observations in a wind-tunnel, many males exposed to disruptants exhibited orientation flights (Yasui et al. 2012). Thus, desensitization or abnormal response are unlikely to be induced by excessive pheromonal input from disruptants, and their efficacy may be particularly attributable to a camouflage effect. It was noteworthy that 2-butanol has a small molecular weight and is very highly volatile. Therefore, it was difficult to release it at a controlled rate for a long period. However, the emergence of adult *D. ishigakiensis* is limited in a few weeks in early spring; therefore, long-term disruption is not necessary. Moreover, adults do not feed after emergence, which implies that their energy resources for mate-finding behavior are limited. Gravid females do not fly for dispersal and consequently do not tend to immigrate into a farm from the outside. These biological features are suitable for application of mating disruption.

5. Case Study: Coccoids

Coccoid insects (Hemiptera: Coccoidea), including scales and mealybugs, are small, plant sap-sucking insects like aphids and whiteflies. They are characterized by their unusual shapes with extreme sexual dimorphism, so much so that it is sometimes hard to recognize them as insects or even as animals (Ross and Shuker 2009). Females are neotenic and almost immobile lacking wings and often even legs. However, they have evolved shield-like structures and a variety of protective secretions including wax or shells to cover their bodies to protect themselves against danger, and as adults can live for several months. In contrast, adult males are winged and mobile, but they are tiny and fragile and have a limited life span of a few days at most. Coccoids are distributed worldwide and are parasites of various plants, and some species are known as serious agricultural pests (Ben-Dov 1994).

The conventional management of coccoid pests involves regular application of insecticides. However, insecticidal chemicals often cannot penetrate because of the protective structures on the body surface in addition to their cryptic behavior and clumped spatial distribution pattern, which render the use of many insecticides ineffective (Franco et al. 2009). Moreover, repeated uses of insecticides effective against a broad range of pests sometimes results in resurgence, which can occur when an insecticide treatment destroys the pest population and kills, repels, irritates, or otherwise deters natural enemies of the pest (Dutcher 2007). It is, therefore, difficult to entirely suppress in populations of mealybugs using these chemicals. Pheromone-based management would potentially give an alternative way to reduce damage caused by coccoid pests.

Because adult males have no mouth to absorb nutrition or water, they can search for mates for only a limited time. Sex pheromones emitted by immobile females are critical to facilitate mating and reproduction by serving as a key navigation tool for males. Mating disruption that interferes this essential chemical communication could have great potential to control coccoid insects. Thus several studies, particularly in mealybugs, challenged to elucidate pheromone structures and apply them for mating disruption.

Mealybug pheromones identified so far are terpenes with unique structures. Currently, 19 pheromone compounds have been isolated and characterized from 17 mealybug species (Tabata et al. 2017), and they are generally carboxylic esters of monoterpene or hemiterpene alcohols. Monoterpenes in mealybug pheromones commonly have an unusual structural feature: most monoterpenes are composed of two isoprene units coupled in a head-to-tail connection, but the two units of the alcohol moieties of mealybug pheromones are linked with irregular non-head-to-tail connections (Tabata and Ichiki 2016). Despite this common motif, all of these pheromones are strictly species specific and no compounds are shared with others. This chemical feature requires to establish a unique synthetic route for each species (Zou and Millar 2015) and is, therefore, an obstacle to formulate pheromones for practical applications in large-scale agricultural fields. However, some mealybug species use pheromones with very similar structures, which are related to lavandulol (5-methyl-2-isopropenyl-4-hexenol) (Tabata and Ichiki 2015; Fig. 2). Lavandulol is a characteristic constituent of lavender essential oils that are widely used

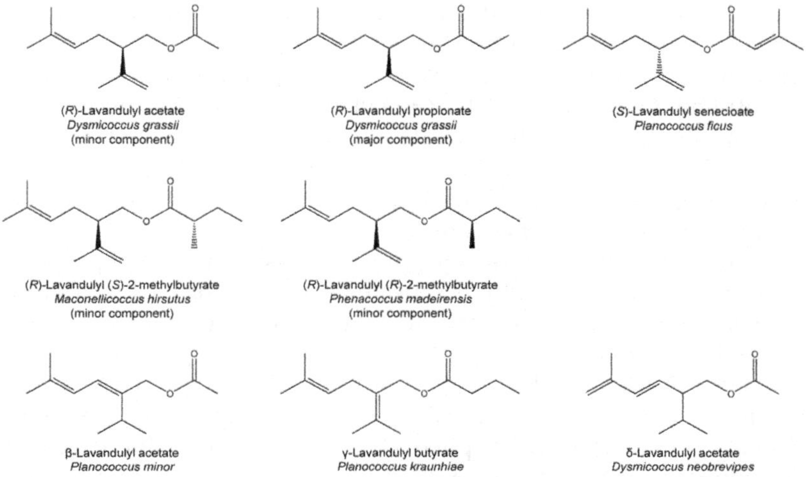

Fig. 2. Lavandulol-related strucures found in mealybug pheromones.

in fragrances of commodities including soaps, colognes, perfumes, skin lotions, food flavors, and aromatherapeutic medicines (Tabata et al. 2015), and, therefore, its synthetic routes, including an industrially preferable route, have been established (e.g. Inomata et al. 1987, Mino et al. 1997). Thus, recent studies, although currently only in limited species, have started to evaluate mating disruption efficacies for mealybug managements.

5.1 Vine Mealybug

The vine mealybug, *Planococcus ficus*, was first described by Hyeres and Nice in France on figs (Ben-Dov 1994). This species is considered to originate from the Mediterranean basin, but it currently invades grapevine yards in many regions including South Africa, California, Mexico, and Brazil (Godfrey et al. 2003, Walton et al. 2004, Pacheco da Silva et al. 2016). In these regions, *P. ficus* is a primary pest in vineyards because it affects crop quality and yield by excreting honeydew, which is a preferred habitat for sooty molds and putrefactive bacteria, which severely reduce the quality of table grapes as well as wine grapes (Chiotta et al. 2010). Moreover, this species is known as a vector of grapevine leafroll-associated viruses (Tsai et al. 2008).

From airborne volatiles released by virgin females of *P. ficus*, (*S*)-lavandulol and (*S*)-lavandulyl senecioate were identified as pheromone candidates that were found in volatiles from mealybug-infested squash but not in volatiles from uninfected squash (Hinkens et al. 2001). However, the former compound showed neither attractiveness nor synergism, and the latter was concluded to be the single pheromone compound of *P. ficus* (Hinkens et al. 2001). Subsequently, Millar et al. (2002) confirmed that lavandulol antagonized the attraction of males at higher doses. (*R*)-Lavandulyl senecioate (the opposite enantiomer) was not inhibitory, and the racemic mixture of lavandulyl senecioate was as attractive as the pure (*S*)-enantiomer (Millar et al. 2002, Zada et al. 2008). Of note, another ester of lavanlulol, (*S*)-lavandulyl isovalerate, was found in a mass-reared culture of *P. ficus* in an Israeli population (Zada et al. 2003). Variations in male response behaviors to (*S*)-lavandulyl isovalerate were also found in this population (Maimon et al. 2010).

The first attempt at mating disruption against *P. ficus* using a racemic mixture of synthetic lavandulyl senecioate was reported by Walton et al. (2006). The sprayable microencapsulated formulation of the racemic lavandulyl senecioate was applied using an air-blast sprayer, with three and four applications in 2003 and 2004, respectively, across 5 to 12 ha of commercial vineyards in California. In 2003, the sprayable pheromone was mixed with water and applied at a rate of 10.69 g/ha across each of the vineyards in May, before any adult male mealybugs emerged, and then in June and in August. In total, 32.07 g/ha of the pheromone was applied in the mating disruption plots. In 2004, the spray was applied at a rate of 10.69 g/ha in April, May, June, and July; in total, 53.45 g/ha of the pheromone was sprayed. As a result, compared with a no-pheromone control plot, there were significantly lower season-long trap catches of adult males, season-long mealybug densities, and crop damage in mating disruption plots. However, mealybug density was only 12.0 ± 15.6% and 31.1 ± 11.6% lower in the mating disruption plots than in control plots in 2003 and 2004, respectively. Mealybug density was significantly reduced after mating disruption treatments in plots with low population sizes whereas it was unchanged in plots with high population sizes (Walton et al. 2006). The observed limitation of mating disruption efficacies was considered to be attributable to the relatively short effective lifetime of the sprayable formulation.

Cocco et al. (2014) carried out *P. ficus* mating disruption in 2008 and 2009 in two commercial vineyards in Sardinia, Italy. They used reservoir dispensers made of solid matrix membranes (4 × 9 cm; CheckMate®; Suterra, Bend, OR, USA) loaded with 100 mg (62.5 g/ha) and 150 mg (93.8 g/ha) of racemic lavandulyl senecioate in 2008 and 2009, respectively. Only one set of the dispensers were set in late-April to mid-May before adult males of the overwintering generation emerged. Consequently, males captured in pheromone traps in mating disruption plots were significantly decreased by 86–95%, and the mealybug density and the proportion of pregnant females were clearly reduced, although crop damage at harvest was very low and did not differ between treatment plots and non-treatment control plots.

Subsequently, the efficacy of the mating disruption method against *P. ficus* when applied during one or two successive years in high and low infestation levels was evaluated in nine commercial vineyards in Israel (Sharon et al. 2016). The CheckMate® reservoir dispensers loaded with 150 mg of racemic lavandulyl senecioate were also used in this study. The disruptants were placed in April at a rate of 620 dispensers (93 g of the active ingredient) per hectare. Males were scarcely captured by monitoring pheromone traps, and a significant reduction in mealybugs was observed in infested vines in plots with low initial infestation levels even with one year of treatment with the mating disruptant. In plots with high initial infestation levels, a gradual reduction in infested vines was observed after two successive years of treatment.

Males of *P. ficus* could invest more time and energy searching for females in the presence of mating disruptants, although it remains unclear whether competitive- or non-competitive effects are primarily imposed. Because mealybug males are morphologically and physiologically fragile, such time-consuming cost should be critical for mating performance. Actually, the reproductively active period for adult males is estimated less than 35 hours after emergence (Zada et al. 2008, Mendel et al. 2012). Moreover, delayed mating may reduce the fecundity of mealybug females as they get older (Ross et al. 2011).

Parasitic wasps of coccoid insects often use their host pheromones as kairomonal cues to forage (Dunkelblum 1999). The sex pheromone of *P. ficus*, (*S*)-lavandulyl senecioate, strongly attracts *Anagyrus* sp. near *pseudococci* (Hymenoptera: Encyrtidae), a solitary koinobiont endoparasitoid of mealybugs (Franco et al. 2009). Thus mating disruption based on lavandulyl senecioate may theoretically affect the behavior of natural enemies including this parasitoid. According to the study in Sardinia (Cocco et al. 2014), however, mating disruption did not negatively affect the parasitism rate, which was even 1.5 times higher in the disrupted plot than in the control plot.

5.2 Japanese Mealybug

The Japanese mealybug, *Planococcus kraunhiae*, is probably of East Asian origin and is currently found in Japan, Korea, Taiwan, and China, and in California as an invader (Kawai 1980, Ben-Dov 1994). This species is one of the most harmful mealybugs that attack many kinds of fruit crops in Japan, including persimmons, grapes, pears, citrus, and figs (Shibao and Tanaka 2000, Ueno 1963). Because frequent resurgence occurs with successive applications of pyrethroid insecticides (Morishita 2005), alternative pest management methods including mating disruption have been explored (Teshiba 2013).

The sex pheromone compound was fully isolated from volatiles emitted by a total of 1,800,000 female day equivalents by means of serial chromatography, and its chemical structure was determined to be

2-isopropyliden-5-methyl-4-hexenyl butyrate by mass spectrometry and nuclear magnetic resonance spectroscopy analyses (Sugie et al. 2008). This compound is a butyrate ester of a double bond-positional isomer of lavandulol (called "γ-lavandulol" in this article), which can be easily synthesized from lavandulol via acid- or base-promoted double-bond migration (Tabata 2013, Kinsho et al. 2015). This opened a way for the practical use of *P. kraunhiae* pheromone in pest management programs.

To evaluate the performance of the *P. kraunhiae* pheromone (γ-lavandulyl butyrate) as a mating disruptant, Teshiba et al. (2009) conducted step-by-step assays. First, they examined frequencies of mating of ≈30 females and ≈50 males in a mesh cage (95 cm^3) in the presence of 1 or 10 rubber septa impregnated with 0.1 mg of γ-lavandulyl butyrate. The results showed clear efficacy, with only 28.7% and 7.9% of mated females observed with the 1 and 10 disruptant-septa, respectively, whereas >90% of females copulated with males in no-disruptant controls. Next, they placed nine potted persimmon trees, each of which had attached 10 septa impregnated with 0.1 mg of γ-lavandulyl butyrate, at an interval of 1 m (3-tree × 3-line) in an open field. Laboratory-reared virgin adult females (≈100 individuals) were released with ≈500 males and recaptured after 2 days. The results of the tests with four replications indicated significant reductions in mating success (0–54.1%) compared to no-treatment controls (75.6–98.4%). Finally, they conducted tests in a commercial persimmon orchard; a 500 m^2 plot with 10 rubber septum dispensers placed at 167 points, each of which contained 3 mg of γ-lavandulyl butyrate (5 g in total). The disruptants were set in April, before adult males of the overwintering generation had emerged. In this plot, no males at all were captured by monitoring pheromone traps, whereas a normal emergence was observed in a neighboring control plot, indicating a strong interference effect on mate-finding behavior by mealybug males. Moreover, wild pregnant females were captured at a rate of only 1.4% in the treatment plot, whereas the rate was 51.0% in the control. The population densities, which were initially similar (0.02–0.03 mealybugs per twig), were significantly different, 0–0.03 in the treatment plot and 0.31–1.30 in the control plot, after treatment. The proportion of fruits with mealybug damage at harvest this season was only 0.2%, which was ≈1/20 of that in the control plot (4.2%).

Similar to the case of *P. ficus*, it is unknown whether competitive attraction to the synthetic pheromone dispensers plays a primary role in the disruption of *P. kraunhiae* mating. In *P. kraunhiae*, male responses to the pheromone source were likely to decrease when the dosage of pheromone was over their optimal range. Considering the amount of natural pheromone released by females (≈ 3 ng per female per day; Sugie et al. 2008), the amount of synthetic pheromone in the dispenser (3 mg × 10) was probably excessive. Thus other non-competitive mechanisms would be, at least partially, effective in the mating disruption of mealybugs.

Another point for successful control of mealybug populations by mating disruption could be attributed to limited mobility of gravid females. The risk of invasion of females who have mated in outside of the treated field can be expected to be very low in coccoid pests.

The sex pheromone of *P. kraunhiae*, as well as that of *P. ficus*, attracts parasitic wasps (Tsueda 2014), and could theoretically confuse kairomonal search by these natural enemies. During field studies of the *P. kraunhiae* pheromone, its cyclization product, (−)-cyclolavandulyl butyrate, was discovered as another attractant for several mealybug parasitic wasps (Tabata et al. 2011). Notably, cyclolavandulyl butyrate strongly attracted *Anagyrus sawadai* Ishii (Hymenoptera: Encyrtidae), which is considered to be synonymous with *A. subalbipes* Ishii (Noyes and Hayat, 1994; Higashiura, 2008) but is not likely to parasitize *P. kraunhiae* in natural conditions. Moreover, Teshiba et al. (2012) demonstrated that *A. sawadai* is a "non-natural" enemy for *P. kraunhiae* but it intensively attacks this pest in the presence of cyclolavandulyl butyrate. A field-scale application of this attractant successfully suppressed mealybug population increase in persimmon orchards by enhancing the parasitism activities of wasps including *A. sawadai* (Teshiba and Tabata 2017). This is a unique example demonstrating that a "non-natural" enemy that does not typically attack the pest under natural conditions can be enrolled in biological control by using its attractant. Because *A. sawadai* is not a natural enemy of *P. kraunhiae* and may not use *P. kraunhiae* pheromones in its foraging behavior, mating disruption could not influence the level of *A. sawadai* parasitism induced by cyclolavandulyl butyrate, and the combination of the mating disruptant with cyclolavandulyl butyrate can be used in a pest management program for *P. kraunhiae*. Furthermore, such a combination of mating disruption and parasitoid recruitment was considered to compensate for their constraints with each other. Mating disruption generally elicits a strong effect when the population density of a target pest is low, but it does not work in the presence of excessive numbers of individuals, because insects will have many opportunities to encounter mates in a high-density population, even if they moved about randomly about in a disrupted environment. In contrast, the performance of parasitoids is expected to improve when the host pest is abundant, because they spend more time feeding and parasitizing in patches with more hosts. Therefore, a combination of these different attractant volatiles with different functions could offer a unique and effective tactic to control mealybugs.

6. Conclusion

As reviewed as the above case studies, both biological features of the target pests and chemical features of their pheromones have greatly influenced projects to develop mating disruptants for practical use. Ideally, more setting points of pheromones with larger dosages across broader field scales are preferred for more efficient mating disruption, but the cost of producing synthetic pheromones is a limiting factor (Rodriguez-Saona et al. 2009). When the chemical structure of the target pheromone is simple to synthesize or it is available inexpensively, mating disruption could be a desirable strategy for pest management, as in the case of the sugarcane wireworm. On the other hand, field-scale application would not be practical when the pheromone has a complicated or unstable structure and is hard to synthesize by industrial means. However, it is almost impossible to judge before the characterization of an unknown pheromone, although it may be possible to estimate to some extent from previous information about the pheromones of closely related species.

From the viewpoint of the biological features of pests, the adult emergence timing and duration are critical for mating disruption treatments. Both the gypsy moth and white grub beetle in the above cases are univoltine and appear as adults for copulation only over a few weeks every year, which reduces constraints on the effective lifetime of disruptants as well as the cost of pheromone synthesis and labor for application. Another essential feature is mobility of target pests, because it is not possible to avoid invasion by individuals from outside of treatment areas even if a disruptant with excellent performance was developed. Therefore, mating disruption of less mobile coccoid pests, including mealybugs, can be potentially effective.

Mating disruption is compatible with the use of natural enemies (Teshiba 2013, Teshiba and Tabata 2017). Because mating disruption can selectively act on pests and can lead to a reduction in the spraying of insecticides that may negatively influence beneficial arthropod fauna, the activities of natural enemies could be promoted along with successive use of mating disruption. Moreover, as described above, these strategies can compensate their constraints with each other; mating disruption is effective with a low pest population density, and natural enemies are more active with a high pest density. Combined uses of mating disruption and attractants for natural enemies or the release of mass-reared natural enemies could hold promise for the protection of crops and forests from pest damage.

Finally, there may well be concerns about the possibility of resistance against mating disruption, although no conclusive evidence of resistance other than the case of tea leafrollers in Japan has so far been reported. For the Japanese tea leafrollers, mating disruptant that contained only one component of the full attractive blend had been used. Such a non-competitive type of disruptant may promote the rapid evolution of resistance, potentially allowing such resistance to evolve via a change in the communication trait of only one of the sexes (Cardé and Haynes 2004). Several previous studies demonstrated that an attractive and more complete pheromone blend is not necessarily more effective for the purposes of mating disruption (e.g., Evenden et al. 1999), and some possible advantages have been postulated when using a single component or a partial blend as the disruptant; for example, no attraction of males from outside the treated area (Flint and Merkle, 1983, 1984), or cost-effectiveness considerations (Evenden et al. 1999). However, studies of the Japanese tea leafrollers support the hypothesis that using only a portion of the natural pheromone blend carries a greater risk of the rapid evolution of resistance (Cardé 1976, Charlton and Cardé 1981, Cardé and Haynes 2004). Although mating disruption based on pheromones is fundamentally considered to have a low risk of resistance evolution, the genetic diversity of insects is vast; therefore, we need to maintain continuous and careful observations of pest populations.

Acknowledgments

The author appreciates advices and comments from Drs. F. Mochizuki, T. Kinsho, T. Fukumoto (Shin-Etsu Chem. Co.), Y. Narai, N. Sawamura (Shimane Agricultural Technology Center), M. Teshiba (Fukuoka Agricultural Research Center), and H. Sugie (National Agriculture and Food Research Organization).

REFERENCES

Ando, T. 2016. Sex pheromones of moths. (http://lepipheromone.sakura.ne.jp/PheromoneList/List_of_Sex_Pheromones.pdf)

Arakaki, N., Y. Sadoyama, M. Kishita, A. Nagayama, A. Oyafuso, M. Ishimine et al. 2004. Mating behavior of the scarab beetle *Dasylepida ishigakiensis* (Coleoptera: Scarabaeidae). Appl. Entomol. Zool. 39: 669–674.

Arakaki, N., A. Nagayama, A. Kobayashi, M. Kishita, Y. Sadoyama, N. Mougi et al. 2008a. Control of the sugarcane click beetle *Melanotus okinawensis* Ohira (Coleoptera: Elateridae) by mass trapping using synthetic sex pheromone on Ikei Island, Okinawa, Japan. Appl. Entomol. Zool. 43: 37–47.

Arakaki, N., A. Nagayama, A. Kobayashi, Y. Hokama, Y. Sadoyama, N. Mogi et al. 2008b. Mating disruption for control of *Melanotus okinawensis* (Coleoptera: Elateridae) with synthetic sex pheromone. J. Econ. Entomol. 101: 1568–1574.

Arakaki, N., Y. Hokama, A. Nagayama, H. Yasui, N. Fujiwara-Tsujii, S. Tanaka et al. 2013. Mating disruption for control of the white grub beetle *Dasylepida ishigakiensis* (Coleoptera: Scarabaeidae) with synthetic sex pheromone in sugarcane fields. Appl. Entomol. Zool. 48: 441–446.

Arakaki, N., A. Nagayama, K. Kijima, H. Yasui, N. Tsujii, S. Tanaka et al. 2017. Ground surface application of mini-dispenser for mating disruption by pheromone to the white grub beetle *Dasylepida ishigakiensis* (Coleoptera: Scarabaeidae). Appl. Entomol. Zool. 52: 159–164.

Atterholt, C.A., M.J. Delwiche, R.E. Rice and J.M. Krochta. 1999. Controlled release of insect sex pheromones from paraffin wax and emulsions. J. Control. Release 57: 233–247.

Baker, T.C., J.J. Zhu and J.G. Millar. 2016. Delivering on the promise of pheromones. J. Chem. Ecol. 42: 553–556.

Bartell, R.J. 1982. Mechanisms of communication disruption by pheromone in the control of Lepidoptera: a review. Physiol. Entomol. 7: 353–364.

Ben-Dov, Y. 1994. A Systematic Catalogue of the Mealybugs of the World. Intercept, Hampshire, UK.

Beroza, M. and E.F. Knipling. 1972. Gypsy moth control with the sex attractant pheromone. Science. 177: 19–27.

Bierl, B.A., M. Beroza, C.W. Collier. 1970. Potent sex attractant of the gypsy moth: its isolation, identification, and synthesis. Science 170: 87–89.

Butenandt, A., R. Beckmann, D. Stamm and E. Hecker. 1959. Über den sexuallockstoff des seidenspinners *Bombyx mori*. Reindarstellung und konstitution. Z. Naturforsch. B 14: 283–284.

Cardé, R.T. 1976. Utilization of pheromones in the population management of moth pests. Environ. Health Perspect. 14: 133–144.

Cardé, R.T. 1990. Principles of mating disruption. pp. 47–71. *In*: R. Ridgeway, R.M. Silverstein and M. Inscoe [eds.]. Behavior-Modifying Chemicals for Insect Management. Marcel Dekker, New York, NY, USA.

Cardé, R.T. and A.K. Minks. 1995. Control of moth pests by mating disruption: successes and constraints. Annu. Rev. Entomol. 40: 559–585.

Cardé, R.T. and H.F. Haynes. 2004. Structure of the pheromone communication channel in moths. pp. 283–332. *In*: R.T. Cardé and J.G. Millar [eds.]. Advances in Insect Chemical Ecology. Cambridge University Press, Cambridge, UK.

Cardé, R.T., C.C. Doane, J. Granett, A.S. Hill, J. Kochansky, W.L. Roelofs. 1977. Attractancy of racemic disparlure and certain analogues to male gypsy moths and the effect of trap placement. Environ. Entomol. 6: 765–767

Charlton, R.E. and R.T. Cardé. 1981. Comparing the effectiveness of sexual communication disruption in the oriental fruit moth (*Grapholitha molesta*) using different combinations and dosages of its pheromone blend. J. Chem. Ecol. 7: 501–508.

Chiotta, M.L., M.L. Ponsone, A.M. Torres, M. Combina and S.N. Chulze. 2010. Influence of *Planococcus ficus* on *Aspergillus* section Nigri and ochratoxin A incidence in vineyards from Argentina. Lett. Appl. Microbiol. 51: 212–218.

Cocco, A., A. Lentini and G. Serra. 2014. Mating disruption of *Planococcus ficus* (Hemiptera: Pseudococcidae) in vineyards using reservoir pheromone dispensers. J. Insect Sci. 14: 144.

Collins, C.W. and S.F. Potts. 1932. Attractants for the flying gipsy moth as an aid in locating new infestation. USDA Tech. Bull. 336.

Dunkelblum, E. 1999. Scale insects. pp. 251–276. *In*: J. Hardie and A.K. Minks [eds.]. Pheromones of Non-lepidopteran Insects Associated with Agricultural Plants. CAB International, Oxfordshire, UK.

Dutcher, J.D. 2007. A review of resurgence and replacement causing pest outbreakes in IPM. pp. 27–43. *In*: A. Ciancio and K.G. Mukerji [eds.]. General Concepts in Integrated Pest and Disease Management. Springer, Dordrecht, Netherland.

Eiter, K., E. Truscheit and M. Boneß. 1967. Neuere Ergebnisse der Chemie von Insektensexuallockstoffen. Synthesen von D,L-10-Acetoxy-hexadecen-(7-*cis*)-ol-(1), 12-Acetoxy-octadecen-(9-*cis*)-ol-(1) ("Gyplure") und 1-Acetoxy-10-propyl-tridecadien-(5-*trans*. 9). Justus Liebigs Ann. Chem. 709: 29–45.

El-Sayed, A.M. 2016. Pherobase. (http://www.pherobase.com/)

Evenden, M.L. 2016. Mating disruption of moth pests in integrated pest management: a mechanistic approach. pp. 365–393. *In*: J.D. Allison and R.T. Cardé [eds.]. Pheromone Communication in Moths: Evolution, Behavior, and Application. University of California Press, Oakland, CA, USA.

Evenden, M.L. and K.F. Haynes. 2001. Potential for the evolution of resistance to pheromone-based mating disruption tested using two pheromone strains of the cabbage looper, *Trichoplusia ni*. Entomol. Exp. Appl. 100: 131–134.

Evenden, M.L., G.J.R. Judd and J.H. Borden. 1999. Pheromone-mediated mating disruption of Choristoneura rosaceana: is the most attractive blend really the most effective? Entomol. Exp. Appl. 90: 37–47.

Flint, H.M. and J.R. Merkle. 1983. Pink bollworm (Lepidoptera: Gelechiidae): communication disruption by pheromone composition imbalance. J. Econ. Entomol. 76: 40–46.

Flint, H.M. and J.R. Merkle. 1984. Pink bollworm (Lepidoptera: Gelechiidae): alteration of male response to gossyplure by release of its component *Z,Z*-isomer. J. Econ. Entomol. 77: 1099–1104.

Franco, J.C., E.B. Silva, E. Cortegano, L. Campos, M. Branco, A. Zada et al. 2008. Kairomonal response of the parasitoid *Anagyrus* spec. nov. near *pseudococci* to the sex pheromone of the vine mealybug. Entomol. Exp. Appl. 126: 122–130.

Franco, J.C., A. Zada and Z. Mendel. 2009. Novel approaches for the management of mealybug pests. pp. 233–278. *In*: I. Ishaaya and A.R. Horowitz [eds.]. Biorational Control of Arthropod Pests. Springer, Dordrecht, Netherland.

Fukumoto, T. and F. Mochizuki. 2007. Utilization of mating disruption products in the world in 2006. Plant Prot. 61: 276–279.

Godfrey, K., J. Ball, D. Gonzalez and E. Reeves. 2003. Biology of the vine mealybug in vineyards in the Coachella valley, California. Southwest. Entomol. 28: 183–196.

Higashiura, Y. 2008. *Anagyrus sawadai* Ishii. *In*: Y. Hirashima and K. Morimoto [eds.]. Iconographia Insectorum Japonicorum Colore Naturali Edita Vol. 3. Hokuryukan, Tokyo, Japan.

Hinkens, D.M., J.S. McElfresh and J.G. Millar. 2001. Identification and synthesis of the sex pheromone of the vine mealybug, *Planococcus ficus*. Tetrahedron Lett. 42: 1619–1621.

Hirai, Y., T. Akino, S. Wakamura and N. Arakaki. 2008. Morphological and chemical comparisons of males of the white grub beetle *Dasylepida ishigakiensis*

(Coleoptera: Scarabaeidae) among four island populations in the Sakishima Islands of Okinawa. Appl. Entomol. Zool. 43: 65–72.

Inomata, K., S. Igarashi, M. Mohri, T. Yamamoto, H. Kinoshita and H. Kotake. 1987. Regio- and stereo-selective synthesis of allylic and homoallylic alcohols by the reductive desulfonylation of allylic sulfone derivatives. Application to the syntheses of (±)-lavandulol and isolavandulol. Chem. Lett. 16: 707–710.

Jacobson, M., M. Beroza and W.A. Jones. 1960. Isolation, identification, and synthesis of the sex attractant of gypsy moth. Science 132: 1011–1012.

Jacobson, M., M. Beroza and W.A. Jones. 1961. Insect sex attractants. I. The isolation, identification, and synthesis of the sex attractants of the gypsy moth. J. Am. Chem. Soc. 83: 4819–4824.

Jacobson, M. and W.A. Jones. 1962. Insect sex attractants. II. The synthesis of a highly potent gypsy moth sex attractant and some related compounds. J. Org. Chem. 27: 2523–2524.

Japan Plant Protection Association. 2014. Guidebook for Biopesticides and Pheromones. JPPA, Tokyo, Japan.

Japan Tea Central Public Interest Incorporated Association, Report, 2015. Tokyo, Japan.

Kawai, S. 1980. Scale Insects of Japan in Colors. Zenkoku Noson Kyouiku Kyoukai Publishing, Tokyo, Japan.

Kinsho, T., N. Ishibashi, M. Yamashita, Y. Miyake, A. Baba and Y. Nagae. 2015. Method to produce 2-isopropyliden-5-methyl-4-hexenyl butyrate. Japan Patent No. P2015-110553A.

Klimetzek, D., G. Loskant, J.P. Vité and K. Mori. 1976. Differences in pheromone perception between gypsy moth and nun moth. Naturwissenschaften 63: 581–582.

Mafra-Neto, A., F.M. de Lame, C.J. Fettig, A.S. Munson, T.M. Perring, L.L. Stelinski et al. 2013. Manipulation of insect behavior with specialized pheromone and lure application technology (SPLAT®). pp. 31–58. *In*: J.J. Beck, J.R. Coats, S.O. Duke and M.E. Koivunen [eds.]. Pest Management with Natural Products, ACS Symposium Series 1141. American Chemical Society, Washington D.C., USA.

Maimon, H.K., A.L. Zada, J.C. Franco, E. Dunkelblum, A. Protasov, M. Eliyaho et al. 2010. Male behaviors reveal multiple pherotypes within vine mealybug *Planococcus ficus* (Signoret) (Hemiptera: Pseudococcidae) populations. Naturwissenschaften 97: 1047–1057.

Meijer, G.M., F.J. Ritter, C.J. Persoons, A.K. Minks and S. Voerman. 1972. Sex pheromones of summer fruit tortrix moth *Adoxophyes orana*: two synergistic isomers. Science 175: 1469–1470.

Mendel, Z., A. Protasov, P. Jasrotia, E.B. Silva, A. Zada and J.C. Franco. 2012. Sexual maturation and aging of adult male mealybugs. Bull. Entomol. Res. 102: 385–394.

Millar, J.G., K.M. Daane, J.S. McElfresh, J.A. Moreira, R. Malakar-Kuenen, M. Guillén et al. 2002. Development and optimization of methods for using sex pheromone for monitoring the mealybug *Planococcus ficus* (Homoptera: Pseudococcidae) in California vineyards. J. Econ. Entomol. 95: 706–714.

Millar, J.G., T.C. Baker and J.J. Zhu. 2016. Delivering on the Promise of Pheromones – Part 2. J. Chem. Ecol. 42: 851–852.

Miller, J.R., L.J. Gut, F.M. de Lame and L.L. Stelinski. 2006a. Differentiation of competitive vs. non-competitive mechanisms mediating disruption of moth sexual communication by point sources of sex pheromone (part 1): theory. J. Chem. Ecol. 32: 2089–2114.

Miller, J.R., L.J. Gut, F.M. de Lame and L.L. Stelinski. 2006b. Differentiation of competitive vs. non-competitive mechanisms mediating disruption of moth sexual communication by point sources of sex pheromone (part 2): case studies. J. Chem. Ecol. 32: 2115–2143.

Mino, T., S. Fukui and M. Yamashita. 1997. Hydrolysis of β,γ-unsaturated aldehyde dimethyllhydrazones with copper dichloride: a new synthesis of lavandulol. J. Org. Chem. 62: 734–735.

Mochizuki, F., T. Fukumoto, H. Noguchi, H. Sugie, T. Morimoto and K. Ohtani. 2002. Resistance to a mating disruption composed of (Z)-11-tetradecenyl acetate in the smaller tea tortrix, *Adoxophyes honmai* (Yasuda) (Lepidoptera: Tortricidae). Appl. Entomol. Zool. 37: 299–304.

Mochizuki, F., H. Noguchi, H. Sugie, J. Tabata and Y. Kainoh. 2008. Sex pheromone of the smaller tea tortrix, *Adoxophyes honmai* (Yasuda) (Lepidoptera: Tortricidae) promotes resistance to a mating disruptant composed of (Z)-11-tetradecenyl acetate. Appl. Entomol. Zool. 43: 293–298.

Morishita, M. 2005. Resurgence of Japanese mealybug, *Planococcus kraunhiae* (Kuwana), in persimmon induced by a synthetic pyrethroid cypermethrin. Annu. Rep. Kansai Plant Prot. Soc. 47: 125–126.

Muraji, M., Y. Hirai, T. Akino, S. Wakamura and N. Arakaki. 2008. Genetic divergence among populations of the white grub beetle, *Dasylepida ishigakiensis* (Coleoptera: Scarabaeidae), distributed in the southern part of the Ryukyu Islands of Japan, detected from the mitochondrial DNA sequences. Appl. Entomol. Zool. 43: 287–292.

Muraji, M., Y. Hirai, T. Akino, S. Wakamura, H. Yasui, N. Arakaki et al. 2010. Relationship among local populations of the white grub beetle, *Dasylepida ishigakiensis* (Coleoptera: Scarabaeidae), detected by phylogenetic analysis based on the mitochondrial DNA sequence. Appl. Entomol. Zool. 45: 289–296.

Nagamine, M. and M. Kinjo. 1990. Report on the Establishment of the Control Technique for the Sugarcane Wireworm *Melanotus okinawensis*. Okinawa Prefectural Sugar Industry Development Association. Haebaru, Okinawa, Japan.

Nakamori, H. and F. Kawamura. 1997. What are the key factors to obstruct ratooning of sugarcane in Okinawa? Present situation of researches and counterplans. J. Okinawa Agric. 32: 36–47.

Noguchi, H. 1991. Geographic variation of sex pheromone of the tea tortrix moth. Plant Prot. 51: 122–124.

Noguchi, H. and H. Sugie. 2004. Geographic variation in the sex pheromone composition of the smaller tea tortrix moth, *Adoxophyes honmai*. Res. Exec. Sum. Natl. Inst. Agro-Environ. Sci. 20: 32–33.

Noguchi, H., Y. Tamaki and K. Yushima. 1979. Sex pheromone of the tea tortrix moth: isolation and identification. Appl. Entomol. Zool. 14: 225–228.

Noyes, J.S. and M. Hayat. 1994. Oriental Mealybug Parasitoids of the Anagyrini (Hymenoptera: Encyrtidae). CAB International, Wallingford, UK

Onufrieva, K.S., A.D. Hickman, D.S. Leonard and P.C. Tobin. 2015. Efficacies and second-year effects of SPLAT GM™ and SPLAT GM™ Organic formulations. Insects 6: 1–12.

Oyafuso, A., N. Arakaki, Y. Sadoyama, M. Kishita, F. Kawamura, M. Ishimine et al. 2002. Life history of the white grub *Dasylepida* sp. (Coleoptera: Scarabaeidae), a new and severe pest on sugarcane on the Miyako Islands, Okinawa. Appl. Entomol. Zool. 37: 595–601.

Pacheco da Silva, V.C., E.C. Galzer, T. Malausa, J.-F. Germain, M.B. Kaydan and M. Botton. 2016. The vine mealybug *Planococcus ficus* (Signoret) (Hemiptera: Pseudococcidae) damaging vineyards in Brazil. Neotrop. Entomol. 45: 449–451.

Phelan, P.L. 1992. Evolution of sex pheromones and the role of asymmetric tracking. pp. 265–314. *In*: B.D. Roitberg and M.B. Isman [eds.]. Insect Chemical Ecology: An Evolutionary Approach. Chapman & Hall, New York, NY, USA.

Pogue, M.G. and P.W. Schaefer. 2007. A Review of Selected Species of *Lymantria* Hübner [1819] Including Three New Species (Lepidoptera: Noctuidae: Lymantriinae) from Subtropical and Temperate Regions of Asia, Some Potentially Invasive to North America. USDA Forest Service, Morgantown, WV, USA.

Rodriguez-Saona, C.R., D.F. Polk and J.D. Barry. 2009. Optimization of pheromone deployment for effective mating disruption of oriental beetle (Coleoptera: Scarabaeidae) in commercial blueberries. J. Econ. Entomol. 102: 659–669.

Ross, L. and D.M. Shuker. 2009. Scale insects. Curr. Biol. 19: R184–R186.

Ross, L., E.J. Dealey, L.W. Beukeboom and D.M. Shuker. 2011. Temperature, age of mating and starvation determine the role of maternal effects on sex allocation in the mealybug *Planococcus citri*. Behav. Ecol. Sociobiol. 65: 909–919.

Rothschild, G.H.L. 1981. Mating disruption of lepidopterous pests: current status and future prospects. pp. 207–228. *In*: E.R. Mitchell [ed.]. Management of Insect Pests with Semiochemicals. Plenum Press, New York, NY, USA.

Sanders, C.J. 1997. Mechanisms of mating disruption in moths. pp. 333–346. *In*: R.T. Cardé and A.K. Minks [eds.]. Insect Pheromone Research New Directions. Chapman & Hall, New York, NY, USA.

Sharon, R., T. Zahavi, T. Sokolsky, C. Sofer-Arad, M. Tomer, R. Kedoshim et al. 2016. Mating disruption method against the vine mealybug, *Planococcus ficus*: effect of sequential treatment on infested vines. Entomol. Exp. Appl. 161: 65–69.

Sharov, A.A., D. Leonard, A.M. Liebhold, E.A. Roberts and W. Dickerson. 2002. "Slow The Spread": a national program to contain the gypsy moth. J. Forest. 100: 30–36.

Shibao, M. and H. Tanaka. 2000. Seasonal occurrence of Japanese mealybug, *Planococcus kraunhiae* (Kuwana), on the fig, *Ficus carica* L., and control of the pest by insecticides. Jpn. J. Appl. Entomol. Zool. Chugoku Branch 42: 1–6.

Silverstein, R.M. 1990. Practical use of pheromones and other behavior-modifying compounds: overview. pp. 1–8. *In*: R. Ridgeway, R.M. Silverstein and M. Inscoe [eds.]. Behavior-Modifying Chemicals for Insect Management. Marcel Dekker, New York, NY, USA.

Stelinski, L.L., L.J. Gut, A.V. Pierzchala and J.R. Miller. 2004. Field observations quantifying attraction of four tortricid moths to high-dosage pheromone dispensers in untreated and pheromone-treated orchards. Entomol. Exp. Appl. 113: 187–196.

Stelinski, L.L., J.R. Miller and L.J. Gut. 2005a. Captures of two leafroller moth species in traps baited with varying dosages of pheromone lures or commercial mating-disruption dispensers in untreated and pheromone-treated orchard plots. Can. Entomol. 137: 98–109.

Stelinski, L.L., L.J. Gut, D. Epstein and J.R. Miller. 2005b. Attraction of four tortricid moth species to high dosage pheromone rope dispensers: observations implicating false plume following as an important factor in mating disruption. IOBC/WPRS Bull. 28: 313–317.

Stelinski, LL, J.R. Miller, R. Ledebuhr, P. Siegert and L.J. Gut. 2007. Season-long mating disruption of *Grapholita molesta* (Lepidoptera: Tortricidae) by one machine application of pheromone in wax drops (SPLAT-OFM). J. Pest Sci. 80: 109–117.

Stevens, L.J. and M. Beroza. 1972. Mating-inhibition field tests using Disparlure, the synthetic gypsy moth sex pheromone. J. Econ. Entomol. 65: 1090–1095.

Suckling, D.M., A.M. El-Sayed and J.T.S. Walker. 2016. Regulatory innovation, mating disruption and 4-Play^{TM} in New Zealand. J. Chem. Ecol. 42: 584–589.

Sugie, H., M. Teshiba, Y. Narai, T. Tsutsumi, N. Sawamura, J. Tabata et al. 2008. Identification of a sex pheromone component of the Japanese mealybug, *Planococcus kraunhiae* (Kuwana). Appl. Entomol. Zool. 43: 369–375.

Tabata, J. 2013. A convenient route for synthesis of 2-isopropyliden-5-methyl-4-hexen-1-yl butyrate, the sex pheromone of *Planococcus kraunhiae* (Hemiptera: Pseudococcidae), using a β,γ- to α,β-double-bond migration in an unsaturated aldehyde. Appl. Entomol. Zool. 48: 229–232.

Tabata, J. and R.T. Ichiki. 2015. A new lavandulol-related monoterpene in the sex pheromone of the grey pineapple mealybug, *Dysmicoccus neobrevipes*. J. Chem. Ecol. 41: 194–201.

Tabata, J. and R.T. Ichiki. 2016. Sex pheromone of the cotton mealybug, *Phenacoccus solenopsis*, with an unusual cyclobutane structure. J. Chem. Ecol. 42: 1193–1200.

Tabata, J., H. Noguchi, Y. Kainoh, F. Mochizuki and H. Sugie. 2007a. Sex pheromone production and perception in the mating disruption-resistant strain of the smaller tea leafroller moth, *Adoxophyes honmai*. Entomol. Exp. Appl. 122: 145–153.

Tabata, J., H. Noguchi, Y. Kainoh, F. Mochizuki and H. Sugie. 2007b. Behavioral response to sex pheromone-component blends in the mating disruption-resistant strain of the smaller tea tortrix, *Adoxophyes honmai* Yasuda (Lepidoptera: Tortricidae), and its mode of inheritance. Appl. Entomol. Zool. 42: 675–683.

Tabata, J., M. Teshiba, S. Hiradate, T. Tsutsumi, N. Shimizu and H. Sugie. 2011. Cyclolavandulyl butyrate: an attractant for a mealybug parasitoid, *Anagyrus sawadai* (Hymenoptera: Encyrtidae). Appl. Entomol. Zool. 46: 117–123.

Tabata, J., M. Teshiba, N. Shimizu and H. Sugie. 2015. Mealybug mating disruption by a sex pheromone derived from lavender essential oil. J. Essent. Oil Res. 27: 232–237.

Tabata, J., R.T. Ichiki, C. Moromizato and K. Mori. 2017. Sex pheromone of a coccoid insect with sexual and asexual lineages: fate of an ancestrally essential sexual signal in parthenogenetic females. J. R. Soc. Interface 14: 20170027.

Tamaki, Y., H. Noguchi, K. Yushima and C. Hirano. 1971. Two sex pheromones of the smaller tea tortrix: isolation, identification, and synthesis. Appl. Entomol. Zool. 6: 139–141.

Tamaki, Y., H. Noguchi, H. Sugie, R. Sato and A. Kariya. 1979. Minor components of the female sex-attractant pheromone of the smaller tea tortrix moth (Lepidoptera: Tortricidae): isolation and identification. Appl. Entomol. Zool. 14: 101–113.

Tamaki, Y., H. Noguchi and H. Sugie. 1983. Selection of disruptants of pheromonal communication in both the smaller tea tortrix moth and tea tortrix moth. Jpn. J. Appl. Entomol. Zool. 27: 124–130.

Tamaki, Y., H. Sugie, M. Nagamine and M. Kinjo. 1986. Female sex pheromone of the sugarcane wireworm *Melanotus okinawensis* Ohira (Coleoptera: Elateridae). Japan Patent No. JP 61-12601.

Tamaki, Y., H. Sugie, M. Nagamine and M. Kinjo. 1990. 9,11-dodecadienyl butyrate and 9,11-dodecadienyl hexanoate female sex pheromone of the sugarcane wireworm *Melanotus sakishimensis* Ohira (Coleoptera: Elateridae). Japan Patent No. JP 2-53753.

Teshiba, M. 2013. Integrated management of *Planococcus kraunhiae* Kuwana (Homoptera: Pseudococcidae) injuring Japanese persimmons. Jpn. J. Appl. Entomol. Zool. 57: 129–135.

Teshiba, M. and J. Tabata. 2017. Suppression of population growth of the Japanese mealybug, *Planococcus kraunhiae* (Hemiptera: Pseudococcidae), by using an attractant for indigenous parasitoids in persimmon orchards. Appl. Entomol. Zool. 52: 153–158.

Teshiba, M., N. Shimizu, N. Sawamura, Y. Narai, H. Sugie, R. Sasaki et al. 2009. Use of a sex pheromone to disrupt the mating of *Planococcus kraunhiae* (Kuwana) (Hemiptera: Pseudococcidae). Jpn. J. Appl. Entomol. Zool. 53: 173–180.

Teshiba, M., H. Sugie, T. Tsutsumi and J. Tabata. 2012. A new approach for mealybug management: recruiting an indigenous, but 'non-natural' enemy for biological control using an attractant. Entomol. Exp. Appl. 142: 211–215.

Tcheslavskaia, K.S., K.W. Thorpe, C.C. Brewster, A.A. Sharov, D.S. Leonard, R.C. Reardon et al. 2005. Optimization of pheromone dosage for gypsy moth mating disruption. Entomol. Exp. Appl. 115: 355–361.

Thorpe, K., R. Reardon, K. Tcheslavskaia, D. Leonard and V. Mastro. 2006. A Review of the Use of Mating Disruption to Manage Gypsy Moth, *Lymantria dispar* (L.). The United States Department of Agriculture, Washington D.C., USA.

Tóth, M. 2013. Pheromones and attractants of click beetles: an overview. J. Pest Sci. 86: 3–17.

Tsai, C.-W., J. Chau, L. Fernandez, D. Bosco, K.M. Daane and R.P. Almeida. 2008. Transmission of Grapevine leafroll-associated virus 3 by the vine mealybug (*Planococcus ficus*). Phytopathology 98: 1093–1098.

Tsueda, H. 2014. Response of two parasitoid wasps to the sex pheromone of Japanese mealybug, *Planococcus kraunhiae*. Jpn. J. Appl. Entomol. Zool. 58: 147–152.

Uchiyama, T. and A. Ozawa. 2014. Rapid development of resistance to diamide insecticides in the smaller tea tortrix, *Adoxophyes honmai* (Lepidoptera: Tortricidae), in the tea fields of Shizuoka Prefecture, Japan. Appl. Entomol. Zool. 49: 529–534.

Ueno, H. 1963. Studies on the scale insects injury to the Japanese persimmon, *Diospyros kaki* L. I. On the overwintered larvae of the mealy bug, *Pseudococcus kraunhiae* Kuwana. Jpn. J. Appl. Entomol. Zool. 7: 85–91.

Vité, J.P., D. Klimetzek, G. Loskant, R. Hedden and K. Mori. 1976. Chirality of insect pheromones: response interruption by inactive antipodes. Naturwissenschaften 63: 582–583.

Wakamura, S., H. Yasui, T. Akino, T. Yasuda, M. Fukaya, S. Tanaka et al. 2009a. Identification of (*R*)-2-butanol as a sex attractant pheromone of the white grub beetle, *Dasylepida ishigakiensis* (Coleoptera: Scarabaeidae), a serious sugarcane pest in the Miyako Islands of Japan. Appl. Entomol. Zool. 44: 231–239.

Wakamura, S., H. Yasui, F. Mochizuki, T. Fukumoto, N. Arakaki, A. Nagayama et al. 2009b. Formulation of highly volatile pheromone of the white grub beetle *Dasylepida ishigakiensis* (Coleoptera: Scarabaeidae) to develop monitoring traps. Appl. Entomol. Zool. 44: 579–586.

Walton, V.M., K.M. Daane and K.L. Pringle. 2004. Utilizing the sex pheromone of *Planococcus ficus* to improve pest management in South African vineyards. Crop Prot. 23: 1089–1096.

Walton, V.M., K.M. Daane, W.J. Bentley, J.G. Millar, T.E. Larsen and R. Malakar-Kuenen. 2006. Pheromone-based mating disruption of *Planococcus ficus* (Hemiptera: Pseudococcidae) in California vineyards. J. Econ. Entomol. 99: 1280–1290.

Witzgall, P., P. Kirsch and A. Cork. 2010. Sex Pheromones and Their Impact on Pest Management. J. Chem. Ecol. 36: 80–100.

Wright, R.H. 1964a. After pesticides—What? Nature 204: 121–125.

Wright, R.H. 1964b. Insect control by nontoxic means. Science 144: 487.

Yasuda, T. 1998. The Japanese species of the genus *Adoxophyes* Meyrick (Lepidoptera, Tortricidae). Trans. Lepid. Soc. Japan 49: 159–173.

Yasui, H., S. Wakamura, N. Fujiwara-Tsujii, N. Arakaki, A. Nagayama, Y. Hokama et al. 2012. Mating disruption by a synthetic sex pheromone in the white grub beetle *Dasylepida ishigakiensis* (Coleoptera: Scarabaeidae) in the laboratory and sugarcane fields. Bull. Entomol. Res. 102: 157–164.

Zada, A., E. Dunkelblum, F. Assael, M. Harel, M. Cojocaru and Z. Mendel. 2003. Sex pheromone of the vine mealybug, *Planococcus ficus* in Israel: occurrence of a second component in a mass-reared population. J. Chem. Ecol. 29: 977–988.

Zada, A., E. Dunkelblum, F.C. Assael, J. Franco, E.B. da Silva, A. Protasov et al. 2008. Attraction of *Planococcus ficus* males to racemic and chiral pheromone baits: flight activity and bait longevity. J. Appl. Entomol. 132: 480–489.

Zou, Y. and J.G. Millar. 2015. Chemistry of the pheromones of scale and mealybug insects. Nat. Prod. Rep. 32: 1067–1113.

CHAPTER

10

Applied Chemical Ecology to Enhance Insect Parasitoid Efficacy in the Biological Control of Crop Pests

Ezio Peri[1*], Rihem Moujahed[1], Eric Wajnberg[2] and Stefano Colazza[1]

[1] Dipartimento di Scienze Agrarie, Alimentari e Forestali, Università degli Studi di Palermo, Viale delle Scienze, 90128 Palermo, Italy
E-mail: ezio.peri@unipa.it (EP)
E-mail: rihem.moujahed@unipa.it (RM)
E-mail: stefano.colazza@unipa.it (SC)

[2] INRIA, Projet Hephaistos, 2004 Route des Lucioles, BP 93, 06902 Sophia Antipolis Cedex, France
E-mail: eric.wajnberg@inra.fr

Abstract

The field application of semiochemicals, used by parasitoids to find mates and to locate their hosts, is a promising environmentally sustainable and highly specific pest control strategy and an attractive alternative to the use of pesticides. In this chapter, we first examine research progress dealing with the effect of semiochemical cues on parasitoid foraging strategy. In the second part, we review the possible field applications of these chemical cues to enhance pest control strategies, either through direct pest control or by manipulating parasitoid behaviour. We then consider novel approaches, such as the "attract and reward" strategy, combining semiochemical application and habitat management to improve the success of pest management. Even if semiochemical-based tactics are promising in pest control, they still are developing and may face several constraints leading to different challenges. Therefore, in the last part of this chapter, we draw attention to the potential

*Corresponding author

limitations and risks of semiochemical applications in the field. Then, we propose potential solutions to overcome these different constraints.

1. Introduction

In recent years, scientists, farmers and citizens have been seeking more sustainable agricultural systems that can combine the production of healthy crops along with the greatest possible protection of the agro-ecosystems. One possible approach is the use of a combination of strategies, such as resistant plants, biological control of pests, habitat manipulation, or modification of cultural practices that allow synergy with the long-term prevention of pests and a reduction in risks to human health and the environment. In this view, the use of semiochemicals appears to be an ecologically sustainable strategy that can reduce the use of pesticides. As a consequence, many studies have been conducted to identify, synthesise and produce semiochemical compounds and to investigate their possible use. Some of these have now been commercialised and become tools in Integrated Pest Management (IPM) strategies (Suckling and Karg 2000, Witzgall et al. 2010).

Semiochemical-based manipulation strategies include either "pheromone-based tactics" or "allelochemical-based tactics". The former tactics are well known and represent successful strategies in IPM programs, mainly in orchards, for both direct and indirect control of pests. In contrast, allelochemical-based tactics represent a relatively new approach and few successful applications have been reported thus far.

In recent years, many researchers have investigated the chemical ecology of parasitoids and predators in order to understand the chemical cues that mediate their host location process. Several studies have demonstrated that the use of semiochemicals can modify the behaviour of insect natural enemies, opening new perspectives in using semiochemicals to conserve and/or improve the efficiency of natural enemies in crop systems (Pickett et al. 1997, Khan et al. 2008). However, despite the several researches conducted, mainly under laboratory conditions, to elucidate the interactions between semiochemicals and natural enemies in multitrophic contexts, the application of semiochemicals in the field is still limited (Blassioli-Moares et al. 2013, Colazza et al. 2013, Kaiser et al. 2016).

In this chapter, we review important advances in research concerning chemical compounds that mediate relationships between parasitoids and their host-plant complex and the possible field applications of these compounds for improving pest control efficiency. We first address basic aspects of parasitoid chemical ecology, focusing on the main chemical cues exploited by wasps during both intra-specific and inter-specific relationships leading them to find a mate or hosts. Then, we review studies on chemical compounds that can be applied in the field to improve insect pest control, addressing both "pheromone-based" tactics and "allelochemical-based" tactics as well as other strategies used to manipulate parasitoid behaviour in order to conserve or augment pest control efficiency. Finally, we highlight the limitations and provide our perspective on the application of these chemical compounds in cropping systems to improve parasitoid efficiency in biological control of insect pests.

2. Semiochemicals and their Effect on Parasitoids: How to Cope with a Complex World

As soon as they emerge in a new environment, adult parasitoid females have to face many challenges in order to search for and find a mate and hosts, which are usually both small and inconspicuous, to produce progeny (Godfray 1994). Female parasitoids live in environments that are highly complex (Ode 2013, Wäschke et al. 2013), in which they are continuously flooded with information involving simultaneously visual, auditory, physical, gustative and olfactory cues and only a small part of these are relevant stimuli to find a mate and then hosts (Vet and Dicke 1992). In the course of evolution and thanks to important evolutionary constraints, females of parasitoid species have progressively developed remarkable and sharp capabilities to recognise, from long and/or short distance, reliable and detectable information about where they are, where potential mates and hosts are and even sometimes, where other competing conspecific or heterospecific females are (Hilker and McNeil 2008). Even if such capabilities are necessarily associated with several morphologic, metabolic and physiologic costs, females are constrained to do so in order to produce progeny. For this, they develop a rich arsenal of different types of receptors (Quicke 1997), leading them to discriminate, at each step of their foraging process, what information to take into account versus stimuli that are not relevant. Several decades of work have led to an understanding that, among all stimuli that can be perceived by parasitic wasp females in their foraging environments, olfactory cues play the most important role (Bernal and Luck 2007). Semiochemicals are chemical cues involved in interactions between two organisms. They are classified into two broad groups, pheromones and allelochemicals. Pheromones mediate interactions between organisms of the same species and are subdivided according to their function, such as sex pheromones, alarm pheromones, aggregation pheromones or host marking pheromones. Allelochemicals, on the other hand, mediate interactions between organisms belonging to different species. They are classified into different categories depending on whether the benefit is for the emitter or the receiver: allomones are favourable to the organism that emits the substance; synomones are favourable to both the emitter and the receiver; and kairomones are favourable to the receiver only.

All these terms are context-specific rather than chemical-specific. For example, the sex pheromone emitted by *Sesamia nonagrioides* (Lepidoptera: Noctuidae) acts as a kairomone for adult females of the egg parasitoid *Telenomus busseolae* (Hymenoptera: Scelionidae), which exploits this cue in their host location process (Colazza et al. 1997). The following paragraphs describe the importance of pheromones, synomones and kairomones in the ability of parasitoid females to locate and attack their hosts. The corresponding sections below will describe, for each of these two types of compounds, the use of long- and short-range semiochemicals cues used by parasitoid females in their foraging behaviour.

2.1 Parasitoid Pheromones

In the last fifteen years, several studies have highlighted pheromonal communication in insect parasitoids, mainly belonging to the Hymenoptera. The identified compounds were classified into four groups (Ruther 2013): (1) sex pheromones that can be released by either sex and are involved in three levels of sexual communication, i.e., attraction to the site of release (emitter individual), mate recognition and elicitation of female receptiveness (i.e., aphrodisiac pheromones released by males); (2) marking pheromones that are released by females, e.g., to avoid superparasitism and competition among larvae inside the host, or to optimise clutch size and offspring sex ratio; (3) putative alarm pheromones identified in some bethylid wasp species and (4) aggregation and anti-aggregation pheromones. Although pheromones have been characterised in parasitoids from numerous families, information about the chemistry involved is still scarce (Li 2006, Ruther 2013). For example, a number of studies demonstrated that Dufour's gland is a source of marking pheromone in some parasitoid species (e.g., Guillot and Vinson 1972, Jaloux et al. 2005). Chemical compounds in this gland often match the cuticular hydrocarbon profiles in the parasitoid species studied, suggesting that these cuticular hydrocarbon compounds play a role (Jaloux et al. 2005).

2.2 Synomones

Upon herbivore attack, plants emit complex mixtures of organic compounds, called synomones that recruit natural enemies of the attacking herbivores. From the plant side, these are an indirect defense mechanism (e.g., "cry for help", see Dicke and Baldwin 2010), induced either by the feeding activities of insect herbivores, i.e., herbivore-induced plant volatiles (HIPVs) and/or by their egg-laying behaviour, i.e., oviposition-induced plant volatiles (OIPVs) (Kessler and Heil 2011, Hilker and Fatouros 2015). Depending on the attacking herbivore species, OIPVs can occur with or without plant wounding caused by females during oviposition (Hilker and Fatouros 2015). Interestingly, plants under insect herbivore attack not only emit volatile defensive compounds to attract parasitoid females, but doing so, they can inform other non-attacked plants. Indeed, volatile cues emitted by plants damaged by herbivores have been showed to be selectively detected by healthy plants that are then induced to also release defensive compounds (Guerrieri 2016, Pickett and Khan 2016).

The emission of HIPVs and OIPVs, which are perceived by parasitoids at long distance (i.e., long-range cues) is a widespread ecological phenomenon that has been recorded for at least 49 plant species belonging to 25 different families (Mumm and Dicke 2010, Meiners and Peri 2013, Hilker and Fatouros 2015, Fatouros et al. 2016). However, parasitoids can also perceive synomones after landing on the host plant (i.e., contact or short-range cues). Some recent studies showed that some plants, such as maize and brassicaceous plants, when they are attacked by insect pests, emit substrate-borne chemical cues (i.e., alteration of leaf chemistry composition), which are exploited by insect parasitoids to increase their ability to find hosts to attack (Fatouros et al. 2005, Conti et al. 2010, Blenn et al. 2012, Salerno et al. 2013). For example, oviposition by *Pieris brassicae* and *P. rapae* (Lepidoptera: Pieridae) on *Brassica oleracea* induces changes in the chemistry of the leaf surface that induce an arrestment response in females of the egg parasitoids *Trichogramma brassicae* and *T. evanescens* (Hymenoptera: Trichogrammatidae) (Fatouros et al. 2009, Pashalidou et al. 2010).

Most of these studies have been conducted in tri-trophic systems, including a host plant, an herbivore and a parasitoid species. These systems differ considerably from field conditions, where plants are often attacked by multiple herbivore species, even sometimes simultaneously, leading to alterations in the synomone blends modifying the response of natural enemies (Soler et al. 2013).

In fact, during the last decade studies on parasitoid foraging behaviour has switched from tri-trophic to multi-trophic interaction investigations (De Rijk et al. 2013, Gols 2014). For example, in multi-herbivores communities, some studies demonstrated that the presence of non-host species could have a strong effect on parasitoid foraging behaviour. Non-host herbivores may either share the same plant with the host herbivores of a parasitoid or be present on neighbouring plants (Dicke et al. 2009). On shared plants, herbivores may all feed on a single plant organ or on different plant organs located either above- or below-ground (van Dam and Heil 2011). Using a system including *Vicia faba* plants, two herbivorous insects, the above-ground herbivore *Nezara viridula* (Hemiptera: Pentatomidae) and the above- and below-ground herbivore *Sitona lineatus* (Coleoptera: Curculionidae), at adult and larval stages, respectively and *Trissolcus basalis* (Hymenoptera: Scelionidae), an egg parasitoid of *N. viridula*, Moujahed et al. (2014) demonstrated that induced plant responses caused by the concurrent infestation of non-host beetle *S. lineatus* reduced the attraction of the egg parasitoid toward OIPVs emitted by plants infested by *N. viridula*. Non-host beetle chewing damage from both adults feeding on leaves and larvae feeding on roots significantly changed the composition of the OIPV blend, resulting in a disruptive effect on *T. basalis* host location (Fig. 1).

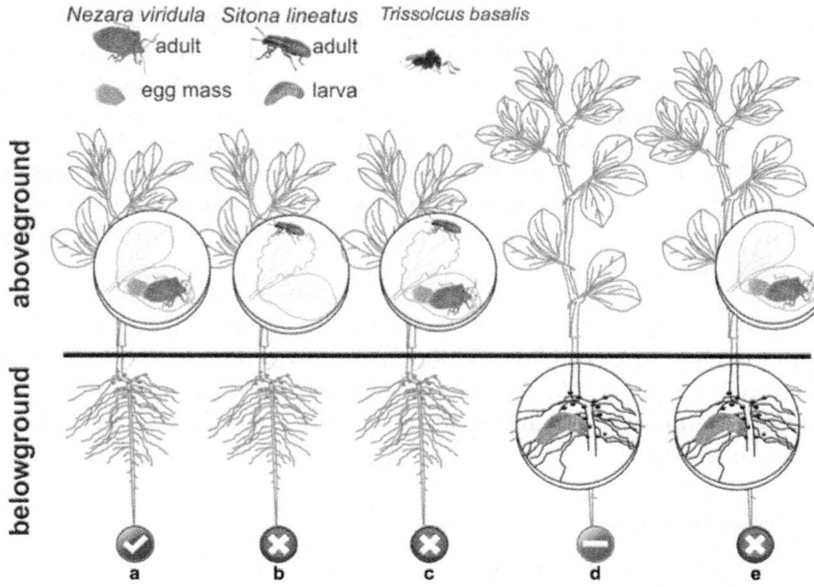

Fig. 1. Effect of non-host herbivore presence (*Sitona lineatus*) on egg parasitoid (*Trissolcus basalis*) attraction towards *Vicia faba* plant attacked by the host, *Nezara viridula*. Above-ground treatment: (a) *Nezara viridula* feeding and oviposition, (b) *Sitona lineatus* adult leaf-feeding, (c) *N. viridula* feeding and oviposition and *S. lineatus* adult leaf-feeding; below-ground treatment: (d) *S. lineatus* larvae root-nodules feeding; above- and below-ground treatment: (e) *N. viridula* feeding and oviposition and *S. lineatus* larvae root-nodules feeding.

Symbols: ✓: positive effect; ×: disruptive effect; –: neutral effect (modified from Moujahed et al. 2014)

Different factors for non-host herbivore attack may induce changes in synomones emission such as the composition of the insect feeding guild (Cusumano et al. 2015), the density of the pest population (Ponzio et al. 2016) or the co-evolutionary history of the plant–insect interaction (Cusumano et al. 2015, Martorana et al. 2017).

The situation is sometimes even more complex than this. In a recent intriguing study, Poelman et al. (2012) demonstrated that cabbage plants attacked by *P. rapae* release volatile compounds that attract primary parasitoids of the genus *Cotesia* (Hymenoptera: Braconidae), e.g., *C. glomerata*, enabling them to find their hosts and to attack and kill them (i.e., a "cry for help" strategy). However, as the authors discovered, volatile compounds released by the plants were also attractive for a hyperparasitoid species, *Lysibia nana* (Hymenoptera: Ichneumonidae) that is then able to find primary parasitoids and kill them. Furthermore, hyperparasitoid females were much more attracted by the volatiles produced by plants attacked by *P. rapae* caterpillars that were previously attacked by their primary parasitoids than by the volatiles produced by plants attacked by unparasitised herbivorous hosts. The difference was due to changes in the oral secretion of the caterpillar hosts when they were parasitised and such changes in oral secretion induced changes in plant volatile emission. Hence, plants that play a "cry for help" strategy attracting natural enemies to protect them actually attract hyperparasitoids that are killing these natural enemies. As a consequence, the overall output is that plant fitness might decrease.

Finally, the response of parasitoids to plant synomones can also be affected by abiotic stresses. For example, *V. faba* plants that were simultaneously in detrimental abiotic conditions (i.e., water stress) and biotic stress (i.e., *N. viridula* attack) showed an alteration in volatile blend emission, resulting in enhanced attraction of the egg parasitoid *T. basalis* (Salerno et al. 2017).

2.3 Kairomones

The ability of parasitoid females to eavesdrop on chemical cues emitted by their hosts and to use them as kairomones in their host location process is a well-documented phenomenon (Powell 1999, Fatouros et al. 2008). Over long distances, parasitoid females can exploit different types of pheromone cues emitted by their hosts to locate their habitat, such as sex (Boo and Yang 2000) or aggregation pheromones (Noldus 1989). Eavesdropping on sex pheromone signals emitted by females to attract males has been demonstrated as a key ability allowing many parasitoid species to find and attack their hosts (Huigens and Fatouros 2013). For example, the egg parasitoid *Telenomus euproctidis* (Hymenoptera: Scelionidae) is attracted by the sex pheromone [(Z)-16-methyl-9-heptadecenyl isobutyrate] emitted by females of its nocturnal host, *Orvasca taiwana* (Lepidoptera: Lymantriidae), before mating occurs (Arakaki et al. 2011). The wasp actually remains able to detect traces of the host sex pheromone even 48 hours after the release. Such pheromone traces are retained on host scale hairs of the anal tufts until female moths start laying their egg clutch on a plant. By doing so, female wasps can extend their host searching time and remain active during the temporal gap between nocturnal host mating and oviposition activities and their diurnal foraging time. Some authors even demonstrated a potential synergism between compounds originating from host plants and host sex pheromones, which may create another "bridge-in-time" strategy improving the overall efficacy of their chemical espionage. Whether the detection of adsorbed pheromones is influenced by variation in the epicuticular wax composition of the plant is still unknown (Wäschke et al. 2013).

Other chemical compounds, for example anti-sex pheromone and aggregation pheromone, can act as kairomone signals for parasitoids in their host searching behaviour. For example, some egg parasitoid species are able to exploit host anti-aphrodisiac compounds transferred post-mating from males to female hosts to render them less attractive to conspecific mates, such as benzyl cyanide, indole and/or methyl salicylate (Fatouros et al. 2005, 2009, Huigens et al. 2009). In addition, several species of tachinid parasitoids use the aggregation pheromone produced by pentatomid bugs as a host-finding kairomone (Nakamura et al. 2013). Finally, parasitoids are also able to spy on allomones. For example, defensive compounds such as (*E*)-2-decenal emitted from the metathoracic gland of *N. viridula* adults acts as a long-range kairomone orienting *T. basalis* females to their hosts (Mattiacci et al. 1993).

After landing on a host plant, parasitoids mainly perceive contact and/or short-range kairomones to find their hosts, such as, for example, alarm pheromone or host marking pheromones (Hoffmeister et al. 2000, Francis et al. 2005). In this phase of parasitoid foraging, a role is played by chemical footprints left by the hosts. For instance, footprints left on the substrate by the southern green stink bug *N. viridula* are involved in the arrestment responses of the parasitoid *T. basalis* (Colazza et al. 1999). Wölfling and Rostás (2009) revealed that the chemical footprints left by caterpillar *Spodoptera frugiperda* (Lepidoptera: Noctuidae), while walking over a plant surface, are detected by female wasps *Cotesia marginiventris* for up to two days after their hosts have left the site.

Footprints left by herbivores consist of chemicals that probably originate from the insect cuticle, such as (mostly linear) alkanes, diglycerides, triglycerides of high molecular weight, long-chain alcohols and fatty acids (Rostás and Wölfling 2009, Geiselhardt et al. 2011, Lo Giudice et al. 2011). These compounds are not only signals used by parasitoids to find their hosts, but also allow them to fine-tune their host searching behaviour. Indeed, Colazza et al. (2007) revealed that *n*-nonadecane (nC19), which is only present in footprints of *N. viridula* males, acts as a "host sex recognition" cue for *T. basalis*, allowing the wasps to discriminate between the chemical traces of females and males host. Such an ability to discriminate host sex through host chemical footprints was later confirmed with other egg parasitoids attacking pentatomid host bugs (Peri et al. 2013). The ability of parasitoid females to discriminate between male and female adults of their hosts helps them to find potential host eggs more efficiently (Puente et al. 2008a). In this respect, some parasitic wasp species were demonstrated to use associative learning. For instance, rewarded experience (i.e., successful oviposition) occurring after host footprints exploitation increased the arrestment response of *T. basalis* and *Trissolcus brochymenae* females to footprints of their hosts (Peri et al. 2006, 2016). Finally, other contact host-derived by-products (e.g., frass, honeydew, exuviae, mandibular gland secretions, scales) have also been demonstrated to be potential sources of host location kairomones (Fatouros et al. 2008, Colazza et al. 2010, Conti and Colazza 2012, Meiners and Peri 2013). For example, the wasp *Psyllaephagus pistaciae* (Hymenoptera: Encyrtidae) exploits, as both volatile and contact kairomones, cues from honeydew released by the common pistachio psylla, *Agonoscena pistaciae* (Hemiptera: Psyllidae) (Mehrnejad and Copland 2006). Obonyo et al. (2010) showed the importance of frass produced by the hosts *Busseola fusca* (Lepidoptera: Noctuidae) and *Chilo partellus* (Lepidoptera: Crambidae) in short-range host recognition by parasitoids, *Cotesia sesamiae* and *C. flavipes*, respectively. Cues from frass can even be used by female parasitoids to discriminate host from non-host species, hosts of different ages/instars and even hosts feeding on different host plants (e.g., Mattiacci and Dicke 1995, Chuche et al. 2006). The capacity to discriminate between host and non-host species, based on host by-product kairomones, was also reported for cues emitted by cocoons (Bekkaoui and Thibout 1993) or exuviae (Battaglia et al. 2000).

3. Use of Semiochemicals for Agricultural Pest Management

Since the first discovery of semiochemicals, scientists have investigated the possibility of applying these compounds to increase the efficacy of biological control or IPM programs (Suckling and Karg 2000). The potential of semiochemicals to manipulate parasitoid behaviour has opened up new possibilities in pest control, with the promise of environmentally sustainable and highly specific pest control tactics. The use of semiochemicals to improve pest control efficacy represents a relatively new approach that mainly uses plant volatiles. The most promising application involves the use of HIPVs to manipulate the natural enemies of pest species in order to attract and conserve them in the vicinity of the crops to be protected. The following sections describe the different tactics used so far.

3.1 Pheromones-based Tactics

In recent years, the identification of pheromones for a number of parasitoid species has opened up new opportunities to apply pheromone-based tactics for a direct manipulation of parasitoid behaviour. Indeed, studies have proved that parasitoid sex pheromones can be applied for monitoring parasitoid populations. For instance, in apple orchards in New Zealand, Suckling et al. (2002) used the sex pheromone of the female wasp *Ascogaster quadridentata* (Hymenoptera: Braconidae) to assess parasitoid establishment, abundance and phenology synchronisation with the host, the codling moth *Cydia pomonella* (Lepidoptera: Tortricidae). The authors concluded that pheromone trapping of insect biological control agents is a valuable tool that can help to determine and improve parasitoid success. In another study, Hardy and Goubault (2007) suggested application of female bethylid wasp sex pheromones to improve their efficiency as biological control agents. Although parasitoid pheromones have been proved to be promising tools in the manipulation of parasitoids, further field and laboratory tests are required to properly develop this tactic in biological control programs.

3.2 Allelochemicals-based Tactics

3.2.1 Host-Associated Volatiles (Kairomones)

Several studies tested the application of host-associated volatiles (HAVs) to enhance the efficacy of natural enemies. Among these, field studies mainly focused on two kinds of long-range HAVs, host sex and aggregation pheromones, due to their greater potential for parasitoid behavioural manipulation. Table 1 summarises current information about their field applications in different agriculture crops. In fact, host sex and aggregation pheromones have been applied either as tools for monitoring parasitoid populations or for enhancing parasitism rate in the field. An initial experiment on the use of host sex pheromone to monitor parasitoid population density was conducted in the UK by Hardie et al. (1991), who reported the attraction of three parasitoid species, *Praon abjectum*, *P. dorsal* and *P. volucre* (Hymenoptera: Braconidae), to water traps baited with synthetic sex pheromone of *Aphis fabae* (Hemiptera: Aphididae), a blend of (4a*S*,7*S*,7a*R*)-(+)-nepetalactone and (1*R*,4a*S*,7*S*,7a*R*)-(−)-nepetalactol.

In subsequent years, tests in cereal fields, in the UK and Germany, showed that (4a*S*,7*S*,7a*R*)-(+)-nepetalactone was the most effective lure with strongest attraction in autumn (Powell et al. 1993). On the basis of these promising results, the authors proposed to use sex pheromone-baited traps in autumn to manipulate parasitoid populations attacking aphids, a strategy that was later tested on commercial farms in the UK (Powell and Pickett 2003). The role of host sex pheromones in recruiting

Table 1. Host-associated volatile (HAV) candidates for improving parasitoid efficacy in agricultural systems

Crop	HAVs			Effect	References
	Chemicals	Origin/Function	Aim		
Apple	Methyl (2*E*,4*Z*)-decadienoate	Aggregation pheromone of *Euschistus conspersus*	Monitoring adult parasitoid populations	No effect	Krupke and Brunner (2003)
			Test for increased egg parasitism rates	Increase in *Gymnoclytia occidentalis* recruitment by host pheromone-baited traps	
Cereals	(4a*S*,7*S*,7a*R*)-Nepetalactone	Sex pheromone of aphids	Test for increased parasitism rates	Attraction of *Praon volucre*, *P. dorsale*, and *P. abjectum*	Powell et al. (1993)
				Increase in *P. volucre* parasitism	Glinwood et al. (1998)
Citrus	(1*R*,3*R*)-(2,2-Dimethyl-3-isopropenylcyclobutyl) methyl acetate (planococcyl acetate)	Sex pheromone of *Planococcus citri*	Monitoring parasitoid populations	No effect on *Anagyrus* sp. *nov.* near *pseudococci*	Franco et al. (2008)
	(*S*)-Lavandulyl senecioate (LS)	Sex pheromone of *P. ficus*		Attraction of *Anagyrus* sp. *nov.* near *pseudococci*	
	Planococcyl acetate	Sex pheromone *of P. citri*	Test for increased parasitism rates	Increase in *Anagyrus* sp. *nov.* near *pseudococci* parasitism for an Italian population	Franco et al. (2011)

(*Contd.*)

	(*S*)-Lavandulyl senecioate	Sex pheromone of *P. ficus*		Increase in *Anagyrus* sp. *nov.* near *pseudococci* parasitism	
Fig and grapevine	(*S*)-Lavandulyl senecioate	Sex pheromone of *P. ficus*	Monitoring parasitoid populations	More attraction of *Anagyrus* sp. *nov.* near *pseudococci* by LS-baited traps than by LI-baited traps	Franco et al. (2008)
	(*S*)-Lavandulyl isovalerate (LI)	Compound produced by *P. ficus* in mass-rearing conditions			
Grapevine	(*S*)-Lavandulyl senecioate	Sex pheromone of *P. ficus*	Test for increased parasitism rates	Increase in *Anagyrus sp. nov.* near *pseudococci* recruitment	Mansour et al. (2011)
			Test for combined effect of parasitoid inundative release + LS application on parasitism rate		
Soybean	Methyl 2,6,10-trimethyltridecanoate	Sex pheromone of *Euschistus heros*	Monitoring egg parasitoids	Attraction of *Telenomus podisi*, *Trissolcus teretis*, *Tr. urichi* and *Tr. brochymenae*	Borges et al. (1998)
	(*E*)-2-Hexenyl (*Z*)-3-hexanoate	Aggregation pheromone of *Riptortus clavatus*	Test for increased parasitism rates	Early attraction of *Ooencyrtus nezarae* and increase in sentinel eggs parasitism	Mizutani (2006)
	(*E*)-2-Hexenyl (*E*)-2-hexenaoate, (*E*)-2-hexenyl (*Z*)-3-hexenoate, tetradecyl isobutyrate	Aggregation pheromone of *Riptortus pedestris*	Test for increased parasitism rates	Attraction of *O. nezarae*, increase of egg parasitism and reduction of stink bug damage on pods	Alim and Lim (2011)

(*Contd.*)

Table 1. (*Contd.*)

Crop	HAVs			Effect	References
	Chemicals	Origin/Function	Aim		
			Test for distribution and guild composition of egg parasitoids	Increase in *O. nezarae* and *Gryon japonicum* abundance. No effect on parasitoid species relative abundance	Lim and Mainali (2013)
Spindle tree	Mixture of (4a*S*,7*S*,7a*R*)-nepetalactone and (1*R*,4a*S*,7*S*,7a*R*)-nepetalactol	Sex pheromone of *Aphis fabae*	Monitoring parasitoid populations	Increase in parasitoid recruitment by host pheromone-baited traps	Hardie et al. (1991)
Mixed crops	(Z)-3,9-Dimethy-6-isopropenyl-3,9-decadienyl propionate	Sex pheromone of *Pseudaulacaspis pentagona*	Monitoring parasitoid populations	Attraction of *Thomsonisca amathus* by host pheromone-baited traps	Bayoumy et al. (2011)
	Mixture of (Z)-3,7-dimethy1-2,7-octadienyl propionate and 7-methyl-3-methylene-7-octenyl propionate	Sex pheromone of *Diaspidiotus perniciosus*		Attraction of *Encarsia perniciosi* by host pheromone-baited traps	
	Methyl (*E*,*E*,*Z*)-2,4,6-decatrienoate	Aggregation pheromone of *Plautia stali*	Monitoring parasitoid populations	Attraction of *Gymnosoma rotundatum* by host pheromone-baited traps	Jang and Park (2010), Jang et al. (2011)

parasitoids and enhancing their activities has also been evaluated in vineyards, citrus and fig orchards in the Mediterranean basin. The system studied includes two mealybug species, citrus mealybug *Planococcus citri* and vine mealybug *P. ficus* (Hemiptera: Pseudococcidae) and the parasitoid *Anagyrus* sp. *nov.* near *pseudococci* (Hymenoptera: Encyrtidae) (Franco et al. 2008, 2011, Mansour et al. 2011). In all the experiments conducted, regardless of the cropping systems used, a significant increase in mealybug parasitism was observed when the sex pheromone of *P. ficus*, (*S*)-(+)-lavandulyl senecioate, was applied in the field. Hence, it was suggested that this new HAV-based strategy could be used efficiently in the control of vine mealybug in vineyards and citrus orchards (Franco et al. 2011, Mansour et al. 2011).

Examples of the use of host aggregation pheromone to manipulate parasitoid behaviour were provided by Jang and Park (2010) and Jang et al. (2011). They obtained successful results in monitoring *Gymnosoma rotundatum* (Diptera: Tachinidae) populations in Korean persimmon orchards, by placing sticky traps baited with methyl (*E*,*E*,*Z*)-2,4,6-decatrienoate, the aggregation pheromone of one of its hosts, the brown-winged green stink bug, *Plautia stali* (Hemiptera: Pentatomidae). Considering that the aggregation pheromone of the brown-winged green stink bug is used to attract both sexes of *P. stali* (Park et al. 2010), the use of appropriate pheromone traps was proposed, with the purpose of both reducing pest population densities and increasing parasitism rates (Jang et al. 2011). In Korean soybeans crop, the use of traps baited with the aggregation pheromone from *Riptortus pedestris* (=*clavatus*) (Hemiptera: Alydidae) has become common practice to monitor and reduce the pest population (Alim and Lim 2011, Mainali and Lim 2012). This host pheromone not only attracts conspecific adults in the field but it also has a kairomonal effect on its egg parasitoids, *Ooencyrtus nezarae* (Hymenoptera: Encyrtidae) and *Gryon japonicum* (Hymenoptera: Scelionidae), increasing the abundance of parasitoids up to 18 m from pheromone locations (Lim and Mainali 2013). Recent research also identified the possibility of developing allelochemical-based tactics with "non-natural" chemical compounds. For example, in persimmon orchards, the synthetic compound (2,4,4-trimethyl-2-cyclohexenyl) methyl butyrate, also known as cyclolavandulyl butyrate (CLB), have been shown to attract two parasitoids, *Anagyrus sawadai* and *Leptomastix dactylopii* (Hymenoptera: Encyrtidae), in order to control the Japanese mealybug *Planococcus kraunhiae* (Hemiptera: Pseudococcidae) (Teshiba et al. 2012, Teshiba and Tabata 2017).

3.2.2 Herbivore-Induced Plant Volatiles (Synomones)

The discovery of the crucial role of HIPVs in indirect plant defense triggered ideas to use them to increase pest control efficacy by manipulating the behaviour of natural enemies (Khan et al. 2008). HIPVs can be applied directly onto plants, by spraying, or as plant hormones, inducing the production of defensive chemical volatiles, which attract parasitoids, or even by using a slow-release dispenser that releases chemical compounds attracting natural enemies directly. Additionally, plants may be genetically engineered to increase their emission of HIPVs (Rodriguez-Saona et al. 2012, Stenberg et al. 2015, Guerrieri 2016).

A potential way to manipulate the behaviours of natural enemies is to identify the natural HIPVs, produce them synthetically and then release them, either separately or in a mixture. One of the early studies that demonstrated the potential of synthetic HIPV lures as direct field attractants for parasitoids was performed by James and Grasswitz (2005). In this study, sticky traps, placed in blocks of grape plants baited with controlled-release dispensers of methyl salicylate (MeSA), methyl jasmonate (MeJA) and (Z)-3-hexenyl acetate, captured more parasitic wasps from the families Encyrtidae and Mymaridae than blocks with unbaited traps. Subsequently, the past decade has witnessed many other laboratory and field studies that demonstrated the capability of synthetic HIPVs to increase the diversity or density of parasitoids within various crop systems (Blassioli-Moraes et al. 2013, Colazza et al. 2013, Simpson et al. 2013). For instance, in *Brassica rapa* crops, the use of dispensers of a synthetic mixture of (Z)-3-hexenyl acetate, n-heptanal, α-pinene and sabinene [i.e., volatile compounds induced by larvae feeding activity of the diamondback moth, *Plutella xylostella* (Lepidoptera: Plutellidae)], is able to recruit the parasitoid *Cotesia vestalis* to uninfested plants and to increase its attack on host larvae on infested plants (Uefune et al. 2012).

Treatment with synthetic HIPVs not only directly attracts natural enemies but also likely elicits the treated plant to produce endogenous HIPVs blends that recruit predator and parasitoid arthropods (Simpson et al. 2011a). In maize fields in Mexico, Von Mérey et al. (2011) demonstrated that, in the presence of glass vial dispensers releasing a mixture of four synthetic green leaf volatiles, maize plants released higher amount of sesquiterpenes than non-exposed plants, but this did not influence the impact of parasitoids on *S. frugiperda*. These encouraging results obtained in terms of parasitoid attraction to synthetic HIPVs need to take into account the effect of these volatiles on all the protagonists involved because the use of synthetic HIPVs may have negative effects, such as increasing the likelihood of intraguild predation on natural enemies, followed by a reduction in the top-down control of pests (Poelman and Kos 2016).

Herbivory-induced changes in plant volatiles are known to be regulated by jasmonic acid (JA), salicylic acid (SA) and ethylene (ET) hormonal signaling pathways (Takabayashi 2006, Ode 2013). In general, SA is triggered by phloem-feeding herbivores, whereas JA/ET are activated by biting/chewing herbivores and these pathways act antagonistically

(Stam et al. 2014). Hence, as an alternative to the use of synthetic HIPVs, plants could be sprayed with plant hormones, such as jasmonates [e.g., JA, MeJA, or *cis*-jasmone (CJ)], which play the role of plant elicitors (=inducers) triggering the production and emissions of their own blend of volatiles and leading to the attraction of natural enemies (Rohwer and Erwin 2008). The effects of jasmonates on the enhancement of parasitoids attraction have been demonstrated within various crops systems (e.g., van Poecke and Dicke 2002, Ozawa et al. 2004). For example, van Poecke and Dicke (2002) showed that *Arabidopsis thaliana* treated with JA increased the attraction of *Cotesia rubecula* compared with untreated plants, whereas treatment with SA did not. Moreover, Lou et al. (2005) reported that JA application on rice plants enhanced more than two-fold the parasitisation of brown planthopper' eggs, *Nilaparvata lugens* (Hemiptera: Delphacidae), by its parasitoid *Anagrus nilaparvatae* (Hymenoptera: Mymaridae). Similarly, treatments with CJ enhanced the foraging behaviour of *Telenomus podisi*, an egg parasitoid of soybean stink bug pests (Moraes et al. 2009) and of *Aphidius ervi* (Hymenoptera: Braconidae), a parasitoid of sweet pepper aphids (Dewhirst et al. 2012). However, recent studies showed no effect of jasmonates in enhancing the parasitism rate in a system composed of soybean/cotton crops, stink bug and egg parasitoids. For instance, CJ application on soybean attracted Platygastridae parasitoids but did not affect the parasitisation rate of eggs of the stink bug *Euschistus heros* (Hemiptera: Pentatomidae) (Vieira et al. 2013).

Apart from jasmonates, SA and its analogues also act as phytohormone elicitors and stimulate the biosynthesis and release of endogenous HIPVs. The effect of SA on parasitoids attraction has been well studied within maize crops (e.g., Rostás and Turlings, 2008, Sobhy et al. 2012, 2014). For instance, treating maize plants with BTH (*S*-methyl 1, 2, 3-benzothiadiazole-7-carbothioate), a mimic of SA, reduced the emission rates of two volatile compounds [indole and (*E*)-caryophyllene] from treated plants. This reduction was expected because of the well-known antagonism between SA and JA pathways (Thaler et al. 2002). Despite this reduction, Rostás and Turlings (2008) recorded a higher attractiveness to the parasitoid *Microplitis rufiventris* (Hymenoptera: Braconidae). This result was confirmed by subsequent tests including another elicitor, laminarin, and was explained by a subtractive hypothesis where the suppression of some volatiles not directly involved in parasitoid attraction may results in a higher attractiveness (Sobhy et al. 2012).

So far, there are no commercial products available using phytohormones. One of the reasons that could prevent the production of these products is the multiple role of these plant hormones, which are also responsible for the regulation of other physiological responses in plants. Therefore, the multifunctionality of potential HIPVs needs to be considered to avoid counter-productive results and to develop cost–

benefit analysis of the products for an efficient manipulation of natural enemies in crops.

3.2.3 Engineering of HIPVs in Crop Plants by Genetic Modification

Genetic modification appeared as an alternative way to increase biological control efficacy. The goal is to manipulate plants traits to produce and release specific volatiles (Åhman et al. 2010, Stenberg et al. 2015). Although plant breeding practices have historically ignored the effects of HIPVs on the third trophic level, this is expected to change with recent advances in molecular technologies (Gurr and You 2016, Pickett and Khan 2016). Generally, variation in the composition of HIPVs can be related to plant species, cultivars, phenological stages, host developmental stage and several other factors (McCormick et al. 2012). In this respect, the first approach that can be applied is selective breeding, which consists in selecting cultivars that can naturally enhance the foraging efficiency of natural enemies in field conditions (Michereff et al. 2015, Tamiru et al. 2015, Mitchell et al. 2016). Michereff et al. (2015) demonstrated that the ability to recruit parasitoids can be cultivar-dependent. Indeed, two pest-resistant soybean cultivars (Dowling and IAC 100) attracted a higher number of Platygastridae wasps compared with a susceptible one (Silvânia), with sufficient abundance to control the stink bug *E. heros*. Therefore, the authors suggested that traits in the resistant cultivars could be useful in breeding programs to obtain new cultivars more resistant to stink bugs and more attractive to natural enemies. The second approach is the use of transgenic plants where specific genes are inserted to prime plants for enhanced HIPV responses (Turlings and Ton 2006, Kos et al. 2009). Most research on transgenic plants has been conducted with arable crops such as maize and cotton, or model organisms such as *Arabidopsis* (Blassioli-Moraes et al. 2013, 2016). For instance, Schnee et al. (2006) transferred a sesquiterpene synthase gene that produces (*E*)-β-farnesene, (*E*)-α-bergamotene and other herbivory-induced sesquiterpenes from maize into *Arabidopsis*, resulting in greater emissions of several sesquiterpenes and enhanced attraction of *C. marginiventris* after wasps learned to associate the presence of hosts with the emissions of these sesquiterpenes.

Genetic engineering research has mainly focused on basic aspects of plant signaling pathways from a tri-tropic perspective, but extension of these laboratory studies to practical field applications remains limited. According to Gurr and You (2016), the new gene-editing tool, CRISPR/Cas9, will offer great power to elucidate and manipulate plant defence mechanisms, including HIPVs that involve natural enemy attraction to attacked plants. Therefore, CRISPR/Cas9 will allow highly accurate changes in genomes, which may reduce potential barriers to regulation and adoption (Voytas and Gao 2014).

3.3 Attract and Reward Strategy

Parasitoid recruitment with semiochemicals can be further improved when combined with habitat management, which involves, for example, the provision of flowering companion plants or wildflower strips in managed landscapes with low biodiversity (Gurr et al. 2004). These plants, by supplying beneficial insects with suitable sugars, increase their lifespan and fecundity and thus improve ecosystem services such as pest suppression. The combination between habitat management and semiochemical-based tactics, as a method to manage natural enemies, has been examined extensively and proved to be efficient (Rodriguez-Saona et al. 2012, Simpson et al. 2013). The ecological basis of such strategy relies on the possibility that crops treated with semiochemicals attract natural enemies that might find other plants in the same field, which provide them with food resources as a reward. To date, the application of the "attract and reward" (A&R) strategy is limited to a few cases, where beneficial insects are "attracted" by synthetically produced HIPVs and are then "rewarded" by providing them with floral resources. Simpson et al. (2011b, c) tested A&R strategy in maize, broccoli and grapevine crops, using some HIPVs, such as methyl anthranilate, MeJA, MeSA, (Z)-3-hexenyl acetate, benzaldehyde and their mixtures, as attractants and plants of buckwheat *Fagopyrum esculentum*, as reward. Indeed, this annual plant induces benefits to parasitoids, because its nectar and pollen resources increase the abundance, fitness and parasitism rates of natural enemies (Lee and Heimpel 2005, Irvin et al. 2006, Foti et al. 2017) (Fig. 2).

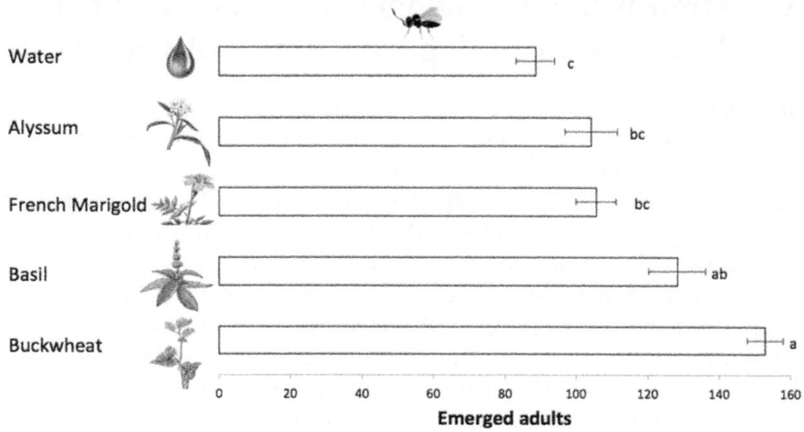

Fig. 2. Effect of floral resources on progeny of the egg parasitoid *Trissolcus basalis*. Significantly more adults emerged from host egg masses parasitised by wasps that had access to buckwheat flowers than to other resources. Means ± SE are reported; different letters indicate where differences among treatments are significant ($P < 0.05$; Based on data from Foti et al. 2017).

Although in some cases parasitoid density significantly increased in treatments with rewards (Simpson et al. 2011b, c), outcomes of these field experiments provided only limited evidence of a positive effect of the A&R strategy, because only one species of scelionids was more abundant in plots with MeSA-treated broccoli and buckwheat plants compared with either MeSA-treated broccoli plots or reward alone (Simpson et al. 2011b). Such non-consistent results obtained so far demonstrate the complexity of the A&R approach and highlight the need for further research to determine the efficacy of such combined strategy in cropping systems. In fact, previous studies showed that A&R could modify the population levels of both herbivorous and carnivorous insects, because each of the two components can determine an unintended recruitment of additional herbivores (Turlings and Ton 2006, Orre et al. 2010) or hyperparasitoids (Baggen et al. 1999, Lavandero et al. 2006, Jonsson et al. 2009), or even a counter-action, as was observed where the reward component may inhibit the attractive one (Orre Gordon et al. 2013). To limit potential negative effects, Orre Gordon et al. (2013) suggested a possible way to improve the A&R strategy and thus the efficacy of biological control of insect pests, by searching for a combination of HIPVs and companion plants that selectively recruit parasitoids, but not herbivorous and/or hyperparasitoid species, because it was demonstrated that there are some HIPVs and food plants that can selectively attract different natural enemy species (Baggen et al. 1999, Begum et al. 2006, Lavandero et al. 2006, Simpson et al. 2011a).

4. Constraints and Challenges of Semiochemical Applications for Biological Control of Crop Pests

As has been discussed in the sections above, the use of semiochemicals to monitor pest density in the field or to disrupt their mating efficiency using pheromones, or to attract natural enemies to control them, has opened up new opportunities leading to efficient, environmentally sustainable pest control strategies. However, the methods and techniques that were – and are still being – developed for this are not necessarily easy and such pest control strategies have always been a subject to controversy regarding the potential limits and risks of using these compounds in the field (Turlings and Ton 2006, Hilker and McNeil 2008, Kaplan 2012, Rodriguez-Saona et al. 2012, Meiners and Peri 2013). More accurately, developing a sound pest control strategy based on the use of semiochemical comes up against several constraints leading to different challenges that are not necessarily easy to solve. The main reason for this is that the use of semiochemicals to manipulate parasitoids' behaviour is done in environments that are, by definition, highly complex, as this has been discussed above (Schröder and Hilker 2008, Wäschke et al. 2013). The processes involved act at the individual, population and ecosystem levels simultaneously and are essentially limitless (Takken and Dicke 2006, Meinwald and Eisner 2008, Colazza and Wajnberg 2013) and, at each level, a handful of both intra- and inter-specific interactions are involved.

Several constraints have been identified and discussed. The most important one is based on the fact that the production and characterisation of the signal molecules used represent a long-lasting and then expensive research effort involving different, and most of the time difficult, steps. Processes involved in the emission and transmission mechanisms of these chemical compounds, the way they are detected by the recipient organisms and the associated neuroendocrine-mediated behavioural and/or physiological changes they trigger must all be accurately understood (Meinwald and Eisner 2008) no matter how these compounds are used in field conditions. It appears that, in order to make increasing strides in our understanding, this entire research effort cannot be done without involving both the genomic and proteomic tools that are now available (Kessler and Baldwin 2002), leading usually to additional costs and the need to solve several technical barriers (Colazza and Wajnberg 2013). Besides this, the consequences of using such chemical compounds in the field at the population level must also be studied, for example in terms of population dynamics, both in time and in space (Meinwald and Eisner 2008). In a nutshell, all of this means that strongly multidisciplinary cooperative partnerships must be developed, essentially between chemical ecologists and genomicists (Meinwald and Eisner 2008), but also involving population and behavioural ecologists (Colazza and Wajnberg 2013). Another constraint is based on the fact that most studies conducted so far have been done under laboratory conditions, which differ considerably from field conditions (Meiners and Peri 2013). The complexity of real field situations and/or the large variation due to unpredicted and fluctuating biotic and abiotic conditions might hamper the efficacy of the chemical compounds used to modify correctly parasitoids' behaviour, even if this has been clearly demonstrated in the laboratory. Hence, the most accurate way to use semiochemicals in different real situations still remains to be discovered in most of the cases.

As a consequence, several attempts at using semiochemicals to improve biological control of crops pest actually led to ineffective pest control or were sometimes even counterproductive (Meiners and Peri 2013). For example, although Roland et al. (1995) demonstrated that releasing borneol in apple orchards can increase the density of *Cyzenis albicans* (Diptera: Tachinidae), a parasitoid of the winter moth *Operophtera brumata* (Lepidoptera: Geometridae), the parasitism rate actually did not increase. One of the possible reasons of such failures might be that semiochemicals could actually decrease parasitoid foraging efficiency because they can attract them to locations without potential hosts to attack, hence leading them to waste time and energy and decreasing their overall pest-killing activity (Powell and Pickett 2003, Puente et al. 2008a). In addition, parasitoid females responding to semiochemicals without finding hosts to attack can rapidly change their foraging behaviour through mechanisms such as habituation (Peri et al. 2006, Abram et al. 2015, Peri et al. 2016), leading also to an overall decrease in their pest control ability.

Indirect negative effects of using semiochemicals with the goal to increase the efficacy of biological control programs can also be observed. Using a semiochemical locally can, for example, attract parasitoids females from remote locations, potentially leading to pest outbreaks in uncontrolled areas with dramatic demographic consequences at larger scales. Additionally, semiochemicals such as plant volatiles that are used to attract parasitoid females can sometimes also attract pests, enhancing overall herbivory levels, for example as shown by Meiners et al. (2005) for the elm leaf beetle *Xanthogaleruca luteola* (Coleoptera: Chrysomelidae). More subtle negative effects can also be observed, for example when semiochemicals attract the primary parasitoid species, which can potentially control the targeted hosts, but also attract hyperparasitoids that can kill the pests' natural enemies, leading to overall greater damage to the plants (Poelman et al. 2012, 2016; see above for a more detailed discussion).

Another challenge might happen when the goal is to use semiochemicals in organic crops, which are becoming more popular nowadays, in order to reduce the use of synthetic pesticides. By definition, with such crops the use of synthetic compounds is prohibited (Simpson et al. 2013). However, the majority of studies involving semiochemicals have been based on synthetic compounds that can be used within conventional agricultural systems only. Such constraints might induce the need to look for products specifically derived from natural sources rather than being synthetically manufactured (Simpson et al. 2013).

Finally, some ecological situations are actually more challenging than others. This is the case, for example, in forest ecosystems in which enhancing pest control by means of using semiochemicals faces very serious constraints. In such ecosystems, the critical pests to control are essentially episodic, complicating the decision about when the treatment must be applied. Additionally, forests are usually grown based on very long rotational harvest cycles (decades or even sometimes centuries) with usual small annual returns, hence limiting the implementation of high-cost pest control strategies. Furthermore, many forest ecosystems are spread over large areas with limited accessibility, further limiting potential pest management decisions (Paine 2013).

Thus, there are still important questions remaining to be solved before an efficient and stable increase in pest control efficacy of insect parasitoids can be achieved by means of using semiochemicals in field conditions

(Meinwald and Eisner 2008, Colazza and Wajnberg 2013). More research is definitely needed into the use of such chemical compounds for pest management (Kaplan 2012), particularly in converting the ever-expanding body of biological knowledge into practicable applications (Simpson et al. 2013). Although quite a lot is already known (see other sections above), the ecological and evolutionary meaning of chemical communication networks, for example, needs to be unraveled more accurately if we want to improve field application. The goal of the end on this section is to propose potential ideas to overcome the different constraints and challenges briefly listed above.

Several ideas have been discussed in the literature. For example, it has been proposed to produce crop varieties, by means of breeding selection or even genetic engineering, with an enhanced ability to emit chemical signals for attracting parasitoids (i.e., synomones) when they are attacked by herbivores (Bottrell et al. 1998), overall leading to a more efficient pest control strategy (see above). In addition, because insect parasitoids are known to change their behaviour according to previous experiences (i.e., learning ability, see above), it has been proposed on several occasions to prime them with semiochemicals prior to release (Hare et al. 1997) in order to enhance their search activity or to retain them in certain areas (Mills and Wajnberg 2008). Insect parasitoids could also potentially be improved by means of breeding selection (Wajnberg 2004), leading them to react more rapidly and more efficiently to semiochemicals when they are released in the field (see above).

There is, however, now a need to address this issue in a more global and formal way, because, as described above, the situation is essentially complex, potentially involving an important number of interacting species both in time and in space. In this respect, the idea here is to make a plea for the development of a modeling framework that is clearly needed if we really want to improve the efficacy of biological control agents to control crop pests with the use of semiochemicals (Mills and Kean 2010, Wajnberg et al. 2016). This is especially true in this case because insects are generally highly time-limited, are also able to modify their behaviour through learning and have to respond to cues that are essentially embedded in a highly complex array of other signals, leading them to experience both costs and benefits. Hence, understanding of the consequences for evolutionary and population dynamics of such features cannot be done without a sound and formal modeling approach (Mills and Wajnberg 2008, Wajnberg et al. 2016). Such modeling work will then provide precious information regarding, for example, how semiochemical compounds can be optimally used, how insect will react (evolve), etc.

Several theoretical works have been published on this topic. For example, Puente et al. (2008a) developed a deterministic model showing that the lag time between herbivory and the emission of plant volatiles in the tri-trophic system *Brassica oleracea–Pieris rapae–Cotesia rubecula* eliminates the benefit for the parasitoids to respond to such chemical compounds. In addition, in the same system, Puente et al. (2008b), using a simulation model, showed that parasitoids do not benefit from responding to plant volatile compounds when their emission continues for a long period of time. These two models demonstrated the importance of both temporal and spatial aspects, likely giving cues about how things should be optimally done. Developing a sound and accurate modeling framework to understand how semiochemicals must ideally be used in field conditions to enhance the efficacy of biological control programs against crop pest is certainly a challenging task. Such a scientific effort needs to be done urgently if we want to obtain efficient, stable and environmentally sustainable pest control strategies without the use of chemical pesticides.

5. Conclusion

Because of the increased use of pesticides worldwide in response to growing human populations and crop production, research on strategies to moderate the negative effects that existing pesticide practices have on human health and the environment is a global challenge in agriculture. One approach to overcome this challenge is to use applied chemical ecology. In recent years, semiochemical-based tactics have emerged as promising tools to enhance the efficacy of natural enemies in the biological control of crop pests. As a consequence, research on the manipulation of insect behaviour by using semiochemicals has increased greatly in the context of pest management. Examples of applications in crop ecosystems of allelochemical-based tactics to reduce insect pest populations by direct (e.g., use of synthetic HIPVs and/or HAVs) or indirect (e.g., use of phytohormonal elicitors or crop varieties, by breeding selection or genetic engineering, enhancing ability to emit attractive chemicals signals) recruitment of natural enemies have increased in number around the world, opening up new opportunities leading to efficient, environmentally sustainable pest control strategies. However, these techniques are often difficult to apply in the field and are subject of controversy regarding the potential limits and risks and the results in terms of enhancing parasitoid efficacy in the biological control of crop pests are sometimes inconsistent. The main reason is that the semiochemical-based tactics aim to manipulate parasitoids behaviour and, therefore, their outcomes can be conditioned by different factors such as, for example, the experience of the parasitoids or the lack of hosts. Moreover, such manipulation is done in highly complex environments, and as a consequence, can act simultaneously at different trophic levels, involving both intra- and inter-specific interactions. Additional comparative studies should be conducted to understand how to limit any negative effects and to investigate, e.g., the effect of landscape complexity on biological control.

For example, the synergy obtained by merging semiochemical-based manipulations and habitat manipulations could be promising, because parasitoids attracted to an allelochemical-treated crop might find some rewards (e.g., food sources such as nectar and pollen or alternative hosts), limiting the possible negative effects of parasitoid recruitment in the absence of hosts. Therefore, an interdisciplinary approach involving different research fields, such as chemical ecology, biological control, insect physiology and behaviour, plant ecology and plant breeding, is needed to open the "chemical ecological toolbox" for developing effective and successful biological control programs.

REFERENCES

Abram, P.K., A. Cusumano, E. Peri, J. Brodeur, G. Boivin and S. Colazza. 2015. Thermal stress affects patch time allocation by preventing forgetting in a parasitoid wasp. Behav. Ecol. 26: 1326–1334.

Åhman, I., G. Robert and N. Velemir. 2010. The potential for modifying plant volatile composition to enhance resistance to arthropod pests. CAB Reviews 5: 1–10.

Alim, M.A. and U.T. Lim. 2011. Refrigerated eggs of *Riptortus pedestris* (Hemiptera: Alydidae) added to aggregation pheromone traps increase field parasitism in soybean. J. Econ. Entomol. 104: 1833–1839.

Arakaki, N., H. Yamazawa and S. Wakamura. 2011. The egg parasitoid *Telenomus euproctidis* (Hymenoptera: Scelionidae) uses sex pheromone released by immobile female tussock moth *Orgyia postica* (Lepidoptera: Lymantriidae) as kairomone. Appl. Entomol. Zool. 46: 195–200.

Atanassov, A., P.W. Shearer and G.C. Hamilton. 2003. Peach pest management programs impact beneficial fauna abundance and *Grapholita molesta* (Lepidoptera: Tortricidae) egg parasitism and predation. Environ. Entomol. 32: 780–788.

Baggen, L.R., G.M. Gurr and A. Meats. 1999. Flowers in tri-trophic systems: mechanisms allowing selective exploitation by insect natural enemies for conservation biological control. Entomol. Exp. Appl. 91: 155–161.

Battaglia, D., G. Poppy, W. Powell, A. Romano, A. Tranfaglia and F. Pennacchio. 2000. Physical and chemical cues influencing the oviposition behaviour of *Aphidius ervi*. Entomol. Exp. Appl. 94: 219–227.

Bayoumy, M.H., B.A. Kaydan and F. Kozár. 2011. Are synthetic pheromone captures predictive of parasitoid densities as a kairomonal attracted tool? J. Entomol. Acarol. Res. 43: 23–31.

Begum, M., G.M. Gurr, S.D. Wratten, P.R. Hedberg and H.I. Nicol. 2006. Using selective food plants to maximise biological control of vineyard pests. J. Appl. Ecol. 43: 547–554.

Bekkaoui, A. and E. Thibout. 1993. Role of the cocoon of *Acrolepiopsis assectella* (Lep., Yponomeutoidae) in host recognition by the parasitoid *Diadromus pulchellus* (Hym., Ichneumonidae). Entomophaga 38: 101–113.

Bernal, J.S. and R.F. Luck. 2007. Mate finding via a trail sex pheromone by *Aphytis melinus* DeBach (Hymenoptera: Aphelinidae) males. J. Insect Behav. 20: 515–525.

Blassioli-Moraes, M.C., M. Borges and R.A. Laumann. 2013. The application of chemical cues in arthropod pest management for arable crops. pp. 225–244. *In*: E. Wajnberg and S. Colazza [eds.]. Chemical Ecology of Insect Parasitoids. Wiley-Blackwell, Oxford, UK.

Blassioli-Moraes, M.C., M. Borges, M.F.F. Michereff, D.A. Magalhães and R.A. Laumann. 2016. Semiochemicals from plants and insects on the foraging behaviour of Platygastridae egg parasitoids. Pesq. Agropec. Bras. 51: 454–464.

Blenn, B., M. Bandoly, A. Kueffner, T. Otte, S. Geiselhardt, N.E. Fatouros et al. 2012. Insect egg deposition induces indirect defense and epicuticular wax changes in *Arabidopsis thaliana*. J. Chem. Ecol. 38: 882–892.

Boo, K.S. and J.P. Yang. 2000. Kairomones used by *Trichogramma chilonis* to find *Helicoverpa assulta* eggs. J. Chem. Ecol. 26: 359–375.

Borges, M., F.G.V. Schmidt, E.R. Sujii, M.A. Medeiros, K. Mori, P.H.G. Zarbin et al. 1998. Field responses of stink bugs to the natural and synthetic pheromone of the Neotropical brown stink bug, *Euschistus heros* (Heteroptera: Pentatomidae). Physiol. Entomol. 23: 202–207.

Bottrell, D.G., P. Barbosa and F. Gould. 1998. Manipulating natural enemies by plant variety selection and modification: a realistic strategy? Annu. Rev. Entomol. 43: 347–367.

Chuche, J., A. Xuéreb and D. Thiéry. 2006. Attraction of *Dibrachys cavus* (Hymenoptera: Pteromalidae) to its host frass volatiles. J. Chem. Ecol. 32: 2721–2731.

Colazza, S., G. Aquila, C. De Pasquale, E. Peri and J. Millar. 2007. The egg parasitoid *Trissolcus basalis* uses n-nonadecane, a cuticular hydrocarbon from its stink bug host *Nezara viridula*, to discriminate between female and male hosts. J. Chem. Ecol. 33: 1405–1420.

Colazza, S., E. Peri and A. Cusumano. 2013. Application of chemical cues in arthropod pest management for orchards and vineyards. pp. 245–265. *In*: E. Wajnberg and S. Colazza [eds.]. Chemical Ecology of Insect Parasitoids. Wiley-Blackwell, Oxford, UK.

Colazza, S., E. Peri, G. Salerno and E. Conti. 2010. Host searching by egg parasitoids: exploitation of host chemical cues. pp. 97-147. *In*: F.L. Cônsoli, J.R.P. Parra and R.A. Zucchi [eds.]. Egg Parasitoids in Agroecosystems with Emphasis on *Trichogramma*. Springer Science and Business Media, Dordrecht, Netherlands.

Colazza, S., C.M. Rosi and A. Clemente. 1997. Response of egg parasitoid *Telenomus busseolae* to sex pheromone of *Sesamia nonagrioides*. J. Chem. Ecol. 23: 2437–2444.

Colazza, S., G. Salerno and E. Wajnberg. 1999. Volatile and contact chemicals released by *Nezara viridula* (Heteroptera: Pentatomidae) have a kairomonal effect on the egg parasitoid *Trissolcus basalis* (Hymenoptera: Scelionidae). Biol. Control 16: 310–317.

Colazza, S. and E. Wajnberg. 2013. Chemical ecology of insect parasitoids: towards a new era. pp. 1–8. *In*: E. Wajnberg and S. Colazza [eds.]. Chemical Ecology of Insect Parasitoids. Wiley-Blackwell, Oxford, UK.

Conti, E. and S. Colazza. 2012. Chemical ecology of egg parasitoids associated with true bugs. Psyche 2012: 1–11.

Conti, E., G. Salerno, B. Leombruni, F. Frati and F. Bin. 2010. Short-range allelochemicals from a plant-herbivore association: a singular case of oviposition-induced synomone for an egg parasitoid. J. Exp. Biol. 213: 3911–3919.

Cusumano, A., B.T. Weldegergis, S. Colazza, M. Dicke and N.E. Fatouros. 2015. Attraction of egg-killing parasitoids toward induced plant volatiles in a multi-herbivore context. Oecologia 179: 163–174.

De Rijk, M., M. Dicke and E.H. Poelman. 2013. Foraging behaviour by parasitoids in multiherbivore communities. Anim. Behav. 85: 1517–1528.

Dewhirst, S.Y., M.A. Birkett, E. Loza-Reyes, J.L. Martin, B.J. Pye, L.E. Smart et al. 2012. Activation of defence in sweet pepper, *Capsicum annuum*, by *cis*-jasmone, and its impact on aphid and aphid parasitoid behaviour. Pest Manag. Sci. 68: 1419–1429.

Dicke, M. and I.T. Baldwin. 2010. The evolutionary context for herbivore-induced plant volatiles: beyond the 'cry for help'. Trend. Plant Sci. 15: 167–175.

Dicke, M. and M.W. Sabelis. 1988. Infochemical terminology: based on cost benefit analysis rather than origin of compounds? Funct. Ecol. 2: 131–139.

Dicke, M., J.J.A. van Loon and R. Soler. 2009. Chemical complexity of volatiles from plants induced by multiple attacks. Nature Chem. Biol. 5: 317–324.

Fatouros, N.E., G. Bukovinszkine'Kiss, L.A. Kalkers, G.R. Soler, M. Dicke and M. Hilker. 2005. Oviposition-induced plant cues: do they arrest *Trichogramma* wasps during host location? Entomol. Exp. Appl. 115: 207–215.

Fatouros, N.E., A. Cusumano, E.G. Danchin and S. Colazza. 2016. Prospects of herbivore egg-killing plant defenses for sustainable crop protection. Ecol. Evol. 6: 6906–6918.

Fatouros, N.E., M. Dicke, R. Mumm, T. Meiners and M. Hilker. 2008. Foraging behaviour of egg parasitoids exploiting chemical information. Behav. Ecol. 19: 677–689.

Fatouros, N.E., F.G. Pashalidou, W.V.A. Cordero, J.J.A. van Loon, R. Mumm, M. Dicke et al. 2009. Anti-aphrodisiac compounds of male butterflies increase the risk of egg parasitoid attack by inducing plant synomone production. J. Chem. Ecol. 35: 1373–1381.

Foti, M.C., M. Rostás, E. Peri, K.C. Park, T. Slimani, S.D. Wratten and S. Colazza. 2017. Chemical ecology meets conservation biological control: identifying plant volatiles as predictors of floral resource suitability for an egg parasitoid of stink bugs. J. Pest Sci. 90: 299–310.

Francis, F., S. Vandermoten, F. Verheggen, G. Lognay and E. Haubruge. 2005. Is the (E)-ß-farnesene only volatile terpenoid in aphids? J. Appl. Entomol. 129: 6–11.

Franco, J.C., E.B. Silva, E. Cortegano, L. Campos, M. Branco, A. Zada et al. 2008. Kairomonal response of the parasitoid *Anagyrus* spec. *nov* near *pseudococci* to the sex pheromone of the vine mealybug. Entomol. Exp. Appl. 126: 122–130.

Franco, J.C., E.B. Silva, T. Fortuna, E. Cortegano, M. Branco, P. Suma et al. 2011. Vine mealybug sex pheromone increases citrus mealybug parasitism by *Anagyrus* sp. near *pseudococci* (Girault). Biol. Control 58: 230–238.

Geiselhardt, S.F., S. Geiselhardt and K. Peschke. 2011. Congruence of epicuticular hydrocarbons and tarsal secretions as a principle in beetles. Chemoecology 21: 181–186.

Glinwood, R.T., W. Powell and C.P.M. Tripathi. 1998. Increased parasitisation of aphids on trap plants alongside vials releasing synthetic aphid sex pheromone and effective range of the pheromone. Biocontrol Sci. Technol. 8: 607–614.

Godfray, H.C.J. 1994. Parasitoids - Behavioural and Evolutionary Ecology. Princeton University Press, Princeton, NJ, USA.

Gols, R. 2014. Direct and indirect chemical defences against insects in a multitrophic framework. Plant Cell Environ. 37: 1741–1752.

Guerrieri, E. 2016. Who's listening to talking plants? pp. 117–136. *In*: J.D. Blande and R. Glinwood [eds.]. Deciphering Chemical Language of Plant Communication, Signaling and Communication in Plants. Springer International Publishing, Gewerbestrasse, Switzerland.

Guillot, F.S. and S.B. Vinson. 1972. Sources of substances which elicit a behavioural response from insect parasitoid, *Campoletis perdistinctus*. Nature 235: 169–170.

Gurr, G.M., S.D. Wratten and M.A. Altieri. 2004. Ecological engineering: a new direction for agricultural pest management. Aust. Farm Bus. Manag. J. 1: 28–35.

Gurr, G.M. and M. You. 2016. Conservation biological control of pests in the molecular era: new opportunities to address old constraints. Front. Plant Sci. 6: 1255.

Hardie, J., S.F. Nottingham, W. Powell and L.J. Wadhams. 1991. Synthetic aphid sex pheromone lures female parasitoids. Entomol. Exp. Appl. 61: 97–99.

Hare, J.D., D.J.W. Morgan and T. Nguyun. 1997. Increased parasitisation of California red scale in the field after exposing its parasitoid, *Aphytis melinus*, to a synthetic kairomone. Entomol. Exp. Appl. 82: 73–81.

Hardy, I.C.W. and M. Goubault. 2007. Wasp fights: understanding and utilising agonistic bethylid behaviour. Biocontrol News Inform. 28: 11–15.

Heil, M. 2014. Herbivore-induced plant volatiles: targets, perception and unanswered questions. New Phytol. 204: 297–306.

Hilker, M. and N.E. Fatouros. 2015. Plant responses to insect egg deposition. Annu. Rev. Entomol. 60: 493–515.

Hilker, M. and J. McNeil. 2008. Chemical and behavioural ecology in insect parasitoids: how to behave optimally in a complex odourous environment? pp. 92–112. *In*: E. Wajnberg, C. Bernstein and J.J.M. van Alphen [eds.]. Behavioural Ecology of Insect Parasitoids: From Theoretical Approaches to Field Applications. Blackwell, Oxford, UK.

Hoffmeister, T.S., B.D. Roitberg and G. Lalonde. 2000. Catching Ariadne by the thread: how a parasitoid exploits the herbivores's marking trails to locate its host. Entomol. Exp. Appl. 95: 77–85.

Huigens, M.E. and N.E. Fatouros. 2013. A hitchhiker's guide to parasitism: chemical ecology of phoretic insect parasitoids. pp. 86–111. *In*: E. Wajnberg and

S. Colazza [eds.]. Chemical Ecology of Insect Parasitoids. Wiley-Blackwell, Oxford, UK.

Huigens, M.E., F.G. Pashalidou, M.H. Qian, T. Bukovinszky, H.M. Smid, J.J.A. van Loon et al. 2009. Hitch-hiking parasitic wasp learns to exploit butterfly antiaphrodisiac. Proc. Natl. Acad. Sci. USA 106: 820–825.

Irvin, N.A., S.L. Scarratt, S.D. Wratten, C.M. Frampton, R.B. Chapman and J.M. Tylianakis. 2006. The effects of floral understoreys on parasitism of leafrollers (Lepidoptera: Tortricidae) on apples in New Zealand. Agric. Forest Entomol. 8: 25–34.

Jaloux, B., C. Errard, N. Mondy, F. Vannier and J.P. Monge. 2005. Sources of chemical signals which enhance multiparasitism preference by a cleptoparasitoid. J. Chem. Ecol. 31: 1325–1337.

James, D.G. and T.R. Grasswitz. 2005. Synthetic herbivore-induced plant volatiles increase field captures of parasitic wasps. BioControl 50: 871–880.

Jang, S.A., J.H. Cho, G.M. Park, H.Y. Choo and C.G. Park. 2011. Attraction of *Gymnosoma rotundatum* (Diptera: Tachinidae) to different amounts of *Plautia stali* (Hemiptera: Pentatomidae) aggregation pheromone and the effect of different pheromone dispensers. J. Asia Pac. Entomol. 14: 119–121.

Jang, S.A. and C.G. Park. 2010. *Gymnosoma rotundatum* (Diptera: Tachinidae) attracted to the aggregation pheromone of *Plautia stali* (Hemiptera: Pentatomidae). J. Asia Pac. Entomol.13: 73–75.

Jonsson, M., S. Wratten, K.A. Robinson and S. Sam. 2009. The impact of floral resources and omnivory on a four trophic level food web. Bull. Entomol. Res. 99: 275–285.

Kaiser, L., P. Ode, S. van Nouhuys, P.-A. Calatayud, S. Colazza, A.-M. Cortesero et al. 2016. The plant as a habitat for entomophagous insects. Adv. Bot. Res. 81: 179–223.

Kaplan, I. 2012. Attracting carnivorous arthropods with plant volatiles: the future of biocontrol or playing with fire? Biol. Control 60: 77–89.

Kessler, A. and I.T. Baldwin. 2002. Plant responses to insect herbivory: the emerging molecular analysis. Annu. Rev. Plant Biol. 53: 299–328.

Kessler, A. and M. Heil. 2011. Evolutionary ecology of plant defences. The multiple faces of indirect defences and their agents of natural selection. Funct. Ecol. 25: 348–357.

Khan, Z.R., C.A.O. Midega, E.M. Njuguna, D.M. Amudavi, J.M. Wanyama and J.A. Pickett. 2008. Economic performance of 'push–pull' technology for stemborer and *Striga* control in smallholder farming systems in western Kenya. Crop Prot. 27: 1084–1097.

Kos, M., J.J.A. van Loon, M. Dicke and L.E.M. Vet. 2009. Transgenic plants as vital components of integrated pest management. Trend. Biotechnol. 27: 621–627.

Krupke, C.H. and J.F. Brunner. 2003. Parasitoids of the consperse stink bug (Hemiptera: Pentatomidae) in North Central Washington and attractiveness of a host-produced pheromone component. J. Entomol. Sci. 38: 84–92.

Lavandero, I.B., S.D. Wratten, R.K. Didham and G. Gurr. 2006. Increasing floral diversity for selective enhancement of biological control agents: a double-edged sward? Basic Appl. Ecol. 7: 236–243.

Lee, J.C. and G.E. Heimpel. 2005. Impact of flowering buckwheat on lepidopteran cabbage pests and their parasitoids at two spatial scales. Biol. Control 34: 290–301.

Li, G. 2006. Host-marking in hymenopterous parasitoids. Acta Entomol. Sinica 49: 504–512.

Lim, U.T. and B.P. Mainali. 2013. Effect of aggregation pheromone trap of *Riptortus pedestris* (Hemiptera: Alydidae) on the distribution and composition of its egg parasitoids. J. Econ. Entomol. 106:1973–1978.

Lo Giudice, D., M. Riedel, M. Rostás, E. Peri and S. Colazza. 2011. Host sex discrimination by an egg parasitoid on brassica leaves. J. Chem. Ecol. 37: 622–628.

Lou, Y.G., M.H. Du, T.C. Turlings, J.A. Cheng and W.F. Shan. 2005. Exogenous application of jasmonic acid induces volatile emissions in rice and enhances parasitism of *Nilaparvata lugens* eggs by the parasitoid *Anagrus nilaparvatae*. J. Chem. Ecol. 31: 1985–2002.

Mainali, B.P. and U.T. Lim. 2012. Annual pattern of occurrence of *Riptortus pedestris* (Hemiptera: Alydidae) and its egg parasitoids *Ooencyrtus nezarae* Ishii and *Gryon japonicum* (Ashmead) in Andong, Korea. Crop Prot. 36: 37–42.

Mansour, R., P. Suma, G. Mazzeo, E. Buonocore, G.K. Lebdi and A. Russo. 2011. Using a kairomone-based attracting system to enhance biological control of mealybugs (Hemiptera: Pseudococcidae) by *Anagyrus* sp. near *pseudococci* (Hymenoptera: Encyrtidae) in Sicilian vineyards. J. Entomol. Acarol. Res. 42: 161–170.

Martorana, L., M.C. Foti, G. Rondoni, E. Conti, S. Colazza and E. Peri. 2017. An invasive insect herbivore disrupts plant volatile-mediated tritrophic signalling. J. Pest Sci. doi:10.1007/s10340-017-0877-5.

Mattiacci, L. and M. Dicke. 1995. The parasitoid *Cotesia glomerata* (Hymenoptera: Braconidae) discriminates between first and fifth larval instars of its host *Pieris brassicae*, on the basis of contact cues from frass, silk, and herbivore-damaged leaf tissue. J. Insect Behav. 8: 485–498.

Mattiacci, L., S.B. Vinson and H.J. Williams. 1993. A long-range attractant kairomone for egg parasitoid *Trissolcus basalis*, isolated from defensive secretion of its host, *Nezara viridula*. J. Chem. Ecol. 19: 1167–1181.

McCormick, A.C., S.B. Unsicker and J. Gershenzon. 2012. The specificity of herbivore-induced plant volatiles in attracting herbivore enemies. Trend. Plant Sci. 17: 303–310.

Mehrnejad, M.R. and M.J.W. Copland. 2006. Behavioural responses of the parasitoid *Psyllaephagus pistaciae* (Hymenoptera: Encyrtidae) to host plant volatiles and honeydew. Entomol. Sci. 9: 31–37.

Meiners, T., N. Hacker, P. Anderson and M. Hilker. 2005. Response of the elm leaf beetle to host plants induced by oviposition and feeding: the infestation rate matters. Entomol. Exp. Appl. 115: 171–177.

Meiners, T. and E. Peri. 2013. Chemical ecology of insect parasitoids: essential elements for developing effective biological control programmes. pp. 193–224. *In*: E. Wajnberg and S. Colazza [eds.]. Chemical Ecology of Insect Parasitoids. Wiley-Blackwell, Oxford, UK.

Meinwald, J. and T. Eisner. 2008. Chemical ecology in retrospect. Proc. Natl. Acad. Sci. USA 105: 4539–4540.

Michereff, M.F.F., M. Michereff-Filho, M.C. Blassioli-Moraes, R.A. Laumann, I.R. Diniz and M. Borges. 2015. Effect of resistant and susceptible soybean cultivars on the attraction of egg parasitoids under field conditions. J. Appl. Entomol. 139: 207–216.

Mills, N.J. and J.M. Kean. 2010. Behavioural studies, molecular approaches, and modeling: Methodological contributions to biological control success. Biol. Control 52: 255–262.

Mills, N.J. and E. Wajnberg. 2008. Optimal foraging behaviour and efficient biological control methods. pp. 3–30. *In*: E. Wajnberg, C. Bernstein and J. van Alphen [eds.]. Behavioural Ecology of Insect Parasitoids: From Theoretical Approaches to Field Application. Blackwell, Oxford, UK.

Mitchell, C., R.M. Brennan, J. Graham and A.J. Karley. 2016. Plant defense against herbivorous pests: Exploiting resistance and tolerance traits for sustainable crop protection. Front. Plant Sci. 7: 1132.

Mizutani, N. 2006. Pheromones of male stink bugs and their attractiveness to their parasitoids. Jpn. J. Appl. Entomol. Zool. 50: 87–99.

Moraes, M.C.B., R.A. Laumann, M. Pareja, F.T.P.S. Sereno, M.F. Michereff, M.A. Birkett et al. 2009. Attraction of the stink bug egg parasitoid, *Telenomus podisi* to defence signals from soybean activated by treatment with *cis*-jasmone. Entomol. Exp. Appl. 131: 178–188.

Moujahed, R., F. Frati, A. Cusumano, G. Salerno, E. Conti, E. Peri et al. 2014. Egg parasitoid attraction toward induced plant volatiles is disrupted by a non-host herbivore attacking above or belowground plant organs. Front. Plant Sci. 5: 601.

Mumm, R. and M. Dicke. 2010. Variation in natural plant products and the attraction of bodyguards involved in direct plant defense. Can. J. Zool. 88: 628–667.

Müller, C. and M. Riederer. 2005. Plant surface properties in chemical ecology. J. Chem. Ecol. 31: 2621–2651.

Nakamura, S., R.T. Ichiki and Y. Kainoh. 2013. Chemical ecology of tachinid parasitoids. pp. 145–167. *In*: E. Wajnberg and S. Colazza [eds.]. Chemical Ecology of Insect Parasitoids. Wiley-Blackwell, Oxford, UK.

Noldus, L.P.J.J. 1989. Semiochemicals, foraging behaviour and quality of entomophagous insects for biological control. J. Appl. Entomol. 108: 425–451.

Obonyo, M., F. Schulthess, B. Le Ru, J. van den Berg, J.F. Silvain and P.A. Calatayud. 2010. Importance of contact chemical cues in host recognition and acceptance by the braconid larval endoparasitoids *Cotesia sesamiae* and *Cotesia flavipes*. Biol. Control 54: 270–275.

Ode, P.J. 2013. Plant defences and parasitoid chemical ecology. pp. 11–36. *In*: E. Wajnberg and S. Colazza [eds.]. Chemical Ecology of Insect Parasitoids. Wiley-Blackwell, Oxford, UK.

Orre, G.U.S., S.D. Wratten, M. Jonsson and R.J. Hale. 2010. Effects of an herbivore-induced plant volatile on arthropods from three trophic levels in brassicas. Biol. Control 53: 62–67.

Orre Gordon, G.U.S., S.D. Wratten, M. Jonsson, M. Simpson and R. Hale. 2013. 'Attract and reward': combining a herbivore-induced plant volatile with floral resource supplementation - multi-trophic level effects. Biol. Control 64: 106–115.

Ozawa, R., K. Shiojiri, M.W. Sabelis, G. Arimura, T. Nishioka and J. Takabayashi. 2004. Corn plants treated with jasmonic acid attract more specialist parasitoids, thereby increasing parasitisation of the common armyworm. J. Chem. Ecol. 30: 1797–1808.

Paine, T.D. 2013. Application of chemical cues in arthropod pest management for forest trees. pp. 282–295. *In*: E. Wajnberg and S. Colazza [eds.]. Chemical Ecology of Insect Parasitoids. Wiley-Blackwell, Oxford, UK.

Park, G.M., S.A. Jang, S.H. Choi and C.G. Park. 2010. Attraction of *Plautia stali* (Hemiptera: Pentatomidae) to different amounts and dispensers of its aggregation pheromone. Korean J. Appl. Entomol. 49: 123–127.

Pashalidou, F.G., M.E. Huigens, M. Dicke and N.E. Fatouros. 2010. The use of oviposition-induced plant cues by *Trichogramma* egg parasitoids. Ecol. Entomol. 35: 748–753.

Peri, E., F. Frati, G. Salerno, E. Conti and S. Colazza. 2013. Host chemical footprints induce host sex discrimination ability in egg parasitoids. PLoS ONE 8: e79054.

Peri, E., G. Salerno, T. Slimani, F. Frati, E. Conti, S. Colazza et al. 2016. The response of an egg parasitoid to substrate-borne semiochemicals is affected by previous experience. Sci. Rep. 6: 27098.

Peri, E., M.A. Sole, E. Wajnberg and S. Colazza. 2006. Effect of host kairomones and oviposition experience on the arrestment behaviour of an egg parasitoid. J. Exp. Biol. 209: 3629–3635.

Pickett, J.A., L.J. Wadhams and C.M. Woodcock. 1997. Developing sustainable pest control from chemical ecology. Agric. Ecosyst. Environ. 64: 149–156.

Pickett, J.A. and Z.R. Khan. 2016. Plant volatile-mediated signalling and its application in agriculture: successes and challenges. New Phytol. 212: 856–870.

Poelman, E.H., M. Bruinsma, F. Zhu, B.T. Weldegergis, A.E. Boursault, Y. Jongema et al. 2012. Hyperparasitoids use herbivore-induced plant volatiles to locate their parasitoid host. PLoS Biol. 10: e1001435.

Poelman, E.H. and M. Kos. 2016. Complexity of plant volatile-mediated interactions beyond the third trophic level. pp. 211–225. *In*: J.D. Blande and R. Glinwood [eds.]. Deciphering Chemical Language of Plant Communication. Springer International Publishing, Gewerbestrasse, Switzerland.

Ponzio, C., P. Cascone, A. Cusumano, B.T. Weldegergis, N.E. Fatouros, E. Guerrieri et al. 2016. Volatile-mediated foraging behaviour of three parasitoid species under conditions of dual insect herbivore attack. Anim. Behav. 111: 197–206.

Powell, W. 1999. Parasitoid hosts. pp. 405–427. *In*: J. Hardie and A.K. Minks [eds.]. Pheromones of Non-Lepidopteran Insects Associated with Agricultural Plants. CABI Publishing, Wallingford, UK.

Powell, W., J. Hardie, A.J. Hick, C. Holler, J. Mann, L. Merritt et al. 1993. Responses of the parasitoid *Praon volucre* (Hymenoptera, Braconidae) to aphid sex-pheromone lures in cereal fields in autumn: implications for parasitoid manipulation. Eur. J. Entomol. 90: 435–438.

Powell, W. and J.A. Pickett. 2003. Manipulation of parasitoids for aphid pest management: progress and prospects. Pest Manag. Sci. 59: 149–155.

Puente, M.E., G.G. Kennedy and F. Gould. 2008a. The impact of herbivore-induced plant volatiles on parasitoid foraging success: a general deterministic model. J. Chem. Ecol. 34: 945–958.

Puente, M., K. Magori, G.G. Kennedy and F. Gould. 2008b. Impact of herbivore-induced plant volatiles on parasitoid foraging success: a spatial simulation of the *Cotesia rubecula, Pieris rapae,* and *Brassica oleracea* system. J. Chem. Ecol. 34: 959–970.

Quicke, D.L.J. 1997. Parasitic Wasps. Chapman and Hall, London, UK.

Reisenman, C.E., H. Lei and P.G. Guerenstein. 2016. Neuroethology of olfactory-guided behaviour and its potential application in the control of harmful insects. Front. Physiol. 7: 271.

Rodriguez-Saona, C., B.R. Blaauw and R. Isaacs. 2012. Manipulation of natural enemies in agroecosystems: habitat and semiochemicals for sustainable insect pest control. pp. 89–126. *In*: M.L. Larramendy and S. Soloneski [eds.]. Integrated Pest Management and Pest Control: Current and Future Tactics. InTech, Rijeka, Croatia.

Rohwer, C.L. and J.E. Erwin. 2008. Horticultural applications of jasmonates: a review. J. Hortic. Sci. Biotechnol. 83: 283–304.

Roland, J., K.E. Denford and L. Jiminez. 1995. Borneol as an attractant for *Cyzenis albicans*, a tachinid parasitoid of the winter moth, *Operophtera brumata* L. (Lepidoptera, Geometridae). Can. Entomol. 127: 413–421.

Rostás, M. and T.C.J. Turlings. 2008. Induction of systemic acquired resistance in *Zea mays* also enhances the plant's attractiveness to parasitoids. Biol. Control 46: 178–186.

Rostás, M. and M. Wölfling. 2009. Caterpillar footprints as host location kairomones for *Cotesia marginiventris*: persistence and chemical nature. J. Chem. Ecol. 35: 20–27.

Ruther, J. 2013. Novel insights into pheromone-mediated communication in parasitic hymenopterans. pp. 112–144. *In*: E. Wajnberg and S. Colazza [eds.]. Chemical Ecology of Insect Parasitoids. Wiley-Blackwell, Oxford, UK.

Salerno, G., F. De Santis, A. Iacovone, F. Bin and E. Conti. 2013. Short-range cues mediate parasitoid searching behaviour on maize: the role of oviposition-induced plant synomones. Biol. Control 64: 247–254.

Salerno, G., F. Frati, G. Marino, L. Ederli, S. Pasqualini, F. Loreto et al. 2017. Effects of water stress on emission of volatile organic compounds by *Vicia faba*, and consequences for attraction of the egg parasitoid *Trissolcus basalis*. J. Pest Sci. 90: 635–647.

Schnee, C., T.G. Köllner, M. Held, T.C.J. Turlings, J. Gershenzon and J. Degenhardt. 2006. The products of a single maize sesquiterpene synthase form a volatile defense signal that attracts natural enemies of maize herbivores. Proc. Natl. Acad. Sci. USA 103: 1129–1134.

Schröder, R. and M. Hilker.2008. The relevance of background odour in resource location by insects: a behavioural approach. BioScience 58: 308–316.

Simpson, M., G.M. Gurr, A.T. Simmons, S.D. Wratten, D.G. James, G. Leeson et al. 2011a. Insect attraction to synthetic herbivore-induced plant volatile treated field crops. Agric. Forest Entomol. 13: 45–57.

Simpson, M., G.M. Gurr, A.T. Simmons, S.D. Wratten, D.G. James, G. Leeson et al. 2011b. Field evaluation of the 'attract and reward' biological control approach in vineyards. Ann. Appl. Biol. 159: 69–78.

Simpson, M., G.M. Gurr, A.T. Simmons, S.D. Wratten, D.G. James, G. Leeson et al. 2011c. Attract and reward: combining chemical ecology and habitat manipulation to enhance biological control in field crops. J. Appl. Ecol. 48: 580–590.

Simpson, M., D.M.Y. Read and G.M. Gurr. 2013. Application of chemical cues in arthropod pest management for organic crops. pp 266–281. *In*: E. Wajnberg and S. Colazza [eds.]. Chemical Ecology of Insect Parasitoids. Wiley-Blackwell, Oxford, UK.

Sobhy, I.S., M. Erb, A.A. Sarhan, M.M. El-Husseini, N.S. Mandour and T.C.J. Turlings. 2012. Less is more: treatment with BTH and laminarin reduces herbivore-induced volatile emissions in maize but increases parasitoid attraction. J. Chem. Ecol. 38: 348–360.

Sobhy, I.S., M. Erb, Y. Lou and T.C.J. Turlings. 2014. The prospect of applying chemical elicitors and plant strengtheners to enhance the biological control of crop pests. Phil. Trans. Roy. Soc. B 369: 20120283.

Soler, R., T.M. Bezemer and J.A. Harvey. 2013. Chemical ecology of insect parasitoids in a multitrophic above and belowground context. pp 64–85. *In*: E. Wajnberg and S. Colazza [eds.]. Chemical Ecology of Insect Parasitoids. Wiley-Blackwell, Oxford, UK.

Stam, J.M., A. Kroes, Y. Li, R. Gols, J.J.A. van Loon, E.H.I. Poelman et al. 2014. Plant interactions with multiple insect herbivores: from community to genes. Annu. Rev. Plant Biol. 65: 689–713.

Steiner, L.F. 1952. Methyl eugenol as an attractant for oriental fruit fly. J. Econ. Entomol. 45: 241–248.

Stenberg, J.A., M. Heil, I. Ahman and C. Bjorkman. 2015. Optimising crops for biocontrol of pests and disease. Trend. Plant Sci. 20: 698–712.

Suckling, D.M. and G. Karg. 2000. Pheromones and semiochemicals. pp. 63–99. *In*: J. Rechcigl and N. Rechcigl [eds.]. Biological and Biotechnical Control of Insect Pests. CRC Press, Boca Raton, FL, USA.

Suckling, D.M., A.R. Gibb, G.M. Burnip and N.C. Delury. 2002. Can parasitoid sex pheromones help in insect biocontrol? A case study of codling moth (Lepidoptera: Tortricidae) and its parasitoid *Ascogaster quadridentata* (Hymenoptera: Braconidae). Environ. Entomol. 31: 947–952.

Takabayashi, J. 2006. Role of the lipoxygenase/lyase pathway of host-food plants in the host searching behaviour of two parasitoid species, *Cotesia glomerata* and *Cotesia plutellae*. J. Chem. Ecol. 32: 969–979.

Takken, W. and M. Dicke. 2006. Chemical ecology: a multidisciplinary approach. pp. 1–8. *In*: M. Dicke and W. Takken [eds.]. Chemical Ecology: From Gene to Ecosystem. Springer, Dordrecht, Netherlands.

Tamiru, A., Z.R. Khan and T.J.A. Bruce. 2015. New direction for improving resistance to insects by breeding for egg induced defense. J. Insect Sci. 9: 51–55.

Teshiba, M. and J. Tabata. 2017. Suppression of population growth of the Japanese mealybug, *Planococcus kraunhiae* (Hemiptera: Pseudococcidae), by using an attractant for indigenous parasitoids in persimmon orchards. Appl. Entomol. Zool. 52: 153–158.

Teshiba, M., H. Sugie, T. Tsutsumi and J. Tabata. 2012. A new approach for mealybug management: recruiting an indigenous, but 'non-natural' enemy for biological control using an attractant. Entomol. Exp. Appl. 142: 211–215.

Thaler, J.S., M.A. Farag, P.W. Paré and M. Dicke. 2002. Jasmonate-deficient plants have reduced direct and indirect defences against herbivores. Ecol. Lett. 5: 764–774.

Turlings, T.C.J. and J. Ton. 2006. Exploiting scents of distress: the prospect of manipulating herbivore-induced plant odours to enhance the control of agricultural pests. Curr. Opin. Plant Biol. 9: 421–427.

Uefune, M., Y. Choh, J. Abe, K. Shiojiri, K. Sano and J. Takabayashi. 2012. Application of synthetic herbivore-induced plant volatiles causes increased parasitism of herbivores in the field. J. Appl. Entomol. 136: 561–567.

van Dam, N.M. and M. Heil. 2011. Multitrophic interactions below and above ground: en route to the next level. J. Ecol. 99: 77–88.

van Poecke, R.M.P. and M. Dicke. 2002. Induced parasitoid attraction by *Arabidopsis thaliana*: involvement of the octadecanoid and the salicylic acid pathway. J. Exp. Bot. 53: 1793–1799.

Vet, L.E.M. and M. Dicke. 1992. Ecology of infochemical use by natural enemies in a tritrophic context. Annu. Rev. Entomol. 37: 141–172.

Vieira, C.R, M.C.B. Moraes, M. Borges, E.R. Sujii and R.A. Laumann. 2013. cis-Jasmone indirect action on egg parasitoids (Hymenoptera: Scelionidae) and its application in biological control of soybean stink bugs (Hemiptera: Pentatomidae). Biol. Control 64: 75–82.

Von Mérey, G., N. Veyrat, G. Mahuku, R.L. Valdez. T.C.J. Turlings and M. D'Alessandro. 2011. Dispensing synthetic green leaf volatiles in maize fields increases the release of sesquiterpenes by the plants, but has little effect on the attraction of pest and beneficial insects. Phytochemistry 72: 1838–1847.

Voytas, D.F. and C. Gao. 2014. Precision genome engineering and agriculture: opportunities and regulatory challenges. PLoS Biol. 12: e1001877.

Wajnberg, E. 2004. Measuring genetic variation in natural enemies used for biological control: Why and how? pp. 19–37. *In*: L. Ehler, R. Sforza and Th. Mateille [eds.]. Genetics, Evolution and Biological Control. CAB International, Wallingford, UK.

Wajnberg, E., B.D. Roitberg and G. Boivin. 2016. Using optimality models to improve the efficacy of parasitoids in biological control programmes. Entomol. Exp. Appl. 158: 2–16.

Wäschke, N., T. Meiners and M. Rostás. 2013. Foraging strategies of parasitoids in complex chemical environments. pp. 37–63. *In*: E. Wajnberg and S. Colazza [eds.]. Chemical Ecology of Insect Parasitoids. Wiley-Blackwell, Oxford, UK.

Witzgall, P., P. Kirsch and A. Cork. 2010. Sex pheromones and their impact on pest management. J. Chem. Ecol. 36: 80–100.

Wölfling, M. and M. Rostás. 2009. Parasitoids use chemical footprints to track down caterpillars. Commun. Integr. Biol. 2: 353–355.

CHAPTER

11

Challenges in Chemical Ecology for the Management of Vector-borne Diseases of Humans and Livestock

Jun Tabata

National Agriculture and Food Research Organization,
3-1-3 Kannondai, Tsukuba City, Ibaraki 305-8604, Japan
E-mail: jtabata@affrc.go.jp

Abstract

It is no exaggeration to say that human history is a fight against infectious diseases. Many arthropod-borne infectious diseases such as malaria and trypanosomiasis still affect people today. Furthermore, emerging infectious diseases that were previously neglected or unknown have been reported frequently, and they are spreading across various regions. Because we currently have no effective vaccine for most of these diseases, the most reliable method for prevention is controlling vector arthropods. In order to establish such pest management strategies, chemical ecological studies would be valuable. In this chapter, studies on representative vector arthropods and semiochemicals that are essential for their lifecycle are reviewed to examine the promotions of applications in chemical ecology.

1. Introduction

Infectious diseases, in particular arthropod-vectored diseases of humans and livestock, have recently attracted considerable attention from both scientists and the public. For example, Zika fever caused by *Zika virus* (ZIKV) emerged on Yap Island, in the Federated States of Micronesia, in 2007 (Duffy et al. 2009), and has subsequently become widespread across the Pacific and the Americas (e.g. Musso et al. 2014, Guerbois et al. 2016, Hennessey et al. 2016). This mosquito-borne flavivirus was first isolated and identified from a sentinel rhesus monkey inhabiting the Zika Forest near Lake Victoria, Uganda, in 1947 (Dick et al. 1952), and it has been reported in Africa and Asia since at least the 1950s (Haddow et al. 2012), but only 14 cases of confirmed human infections were reported from Africa and Southeast Asia before the recent outbreak (Duffy et al. 2009). A study of the nucleotide sequences of the open reading frame of ZIKV revealed the existence of two main virus lineages (African and Asian) and demonstrated that the strain responsible for the Yap epidemic case most likely originated from Southeast Asia (Haddow et al. 2012).

In the summer of 2014, infections from another mosquito-borne flavivirus, *Dengue virus* (DENV), were reported in approximately 160 patients in Japan, where no domestic DENV infections had been found since the early 1940s (Miura et al. 2015). DENV is transmitted by *Aedes* mosquitoes, mainly *A. aegypti*, which is widespread in tropical and subtropical regions but is currently not distributed in Japan (Kamimura et al. 2001). Because some Japanese patients in 2014 had not been overseas before their symptoms, they could have been infected by autochthonous DENV carried by domestic mosquitoes including *A. albopictus* (Kutsuna et al. 2015, Seki et al. 2015). This recent spread of infectious diseases probably coincides with increasing international exchanges of humans and materials and potentially from climate change, including global warming, which can promote the establishment of invading vectors.

Many pathogens, including viruses, bacteria, protozoa, and nematodes, are known to be carried and transmitted by arthropod pests (Table 1). One of the most serious problems of these pathogens is the limited availability of effective vaccines or antibodies. Currently, there are no vaccines established for ZIKV (Chen and Hamer 2016) or DENV (Whitehorn and Farrar 2010). Many antibiotics are available and can be used against bacterial pathogens, but antibiotic misuse and/or overuse often leads to the evolution of antibiotic-resistant bacteria. Pathogens of protozoa, which are unicellular eukaryotic organisms and are currently treated as an informal taxon, are more difficult to manage with medicines such as vaccines or antibodies, because they have complicated lifecycles with ingenious mechanisms to escape from host immune reactions (Miller et al. 2002). One of the most reliable tactics to prevent infections is to reduce the risk of encountering vectors. Hence, chemical ecological approaches that aim to control vectors are essential (Pickett et al. 2010), as well as clinical and medical approaches that directly treat infections in patients. This chapter reviews the challenges for controlling pests vectoring pathogens for humans and pathogens by applying knowledge from chemical ecology and natural products to arthropods, plants, and microbes.

Table 1. Examples of arthropod-borne infectious diseases for humans and livestock

Disease	Pathogen	Recipient	Vector
Viruses			
Dengue fever	*Dengue virus* (DENV; *Flaviviridae*: *Flavivirus*)	Humans, monkeys	Mosquitoes *Aedes aegypti*, *A. albopictus*, *A. polynesiensis*, *A. scutellaris*, etc.
Zika fever	*Zika virus* (ZIKV; *Flaviviridae*: *Flavivirus*)	Humans, monkeys	Mosquitoes *Aedes aegypti*, *A. albopictus*, etc.
Chikungunya fever	*Chikungunya virus* (CHIKV; *Togaviridae*: *Alphavirus*)	Humans, monkeys	Mosquitoes *Aedes aegypti*, *A. albopictus*, etc.
West Nile fever	*West Nile virus* (WNV; *Flaviviridae*: *Flavivirus*)	Humans, birds, bats, horses, etc.	Mosquitoes *Culex pipiens*, *C. tarsalis*, *C. quinquefasciatus*, etc.
Yellow fever	*Yellow fever virus* (YFV; *Flaviviridae*: *Flavivirus*)	Humans, monkeys	Mosquitoes *Aedes aegypti*, *A. albopictus*, etc.
Japanese encephalitis	*Japanese encephalitis virus* (JEV; *Flaviviridae*: *Flavivirus*)	Humans, birds, pigs, dogs, horses, etc.	Mosquitoes *Culex tritaeniorhynchus*, *C. vishnui*, etc.
Akabane disease	*Akabane virus* (*Bunyaviridae*: *Phlebovirus*)	Cattle, sheep, goat, etc.	Midges *Culicoides brevitarsis*, *C. oxystoma*
Severe fever with thrombocytopenia syndrome	*Severe fever with thrombocytopenia syndrome virus* (SFTSV; *Flaviviridae*: *Flavivirus*)	Humans, mammals	Ticks *Haemaphysalis longicornis*, *Rhipicephalus microplus*

(*Contd.*)

Bacteria			
Epidemic typhus	*Rickettsia prowazekii*	Humans, wildlife	Lice *Pediculus humanus*
Scrub typhus	*Orientia tsutsugamushi*	Humans, wildlife	Mites *Leptotrombidium akamusi*, *L. scutellare*, *L. pallidum*, etc.
Plague	*Yersinia* (*Pasteurella*) *pestis*	Humans, rodents, etc.	Fleas *Xenopsylla cheopsi*, etc.
Tularemia	*Francisella tularensis*	Humans, rodents, rabbits, etc.	Ticks *Haemaphysalis* spp., *Ixodes* spp., etc. Biting flies *Chrysops discalis*, etc.
Relapsing fever	*Borrelia* spp.	Humans, rodents, etc.	Lice *Pediculus humanus* Ticks *Ornithodoros* spp., *Ixodes* spp.
Lyme disease	*Borrelia burgdorferi*, *B. garinii*, *B. afzelii*	Human, rodents, deer, birds, etc.	Ticks *Ixodes ricinus*, *I. ricinus*, *I. scapularis*, etc.
Protozoa			
African Trypanosomiasis	*Trypanosoma brucei*	Humans, mammals	Tsetse flies *Glossina* spp.
Malaria	*Plasmodium* spp.	Humans, wildlife, livestock	Mosquitoes *Anopheles* spp.
Leishmania	*Leishmania* spp.	Humans, rodents, dogs, etc.	Sand flies *Lutzomyia* spp., *Phlebotomus* spp.

(*Contd.*)

Table 1. (*Contd.*)

Disease	Pathogen	Recipient	Vector
Chagas' disease	*Trypanosoma cruzi*	Humans, mammals	Assassin bugs *Rhodnius prolixus*, *Triatoma infestans*, etc.
Babesiosis	*Babesia* spp.	Cattle, horses, sheep, dogs, etc.	Ticks *Ixodes scapularis*, *Rhipicephalus microplus*, etc.
East coast fever	*Theileria parva*	Cattle, sheep, goat	Ticks *Rhipicephalus appendiculatus*
Nematodes			
Filariasis	*Dirofilaria immitis*	Dogs, cats, etc.	Mosquitoes *Culex* spp., *Aedes* spp., *Anopheles* spp., etc.
	Wuchereria bancrofti	Humans	Mosquitoes *Culex* spp., *Aedes* spp., *Anopheles* spp., etc.
	Brugia malayi	Humans	Mosquitoes *Culex* spp., *Aedes* spp., *Anopheles* spp., etc.
	Onchocerca volvulus	Humans	Gnats *Simulium* spp.

2. Natural Products: Old but Promising for the Management of Vectors

Arthropods that transmit vertebrate-pathogens are mostly bloodsuckers, including mosquitoes, biting midges and flies, mites and ticks, and assassin bugs, although some necrophagous insects, such as calliphorid flies, can mediate the transmission of some pathogens with high infectiousness by mechanical contact. Natural products that have pesticidal or repellent properties have long been examined to battle against these pests for sanitation. Pyrethroids, which were isolated from pyrethrum plants, *Chrysanthemum cinerariaefolium* (currently, *Tanacetum cinerariifolium*) (Asteraceae), are one of the oldest examples. The first record of the pyrethrum daisy was approximately 2000 years ago at the time of China's Chou Dynasty; the flower was grown in the Dalmatian region (presently Croatia) and traded along the Silk Route (Warmund 2008). In the 19th century, pyrethrum cultivation was introduced to several countries including the United States and Japan, and powder made of crushed flowers was used by soldiers to control fleas and lice on the battlefields. The isolation and characterization of pyrethroids from pyrethrum plants were, however, achieved relatively recently; the structure of the first natural pyrethroid, Pyrethrin I and II, which showed insecticidal activity, were determined in 1924 by Staudinger and Ruzicka (1924a, b, c). Subsequently, additional pyrethroids with insecticidal properties, Cinerin I and II, were discovered from pyrethrum plants and identified by La Forge and Barthel (1945a, b, c). The third pyrethroids isolated were Jasmolin I and II, which were found by Godin et al. (1965, 1966).

These natural pyrethroids have a common motif in their structure; they are esters of chrysanthemum acid. More strictly, Pyrethrin I, Cinerin I, and Jasmolin I are esters of chrysanthemum monocarboxylic acid, and Pyrethrin II, Cinerin II, and Jasmolin II are esters of chrysanthemum dicarboxylic acid monomethyl ester. Chrysanthemum acid [2,2-dimethyl-3-(2-methylprop-1-enyl)cyclopropanecarboxylic acid] is a cyclic monoterpene including a cyclopropane ring, which is relatively unstable and broken by photic stimulation or air oxidation. Because of this physical feature, natural pyrethroids do not remain in the environment and have a low risk of pollution. Chrysanthemum acid has two asymmetric carbons and four isomers; (+)-*trans* [(1*R*,3*R*)], (−)-*trans* [(1*S*,3*S*)], (+)-*cis* [(1*R*,3*S*)], and (−)-*cis* [(1*S*,3*R*)]. Most of natural pyrethroids have a (1*R*,3*R*)-*trans* configuration, which generally shows more pesticidal activity than the corresponding *cis*-isomer (Khambay and Jewess 2010). Pyrethroids are axonic excitoxins, which act on nerve cell receptors of arthropods by manipulating voltage-gated sodium channels in the axonal membranes, but they have less effect on mammals or birds (Khambay and Jewess 2010).

N,N-Diethyl-3-methylbenzamide, also called DEET, is not a naturally occurring substance but should be mentioned because this is currently the most prevalent insect repellent. DEET was developed by screening over 20,000 potential mosquito repellents by the United States Department Agriculture for military uses to protect troops from pests vectoring pathogens during World War II. DEET is one of the best repellents and has been the gold standard for public health threats and emergencies as well as for use in military operations (Novak and Gerberg 2005). The mode of action of DEET as an insect repellency has been well studied. DEET was previously thought to block olfactory receptors for lactic acid or 1-octen-3-ol, which are present in odors, breath, and sweat of mammals and birds and are attractive to female mosquitoes, and to abolish the upwind flight of pests, resulting in "losing" their host (Davis and Sokolove 1976, Peterson and Coats 2001). However, subsequent experiments demonstrated mosquitoes can smell DEET directly and avoid it; mosquitoes have an olfactory receptor neuron that perceives DEET as well as insect repellent terpenoids, including linalool, and both female and male mosquitoes showed avoidance of DEET in a sugar-feeding assay (Syed and Leal 2008). Moreover, an odorant binding protein, *Anopheles gambiae* odorant binding protein 1, was structurally characterized to bind to DEET with high affinity, which provides a way for structure-based modeling to facilitate the design of novel repellents with enhanced binding affinity and selectivity (Tsitsanou et al. 2012).

Although DEET has generally been regarded as safe, some toxic effects have been recorded, including encephalopathy in children, urticaria syndrome, anaphylaxis, hypotension, and decreased heart rate (Peterson and Coats 2001). Thus many studies have searched for natural products, particularly materials derived from plants traditionally used for pest control, which have similar efficacy to DEET (Maia and Moore 2011). For example, extracts of neem (*Azadirachta indica*) (Meliaceae), an evergreen tree native to the Indian subcontinent, have been used in India for several thousand years for medicinal purposes including pest control (Kumar and Navaratnam 2013). According to Sharma et al. (1993), complete protection for 12 hours from *Anopheles* mosquitoes was achieved by application of 2% neem oil mixed in coconut oil to the human body surface. Neem oil, an active ingredient of which is known as azadirachtin (for a recent review, Morgan 2009), is extracted from fruits and seeds of neem trees and is used as a pesticide for agricultural crops as well as a repellent against pests for public health (Isman 2006). Eucalypt plants (*Eucalyptus*, *Corymbia*, and *Angophora* spp.) (Myrtaceae), which are mostly native to Australia, have also long been known as medicinal materials (MacPherson 1939). Some reports showed that eucalyptus oils had equivalent of better effectiveness than DEET (Trigg 1996, Moore et al. 2002). The principal active ingredient of eucalyptus oils was demonstrated to be *p*-menthane-3,8-diol, which was suggested as a potential DEET-alternative for public health applications (Carroll and Loye 2006).

3. Case Studies

3.1 Mosquitoes

Mosquitoes (Diptera: Culicidae) are ubiquitous worldwide and undoubtedly one of the most important vectors that carries many kinds of pathogens; mosquitoes — or more accurately the viruses they carry — kill about 725,000 people a year, mostly in tropical and subtropical regions where the insects thrive (Greenblatt 2016). There are 3,500 named species of mosquito, but only a couple of hundred species feed on human blood (Fang 2010). Moreover, there are specificities in the interactions between mosquito species and pathogens (e.g. Billingsley and Sinden 1997, Alavi et al. 2003, Abraham and Jacobs-Lorena 2004, Sinden et al. 2004), and the species that can transmit pathogens are limited. Three types of mosquito-borne human diseases are known: (i) diseases that are transmitted from other animals to humans but not from humans to humans (e.g. Japanese encephalitis); (ii) diseases that are transmitted from other animals to humans as well as from humans to humans (e.g. dengue fever); and (iii) diseases that are transmitted only from humans to humans (e.g. malaria).

Adult female mosquitoes of blood-feeding species require blood for the maturation of their oocytes. They often use chemical cues to find and approach their forage sources (Takken and Knols 1999). It is well-known that mosquitoes are attracted to carbon dioxide included in breath, lactic acid in sweat (Acree et al. 1968), or 1-octen-3-ol (Hoel et al. 2007, Vezenegho et al. 2014). Notably, human eccrine sweat is attractive to the malaria vector mosquito, *Anopheles gambiae*, after incubation with skin-surface bacteria for several days (Braks and Takken 1999). Moreover, antibacterial treatment of human feet was demonstrated to significantly alter the biting behavior of female mosquitoes (De Jong and Knols 1995). These studies strongly indicated that microbiota on human skin play a major role in body odor production that mediates the attraction of mosquitoes (Verhulst et al. 2010). Moreover, odors from blood agar plates incubated with skin microbiota from human feet elicited significant behavioral responses of mosquitoes in indoor trapping experiments (Verhulst et al. 2009). Fourteen putative attractants, including isovaleric acid (3-methylbutanoic acid) and related compounds, were found and identified in the headspace of the skin bacteria (Verhulst et al. 2009), although it was not yet clear whether the production of these attractive volatiles was specific to bacterial species present only on human skin (Verhulst et al. 2010).

Recently, Homan et al. (2016) assessed the effects of mass deployment of synthetic human-odor-baited traps for *Anopheles* mosquitoes on malaria transmission and disease burden. This assessment was conducted on Rusinga Island in Lake Victoria, western Kenya. All residents with completed health and demographic surveillance system records were eligible to participate. The trap apparatus, Suna trap (suna means mosquito in the Dholuo word) consisted of a funnel and ventilator section, carbon dioxide release pipe, perforated plastic base, netting catch bag, hanging tripod, and conical plastic cover, was operated by solar power; the ventilator rotated to suck air up through the funnel at a rate of 3.1 m/s, and, as air circulates under the conical cover of the trap, volatiles from a synthetic chemical blend of attractants were released from the nylon strips suspended from the hanging tripod (Hiscox et al. 2014). The synthetic blend consisted of ammonia, L-lactic acid, tetradecanoic acid, 3-methylbutanol, and *n*-butylamine (Menger et al. 2014). Attractants included in odor-saturated air were forced out of the trap through holes in the plastic base at a rate of 0.5 m/s. This generated a flow of attractants, which were carried away from the trap. A total of 34,041 participants were enrolled, and 4,358 households were provided with the trap. As a result, people living in households with traps had a significant and approximately 30% lower prevalence of malaria than did those living in households without traps. This was the first evaluation of mass trapping of mosquitoes for malaria control (Homan et al. 2016).

Jaleta et al. (2016) reported a unique screening of repellents for *Anopheles arabiensis*, a dominant vector for malaria in sub-Saharan Africa, from a non-preferred host vertebrate. According to a host census and blood meal analyses in western Ethiopian villages, *A. arabiensis* preferred blood of humans, cattle, goats, and sheep but avoided chickens despite their relatively high abundance, which indicated that chickens were a non-host species of this mosquito. Jaleta et al. (2016) compared volatiles released from hair, wool, and feathers of these livestock and discovered chicken-specific compounds including isobutyl butyrate, naphthalene, hexadecane, and trans-limonene oxide. These compounds elicited electrophysiological responses in the mosquito antennae. Moreover, mosquito-capture in traps was significantly reduced when these chemicals were attached compared with a negative control in field trapping experiments (Jaleta et al. 2016), although *A. arabiensis* appeared to show opportunistic foraging behavior and was shown to be attracted to chicken odors in screenhouse experiments (Busula et al. 2015). Jaleta et al. (2016) pointed out that the selective advantage of host-preference of mosquitoes may be explained by variation in nutritional rewards and corresponding fitness accruing from feeding on different host types, as well as the physical barrier provided by the chicken's feathers or the chicken's prey behavior, because the birds will actively feed on mosquitoes.

Disease infections often alter host physiology and biochemistry, which can influence the host-finding behavior of mosquitoes. Many studies have demonstrated that manipulation of host odors and vector attraction to enhance pathogen transmission in malaria–mosquito systems. Children who harbored gametocytes (the stage transmissible from humans to mosquitoes) of human malaria, *Plasmodium falciparum*, attracted about twice as many *A. gambiae* mosquitoes as other children who harbored malaria of the non-infective stage (oocysts) or no malaria in a semi-natural situations in Kenya (Lacroix et al. 2005). The stage-specific manipulation of host odors were also substantiated using a rodent malaria system; the odor profiles of mice infected with rodent malaria *Plasmodium chabaudii* fluctuated more than healthy mice (De Moraes et al. 2014). Metabolome analyses of volatile compounds of these mice showed an overall elevation in volatile emissions from infected mice after the subsidence of acute malaria symptoms, but during which mice remained highly infectious with gametocytes, and 11 compounds including *N,N*-dibutylformamide, tridecane, 2-pyrrolidone, prenol, isovaleric acid, 2-hexanone, and benzaldehyde, were identified as important predictors of infection status during this phase. Consistent with this volatile analyses, mosquitoes appeared to prefer to the odors of mice during this malaria-transmissible phase (De Moraes et al. 2014). Moreover, mosquitoes harboring malaria changed their biting behavior; *Anopheles* mosquitoes with non-transmissible oocysts decreased host-finding behavior and thereby increased the probability that the mosquito survives the parasite's development, whereas mosquitoes with sporozoites (the stage transmissible from mosquitoes to humans) increase both the motivation to bite and biting frequency (Koella et al. 1998, Anderson et al. 1999, Koella et al. 2002, Smallegange et al. 2013). These stage-specific changes in behavior were paralleled by changes in the responsiveness of mosquito odorant receptors, providing a possible neurophysiological mechanism for the responses (Cator et al. 2013).

3.2 Midges, Sandflies, and Other Biting Flies

Many dipteran insects other than mosquitoes, including midges, sandflies, and gnats, feed on blood or secretions of vertebrates. For example, adult females of biting midges, *Culicoides* spp. (Ceratopogonidae), are blood suckers similar to mosquitoes and transmit many pathogens. In particular, *Culicoides* spp. are known as an important vector of cattle pathogens such as Akabane virus or bluetongue virus. The distribution of *Culicoides* spp. and its related pathogens appears to be expanding; bluetongue virus, for example, was first recorded in 2006 in northern Europe, successfully overwintered with vector midges and subsequently caused substantial losses to the farming sector in 2007 and 2008 (Carpenter et al. 2009). Like mosquitoes and other haematophagous insects, *Culicoides* spp. are attracted to carbon dioxide (Mands et al. 2004) and some generic vertebrate-derived volatiles including 1-octen-3-ol (e.g. Kline et al. 1994,

Blackwell et al. 1996, Bhasin et al. 2001, Harrup et al. 2012). In addition, several aliphatic aldehydes and phenols were discovered in headspace volatile extracts from cattle hair and urine, respectively (Isberg et al. 2016). Curiously, decanal or 3-ethylphenol elicited the attraction of host seeking *C. nubeculosus*, a vector of bluetongue and Schmallenberg viruses in Europe, whereas nonanal or 3-propylphenol similar to these compounds inhibited the attraction of midges; it remains unclear whether these related compounds elicit distinct responses in different functional types of olfactory sensory neurons or how different behavioral circuits are activated (Isberg et al. 2016).

Leishmaniasis caused by protozoan parasites of *Leishmania* spp. (Trypanosomatidae) is also a serious zoonosis carried by haematophagous dipteran insects of the subfamily Phlebotominae (Psychodidae). These tiny blood feeders are sometimes called by the common name "sand fly (or sandfly)", although this name is confusing, because some other unrelated dipteran insects including biting midges (Ceratopogonidae), horse flies (Tabanidae), and black flies (Simuliidae) are occasionally referred to as sand flies. Leishmaniasis is endemic in 98 countries in Asia, Africa, South and Central America, and southern Europe (Barrett and Croft 2012). However, the causative vectors are likely to be limited species; approximately 600 species are known in Phlebotominae, but only 10% of them are leishmanial vectors (Sharma and Singh 2008). More specifically, they are found in two genera: *Lutzomyia* spp. distributed only in the New World, generally in forest dwellings from southern areas of the Nearctic to throughout the Neotropic zone (Beati et al. 2004), and *Phlebotomus* spp. in desert or semi-arid environments in the Old World (Sharma and Singh 2008). They generally breed in organic waste such as rotten wood or leaf litter as well as feces and manure, whereas only adult females require blood to obtain proteins and other nutrients for egg development and harbor *Leishmania* parasites. By means of stable carbon and nitrogen isotope analysis which was highly effective to link adult flies with their diets during larval stages, Mascari et al. (2013) demonstrated that *Phlebotomus papatasi*, a vector of *Leishmania major* (a causative agent of zoonotic cutaneous leishmaniasis), feeds on the feces of rodents in the field. These rodents can be reservoirs of *Leishmania* parasites. Furthermore, Mascari et al. (2013) conducted a field trial using rodent bait containing an insecticide that circulated in the blood of baited rodents and showed that adult females feeding on baited rodents were killed. Gravid females of *P. papatasi* were also reported to be attracted to rodent feces with their larval digestion, suggesting that larval digestion and possibly the larval gut microbial community contribute to the production of oviposition attractants (Marayati et al. 2015). Adult females of another sand fly, *Lutzomyia longipalpis*, were similarly shown to respond to the odors from host vertebrates. At least 16 compounds that elicited electrophysiological responses in adult female antennae were identified from fox odors, and synthetic blends of these chemicals induced significant behavioral responses in female flies (Dougherty et al. 1999), and extracts of volatiles from rabbit and chicken feces including hexanal and 2-methyl-2-butanol preferentially attracted gravid females of this species in an oviposition bioassay (Dougherty et al. 1995). In addition, adult males of this species were shown to produce female-attractant pheromones; these pheromones included two homosesquiterpenes (C16), 9-methylgermacrene-B (Hamilton et al. 1996a) and 3-methyl-α-himachalene (Hamilton et al. 1996b), and two diterpenes (C20) of cembrene isomers (Hamilton et al. 2004, 2005). These pheromones were different among populations, which were considered to be sibling species consisting of *L. longipalpis* species complex (Hamilton et al. 2005, Brazil et al. 2009). Among these pheromones, synthetic (±)-9-methylgermacrene-B, produced from a low-cost plant intermediate, was reported to attract both sexes in a field trap experiment in Brazil by formulating dispensers that released this pheromone at a rate similar to that released by aggregating males (Bray et al. 2009).

Trypanosoma parasites (Trypanosomatidae) are causative agents of another notorious zoonosis, which is known as human African trypanosomiasis (sleeping sickness) or animal African trypanosomiasis (nagana). These trypanosomes are transmitted by tsetse flies and are endemic to a relatively limited area of sub-Saharan Africa, but a total of 70 million people are estimated to be at risk of contracting sleeping sickness in Africa (Simarro et al. 2012). All species of tsetse flies are included in the genus *Glossina*, which solely consisted of their own family, Glossinidae. Reproduction systems of tsetse flies are unusual, being termed adenotrophic viviparity. An adult female fertilized only one egg at once, and the larva hatched and developed within its mother's uterus during the entirety of three larval stages, generally for ≈7 days. Gravid females fed the developing offspring with a nutritious substance secreted by a modified gland in the uterus. A matured larva leaves the uterus, crawls into the soil, and forms a puparial case, where it becomes a pupa. An adult fly emerges from its pupa after 30–40 days in general. An adult female can reproduce for 20 to 40 days and can perform parturitions about every 7–11 days (Krinsky 2002), although only one copulation was enough to fertile eggs during its whole lifespan. Gravid females require considerable nutrition, which is supplied by vertebrate blood, for feeding offspring in their bodies, because all development of offspring out of their mother occurs without feeding. In addition, adult males of tsetse flies also feed on blood. Because of the lack of an effective vaccine for trypanosomiasis (Berriman et al. 2005), controlling tsetse flies is essential for the prevention of these diseases. In Ethiopia, a sterile insect technique, that aims to reduce reproduction of tsetse flies by releasing many male flies sterilized by exposure to radiation, was attempted, although it had a high cost and was considered to provide an uncertain outcome (Enserink 2007). In Zambia, host odor-baited insecticidal targets were applied across a 300 km^2 area between 1989 and 1991, and successfully reduced the tsetse population and trypanosomiasis incidence (van den Bossche 1997). As well as other haematophagous dipteran insects, tsetse flies are attracted to odors of hosts including cattle, pigs, and humans, which are composed of 1-octen-3-ol, phenols, acetone, carbon dioxide, etc. (Jaenson et al. 1991, Späth 1995, Sutton et al. 1997, Ahmed et al. 1999, Ndegwa et al. 1999, Mihok et al. 2007, Rayaisse et al. 2012).

3.3 Fleas

Fleas (Siphonaptera) are holometabolous, wingless insects. Their larvae are a tiny worm-like form and they feed on any available organic substances. After three larval stages, they weave silken cocoons to pupate. Within the cocoon, the pre-emergent adult awaits a suitable opportunity to emerge after metamorphosis into the adult form is completed. Cues for emergence include vibrations, heat, and carbon dioxide from hosts. All flea species are ectoparasitoids of warm-blooded vertebrates, and their adults suck fresh blood of host animals. Although adult fleas have no wings, they have long hind legs adapted for jumping and can move among host individuals. Fleas are generally specialized to parasite their own host species or group of species but they often feed on blood of other non-host animals. As a result, fleas can transmit some pathogens of zoonosis from pet animals or livestock to humans. Fleas are a known vector of plague (*Yersinia pestis*), which is transmitted from rodents to humans and has historically swept the world multiple times causing massive pandemics with high death rates.

Little is known about chemical ecology of fleas, but they definitely use chemical information in their lifecycle. For example, fleas discriminate their host animals with their odors. Two rodent flea species, *Xenopsylla dipodilli* and *Parapulex chephrenis*, can discriminate the odors from their own hosts, *Gerbillus dasyurus* and *Acomys cahirinus*, which are similar rodents inhabiting deserts, even with simultaneous exposure (Krasnov et al. 2002). The ability to select an appropriate host species did not differ significantly either between flea species or between individuals of different sex or age classes within flea species (Krasnov et al. 2002). Another flea species, *X. ramesis*, appeared to choose the sex of its host rodent, *Meriones crassus*, using their odors; male fleas chose randomly between a male and a female rodent, whereas female fleas chose a male rodent significantly more often than a female rodent in an olfaction bioassay (Khokhlova et al. 2011). According to regression analyses using a large data set and phylogenetic information on 297 species of flea that parasitize small mammals, such host discriminations was likely to have evolved with a trend for decreasing host specificity, which may be favorable to limit the risk of extinction and to provide more immediate fitness benefits to parasites (Poulin et al. 2006).

3.4 Lice

Lice (Psocodea: Anoplura) are hemimetabolous insects that are complete parasitoids of mammals. They are adapted to parasite their own hosts and generally show a strict species specificity. This feature was in contrast to those of other blood-feeding insects described above. Actually, lice that inflict humans are limited only two species, *Pediculus humanus* (Pediculidae) and *Phthirus pubis* (Pthiridae) (Durden and Lloyd 2002). The former includes two subspecies: *P. h. humanus*, the body louse, and *P. h. capitis*, the head louse. These species are transmitted by human-to-human contact and show a cosmopolitan distribution all over the world. Lice are known as a vector of *Rickettsia prowazekii* (a causative pathogen of endemic typhus) or *Borrelia recurrentis* (a kind of spirochaeta causative to relapsing fever). These lice pests are relatively easy to control because they cannot survive in an environment other than human body surfaces. However, they can cause outbreaks in a relatively small residence, such as in a refuge, which is crowded with people. The annual cost for lice management in the USA is estimated to be one billion USD (Hansen and O'Haver 2004).

Materials that discourage the transmission of lice are valuable, but such studies are relatively limited (Greive and Barnes 2012). Pyrethroids and DEET were used to protect soldiers from body lice (Eldridge 1973). The repellent activities of some plant essential oils against human lice have been reported. For example, lavender oil, which is probably one of the most prevalent plant essential oils (Tabata et al. 2015), was examined to prevent lice from attaching to children (Burgess 1993). Mumcuoglu et al. (1996) compared the repellent activities of essential oils including citronella, rosemary, eucalyptus, and lavender, and their components to that of DEET. Citronella oil showed similar effectiveness and long-lasting repellency as DEET. Citronellal and geraniol were the ingredients that seemed responsible for the repellent activity of citronella oil (Mumcuoglu et al. 1996). On the other hand, human body lice aggregated on filter paper impregnated with an aqueous extract of louse feces (Mumcuoglu et al. 1986). Ammonium salts in the feces were likely to elicit aggregation behavior in lice, but excretory products of other insects and ticks did not induce aggregation (Mumcuoglu et al. 1986). This may indicate the presence of aggregation pheromones in the excretions, although there are no reports so far.

3.5 Assassin Bugs

The great majority of hemipteran insects are phytophagous bugs that sucking plant sap, but a proportion are carnivorous and suck other animals using their long rostrum to inject a lethal saliva that contains enzymes for digesting and liquefying prey tissues, i.e. extraoral digestion. They mostly attack other invertebrates, including agricultural pests, and as such are beneficial for humans. However, some members of the family Reduviidae, in the subfamily Triatominae, have evolved to feed on blood of vertebrates including humans rather than extraoral digestion. Moreover, Central and South American triatomine species transmit the potentially fatal trypanosomal zoonosis Chagas disease, or American trypanosomiasis. The causative agent is *Trypanosoma cruzi*, which infects a broad range of mammals, including humans, dogs, cats, etc. Many triatomine species, sometimes called as assassin bugs or kissing bugs, harbor and deposit *T. cruzi* in their feces, which can be transferred to vertebrates from injuries caused by blood sucking.

As well as other blood feeders, assassin bugs are attracted by some chemical cues emitted by vertebrates. Isobutyric acid, which is prevalently found in odors from humans or livestock, induces odor-conditioned anemotaxis of fifth-instar nymphs of the assassin bug *Triatoma infestans* (Guerenstein and Guerin 2001). Their antennal responses in basiconic and grooved-peg sensilla were shown to be higher to isobutyric acid than other short-chain branched and unbranched acids. Nymphs of *T. infestans* were also attracted to carbon dioxide and 1-octen-3-ol similar to other blood suckers (Barrozo and Lazzari 2004). Of note, they display a clear diel rhythm of behavioral orientation towards CO_2; they had an endogenous circadian control of olfactory-based responses only at the beginning of the scotophase (Barrozo et al. 2004). This was consistent with their habit, because they are mostly inactive and are usually found in a quiescent state or akinesis, aggregated inside their refuges in close contact with other members of the population during daytime, whereas they show most of their activity i.e. searching for food, mates, oviposition, during the night (Lazzari 1992, Lorenzo and Lazzari 1998). In addition, nymphs of *T. infestans* were shown to aggregate around their own feces, which released attractive odorants including ammonia (Taneja and Guerin 1997). Another triatomine bug, *Rhodnius prolixus* responds similarly to CO_2 and ammonia (Taneja and Guerin 1995, Otálora-Luna et al. 2004).

Some pheromonal compounds have been reported in assassin bugs. For example, isobutyric acid, which is an attractive host volatile for *T. infestans* nymphs, was present in secretions of reduviid-specific dosal glands in abdominal segments of triatomine bugs, including *T. infestans* and *R. prolixus* (Games et al. 1974, Kälin and Barrett 1975, Juárez and Brenner 1981, Seigler and Lampman 2000, Manrique et al. 2006, González Audino et al. 2007). These glands, called Brindley's glands, are considered to be the source of a defensive secretion (Remold 1962). Actually, the introduction of synthetic isobutyric acid vapor to adult *R. prolixus* caused rapid arousal and disruption of an equilibrium aggregation, whereas no such effect was observed in a similar equilibrated aggregation of nymphs (Kälin and Barrett 1975), indicating this compound had an alarm pheromone-like dispersal property (Manrique et al. 2006). The opposing functions of isobutyric acid in triatomine bugs seem apparently inconsistent, but they may reflect a key role of this chemical for behaviors of these bugs with the presence of conserved reception nervous systems. Moreover, some low molecule weight ketones and alcohols, including 3-pentanone, 2-methylbutanol, and 3-pentanol in *T. infestans* (Manrique et al. 2006) and 2-methyl-3-buten-2-ol, 2-pentanol, *trans*-2-methyl-3-penten-2-ol, and 4-methyl-3-penten-2-ol in *R. prolixus* (Pontes et al. 2008), were discovered in metasternal glands of each species and were suggested to their mediate sexual communications. In addition, epicuticular lipids including long-chain aliphatic compounds such as octadecanoic acid or eicosanol were identified as pheromones for aggregation or mate-recognition in *T. infestans* (Figueiras et al. 2009, Cocchiararo-Bastias et al. 2011).

3.6 Mites and Ticks

Mites and ticks (Acari) include numerous species that show a great diversity in their biological and ecological features. Only a part of them, such as members of the family Ixodoidae, Trombiculidae, or Dermanyssidae, are ectoparasitoids of vertebrates including humans, but they often transmit serious pathogens, particularly rickettsia. Recently, an emerging viral disease, which has been referred to as severe fever with thrombocytopenia syndrome (SFTS), was discovered in East Asia and was proven to be vectored by ticks. This infectious disease was first reported in rural areas of China in summers in the 2000s, and it currently occurs in China, Korea, and Japan. By early 2007, scientists at the Chinese Center for Disease Control and Prevention identified the causative agent as human granulocytic anaplasmosis, a bacterial infection from tick bites (Stone 2010, Liu et al. 2014), but subsequently, a novel bunyavirus was identified from the patients and ticks via RNA sequences (Yu et al. 2011). This virus was estimated to have originated 50–150 years ago and has undergone a very recent population expansion (Lam et al. 2013).

Acari species are tiny and generally show a limited ability to disperse by themselves. Accordingly, chemical cues often play an important role in their behavior. For example, many ticks produce pheromones for aggregation, attraction, attachment, and mating, as reviewed by Sonenshine (2004, 2006). These chemicals are structurally divergent. Male ticks of the genus *Amblyomma*, in which the adults feed on ungulates, produce a mixture of different classes of organic volatile compounds including 2-nitrophenol, methyl salicylate, benzaldehyde, or 2,6-dichlorophenol, for inducing attraction, aggregation, and attachment of both male and female cohorts (Schöni et al. 1984, Diehl et al. 1991, Lusby et al. 1991, Price et al. 1994). 2-Nitrophenol was commonly emitted in large quantities and served as a primary long-range attractant in two allies, *A. variegatum* and *A. hebraeum* (Norval et al. 1991a, b). However, the amounts of methyl salicylate and benzaldehyde were different among the two species, which may indicate species specificity in their attraction-aggregation-attachment pheromones and may explain species-specific aggregation that is observed in nature where the two ticks are sympatric (Price et al. 1994). 2,6-Dichlorophenol was also reported as a sex pheromone compound that is produced by adult females to attract adult males from many species of several different genera (Sonenshine 2004, 2006). In addition to pheromones, these mites and ticks use host-derived volatiles such as carbon dioxide, ammonia, and 1-octen-3-ol (McMahon et al. 2001, Carr et al. 2012), as well as other haematophagous insects.

4. Conclusion

As reviewed above, vector arthropods that transmit infectious diseases to humans and livestock are mostly tiny and, therefore, use chemical cues to optimize their behavioral efficiency. Thus, such chemicals have great potential for application in controlling these pests with a low risk of environmental pollution or resistance development. Chemicals with irritant or insecticidal properties can be used as repellents, whereas chemicals with attractant properties can serve as a tool for mass trapping, decoy, or "attract-and-kill" strategies. Trapping apparatuses, such as the Suna trap used for malaria–vector mosquitoes (Hiscox et al. 2014), operated by solar power could be particularly promising, because they can be placed anywhere without a requirement for electricity. Further improvement of such "hardware" for applying semiochemicals, along with studying these chemicals ("software") themselves, may promote the application of chemical ecological knowledge for disease-vector management programs.

Vector arthropods often use pheromones and host-derived chemicals (kairomones) for attraction, aggregation, or foraging. The formers are generally designed with species specificity. This feature is an obstacle for practical uses, because we have to develop specific chemicals for each pest species. On the other hand, some of the latter chemicals are likely to be more broadly effective. Carbon dioxide, ammonia, lactic acid, or 1-octen-3-ol are typical volatiles emitted from vertebrates and are attractive to a broad range of haematophagous arthropods. Moreover, most of these compounds are structurally simple and are available relatively inexpensively, and, therefore, may be more valuable for practical applications.

REFERENCES

Abraham, E.G. and M. Jacobs-Lorena. 2004. Mosquito midgut barriers to malaria parasite development. Insect Biochem. Mol. Biol. 34: 667–671.

Acree, F., R.B. Turner, H.K. Gouck, M. Beroza and N. Smith.1968. L-Lactic acid: a mosquito attractant isolated from humans. Science 161: 1346–1347.

Ahmed, M.M. and S. Mihok. 1999. Responses of *Glossina fuscipes fuscipes* (Diptera: Glossinidae) and other Diptera to carbon dioxide in linear and dense forests. Bull. Entomol. Res. 89: 177–184.

Alavi, Y., M. Arai, J. Mendoza, M. Tufet-Bayona, R. Sinha, K. Fowler et al. 2003. The dynamics of interactions between *Plasmodium* and the mosquito: a study of the infectivity of *Plasmodium berghei* and *Plasmodium gallinaceum*, and their transmission by *Anopheles stephensi*, *Anopheles gambiae* and *Aedes aegypti*. Int. J. Parasitol. 33: 933–943.

Anderson, R.A., J.C. Koella and H. Hurd. 1999. The effect of *Plasmodium yoelii nigeriensis* infection on the feeding persistence of *Anopheles stephensi* Liston throughout the sporogonic cycle. Proc. R. Soc. Lond. B. 266: 1729–1733.

Barrett, M.P. and S.L. Croft. 2012. Management of trypanosomiasis and leishmaniasis. Br. Med. Bull. 104: 175–196.

Barrozo, R.B. and C.R. Lazzari. 2004. The response of the blood-sucking bug *Triatoma infestans* to carbon dioxide and other host odours. Chem. Senses. 29: 319–329.

Barrozo, R.B., S.A. Minoli and C.R. Lazzari. 2004. Circadian rhythm of behavioural responsiveness to carbon dioxide in the blood-sucking bug *Triatoma infestans* (Heteroptera: Reduviidae). J. Insect Physiol. 50: 249–254.

Beati, L., A.G. Cáceres, J.A. Lee and L.E. Munstermann. 2004. Systematic relationships among *Lutzomyia* sand flies (Diptera: Psychodidae) of Peru and Colombia based on the analysis of 12S and 28S ribosomal DNA sequences. Int. J. Parasitol. 34: 225–234.

Berriman, M., E. Ghedin, C. Hertz-Fowler, G. Blandin, H. Renauld, D.C. Bartholomeu et al. 2005. The genome of the African trypanosome *Trypanosoma brucei*. Science 309: 416–422.

Bhasin, A., A.J. Mordue and W. Mordue. 2001. Field studies on efficacy of host odour baits for the biting midge *Culicoides impunctatus* in Scotland. Med. Vet. Entomol. 15: 147–156.

Billingsley, P.F. and R.E. Sinden. 1997. Determinants of malaria-mosquito specificity. Parasitol. Today 13: 297–301.

Blackwell, A., C. Dyer, A.J. Mordue, L.J. Wadhams and W. Mordue. 1996. The role of 1-octen-3-ol as a host-odour attractant for the biting midge, *Culicoides impunctatus* Goetghebuer, and interactions of 1-octen-3-ol with a volatile pheromone produced by parous female midges. Physiol. Entomol. 21: 15–19.

Braks, M.A.H. and W. Takken. 1999. Incubated human sweat but not fresh sweat attracts the malaria mosquito *Anopheles gambiae* sensu stricto. J. Chem. Ecol. 25: 663–672.

Bray, D.P., K.K. Bandi, R.P. Brazil, A.G. Oliveira and J.G.C. Hamilton. 2009. Synthetic sex pheromone attracts the leishmaniasis vector *Lutzomyia longipalpis* (Diptera: Psychodidae) to traps in the field. J. Med. Entomol. 46: 428–434.

Brazil, R.P., N.N. Caballero and J.G.C. Hamilton. 2009. Identification of the sex pheromone of *Lutzomyia longipalpis* (Lutz & Neiva, 1912) (Diptera: Psychodidae) from Asunción, Paraguay. Parasit. Vectors 2: 51.

Burgess, I. 1993. New head louse repellent. Br. J. Dermatol. 128: 357–360.

Busula, A.O., W. Takken, D.E. Loy, B.H. Hahn, W.R. Mukabana and N.O. Verhulst. 2015. Mosquito host preferences affect their response to synthetic and natural odour blends. Malaria J. 14: 133.

Carpenter, S., A. Wilson and P.S. Mellor. 2009. Culicoides and the emergence of bluetongue virus in northern Europe. Trends Microbiol. 17: 172–178.

Carr, A.L., R.M. Roe, C. Arellano, D.E. Sonenshine, C. Schal and C.S. Apperson. 2013. Responses of *Amblyomma americanum* and *Dermacentor variabilis* to odorants that attract haematophagous insects. Med. Vet. Entomol. 27: 86–95.

Carroll, S.P. and J. Loye. 2006. PMD, a registered botanical mosquito repellent with deet-like efficacy. J. Am. Mosq. Control Assoc. 22: 507–514.

Cator, L.J., J. George, S. Blanford, C.C. Murdock, T.C. Baker, A.F. Read et al. 2013. 'Manipulation' without the parasite: altered feeding behaviour of mosquitoes is not dependent on infection with malaria parasites. Proc. R. Soc. B 280: 20130711.

Chen, L.H. and D.H. Hamer. 2016. Zika virus: rapid spread in the Western hemisphere. Ann. Intern. Med. 164: 613–615.

Cocchiararo-Bastias, L.M., S.J. Mijailovsky, G.M. Calderon-Fernández, A.N.L. Figueiras and M.P. Juárez. 2011. Epicuticle lipids mediate mate recognition in *Triatoma infestans*. J. Chem. Ecol. 37: 246–252.

Davis, E.E. and P. Sokolove. 1976. Lactic acid-sensitive receptors on the antennae of the mosquito, *Aedes aegypti*. J. Comp. Physiol. A 105: 43–54.

De Jong, R. and B.G.J. Knols. 1995. Selection of biting sites on man by two malaria mosquito species. Experientia 51: 80–84.

De Moraes, C.M., N.M. Stanczyk, H.S. Betz, H. Pulido, D.G. Sim, A.F. Read et al. 2014. Malaria-induced changes in host odors enhance mosquito attraction. Proc. Natl. Acad. Sci. USA 111: 11079–11084.

Dick, G.W., S.F. Kitchen and A.J. Haddow. 1952. Zika virus. I. Isolations and serological specificity. Trans. R. Soc. Trop. Med. Hyg. 46: 509–520.

Diehl, P.A., P. Guerin, M. Vlimant and P. Steullet. 1991. Biosynthesis, production site, and emission rates of aggregation-attachment pheromone in males of two *Amblyomma* ticks. J. Chem. Ecol. 17: 833–847.

Dougherty, M.J., P.M. Guerin and R.D. Ward. 1995. Identification of oviposition attractants for the sandfly *Lutzomyia longipalpis* (Diptera: Psychodidae) in volatiles of faeces from vertebrates. Physiol. Entomol. 20: 23–32.

Dougherty, M.J., P.M. Guerin, R.D. Ward and J.G.C. Hamilton. 1999. Behavioural and electrophysiological responses of the phlebotomine sandfly *Lutzomyia longipalpis* (Diptera: Psychodidae) when exposed to canid host odour kairomones. Physiol. Entomol. 24: 251–262.

Duffy, M.R., T.-H. Chen, W.T. Hancock, A.M. Powers, J.L. Kool, R.S. Lanciotti et al. 2009. Zika virus outbreak on Yap Island, Federated States of Micronesia. N. Engl. J. Med. 360: 2536–2543.

Durden, L.A. and J.E. Lloyd. 2002. Lice (Phthiraptera). pp. 56–79. *In*: G. Mullen and L. Durden [eds.]. Medical and Veterinary Entomology. Academic Press, New York, NY, USA.

Eldridge, B.F. 1973. Repellents and impregnants for the control of body lice. pp. 177–178. *In*: PAHO/WHO [eds.]. The Control of Lice and Louse-borne Diseases. Pan American Health Organization Scientific Publication 263, Washington D.C., USA.

Enserink, M. 2007. Welcome to Ethiopia's fly factory. Science 317: 310–313.

Fang, J. 2010. A world without mosquitoes. Nature 466: 432–434.

Figueiras, A.N.L., J.R. Girotti, S.J. Mijailovsky and M.P. Juárez. 2009. Epicuticular lipids induce aggregation in Chagas disease vectors. Parasit. Vect. 2: 8.

Games, D.E., C.J. Schofield and B.W. Staddon. 1974. The secretion from Brindley's scent glands in Triatominae. Ann. Entomol. Soc. Am. 67: 820.

Godin, P.J., J.H. Stevenson and R.M. Sawicki. 1965. The insecticidal activity of Jasmolin II and its isolation from pyrethrum (*Chrysanthemum cinerariaefolium* Vis.). J. Econ. Entomol. 58: 548–551.

Godin P.J., R.J. Sleeman, M. Snarey and E.M. Thain. 1966. The jasmolins, new insecticidally active constituents of *Chrysanthemum cinerariaefolium* Vis. J. Chem. Soc. C 1966: 332–334.

González Audino, P., R.A. Alzogaray, C. Vassena, H. Masuh, A. Fontan, P. Gatti et al. 2007. Volatile compounds secreted by Brindley's glands of adult *Triatoma infestans*: identification and biological activity of previously unidentified compounds. J. Vect. Ecol. 32: 75–82.

Greenblatt, A. 2016. Mosquito-borne disease. CQ Res. 26: 601–624.

Greive, K.A. and T.M. Barnes. 2012. *In vitro* comparison of four treatments which discourage infestation by head lice. Parasitol. Res. 110: 1695–1699.

Guerbois, M., I. Fernandez-Salas, S.R. Azar, R. Danis-Lozano, C.M. Alpuche-Aranda, G. Leal et al. 2016. Outbreak of Zika virus infection, Chiapas State, Mexico, 2015, and first confirmed transmission by *Aedes aegypti* mosquitoes in the Americas. J. Infect. Dis. 214: 1349–1356.

Guerenstein, P.G. and P.M. Guerin. 2001. Olfactory and behavioural responses of the blood-sucking bug *Triatoma infestans* to odours of vertebrates hosts. J. Exp. Biol. 204: 585–597.

Haddow, A.D., A.J. Schuh, C.Y. Yasuda, M.R. Kasper, V. Heang, R. Huy et al. 2012. Genetic characterization of Zika virus strains: geographic expansion of the Asian lineage. PLoS Negl. Trop. Dis. 6: e1477.

Hamilton, J.G.C., G.W. Dawson and J.A. Pickett. 1996a. 9-Methylgermacrene-B: Proposed structure for novel homosesquiterpene from the sex pheromone glands of *Lutzomyia longipalpis* (Diptera: Psychodidae) from Lapinha, Brazil. J. Chem. Ecol. 22: 1477–1491.

Hamilton, J.G.C., G.W. Dawson and J.A. Pickett. 1996b. 3-Methyl-α-himachalene: Proposed structure for novel homosesquiterpene sex pheromone of *Lutzomyia longipalpis* (Diptera: Psychodidae) from Jacobina, Brazil. J. Chem. Ecol. 22: 2331–2340.

Hamilton, J.G.C., R.P. Brazil and R. Maingon. 2004. A fourth chemotype of *Lutzomyia longipalpis* (Diptera: Psychodidae) from Jaíbas, Minas Gerais State, Brazil. J. Med. Entomol. 41: 1021–1026.

Hamilton, J.G.C., R.D.C. Maingon, B. Alexander, R.D. Ward and R.P. Brazil. 2005. Analysis of the sex pheromone extract of individual male *Lutzomyia longipalpis* sandflies from six regions in Brazil. Med. Vet. Entomol. 19: 480–488.

Hansen, R.C. and J. O'Haver. 2004. Economic considerations associated with *Pediculosis humanus capitis* infestation. Clin. Pediatr. 43: 523–527.

Harrup, L.E., J.G. Logan, J.I. Cook, N. Golding, M.A. Birkett, J.A. Pickett et al. 2012. Collection of *Culicoides* (Diptera: Ceratopogonidae) using CO_2 and enantiomers of 1-octen-3-ol in the United Kingdom. J. Med. Entomol. 49: 112–121.

Hennessey, M., M. Fischer and J.E. Staples. 2016. Zika virus spreads to new areas—region of the Americas, May 2015–January 2016. Am. J. Transplant. 16: 1031–1034.

Hiscox, A., B. Otieno, A. Kibet, C.K. Mweresa, P. Omusula, M. Geier et al. 2014. Development and optimization of the Suna trap as a tool for mosquito monitoring and control. Malaria J. 13: 257.

Hoel, D.F., D.L. Kline, S.A. Allan and A. Grant 2007. Evaluation of carbon dioxide, 1-octen-3-ol, and lactic acid as baits in Mosquito Magnet Pro traps for *Aedes albopictus* in north central Florida. J. Am. Mosq. Control Assoc. 23: 11–17.

Homan, T., A. Hiscox, C.K. Mweresa, D. Masiga, W.R. Mukabana, P. Oria et al. 2016. The effect of mass mosquito trapping on malaria transmission and disease burden (SolarMal): a stepped-wedge cluster-randomised trial. Lancet 388: 1193–1201.

Isberg, E., D.P. Bray, G. Birgersson, Y. Hillbur and R. Ignell. 2016. Identification of cattle-derived volatiles that modulate the behavioral response of the biting midge *Culicoides nubeculosus*. J. Chem. Ecol. 42: 24–32.

Isman, M.B. 2006. Botanical insecticides, deterrents, and repellents in modern agriculture and an increasingly regulated world. Annu. Rev. Entomol. 51: 45–66.

Jaenson, T.G.T., R.C. Barreto Dos Santos and D.R. Hall. 1991. Attraction of *Glossina longipalpis* (Diptera: Glossinidae) in Guinea-Bissau to odor-baited biconical traps. J. Med. Entomol. 28: 284–286.

Jaleta, K.T., S.R. Hill, G. Birgersson, H. Tekie and R. Ignell. 2016. Chicken volatiles repel host-seeking malaria mosquitoes. Malaria J. 15: 354.

Juárez, P. and R.R. Brenner. 1981. Bioquimica del ciclo evolutivo del *Triatoma infestans* (Vinchuca). V. Emision de acidos grasos volatiles. Acta Physiol. Lat. Am. 31: 113–118.

Kälin, M. and F.M. Barrett. 1975. Observations on the anatomy, histology, release site, and function of brindley's glands in the blood-sucking bug *Rhodnius prolixus* (Heteroptera: Reduviidae). Ann. Entomol. Soc. Am. 68: 126–134.

Kamimura, K., I. Matsuse, Y. Shirai, H. Takahashi, T. Fukuda, J. Komukai et al. 2001. Possibility of settlement of *Aedes aegypti* in Japan. Med. Entomol. Zool. 52: 160.

Khambay, B.P.S. and P.J. Jewess. 2010. Pyrethroids. pp. 1–29. *In*: L.I. Gilbert and S.S. Gill [eds.]. Insect Control. Academic Press, London, UK.

Khokhlova, I.S., V. Serobyan, A.A. Degen and B.R. Krasnov. 2011. Discrimination of host sex by a haematophagous ectoparasite. Anim. Behav. 81: 275–281.

Kline, D.L., D.V. Hagan and J.R. Wood. 1994. *Culicoides* responses to 1-octen-3-ol and carbon dioxide in salt marshes near Sea Island, Georgia, USA. Med. Vet. Entomol. 8: 25–30.

Koella, J.C., F.L. Sørensen and R.A. Anderson. 1998. The malaria parasite, *Plasmodium falciparum*, increases the frequency of multiple feeding of its mosquito vector, *Anopheles gambiae*. Proc. R. Soc. Lond. B 265: 763–768.

Koella, J.C., L. Rieu and R.E.L. Paul. 2002. Stage-specific manipulation of a mosquito's host-seeking behavior by the malaria parasite *Plasmodium gallinaceum*. Behav. Ecol. 13: 816–820.

Krasnov, B.R., I.S. Khokhlova, I. Oguzoglu and N.V. Burdelova. 2002. Host discrimination by two desert fleas using an odour cue. Anim. Behav. 64: 33–40.

Krinsky, W.L. 2002. Tsetse flies (Glossinidae). pp. 303–316. *In*: G. Mullen and L. Durden [eds.]. Medical and Veterinary Entomology. Academic Press, New York, NY, USA.

Kumar, V.S. and V. Navaratnam. 2013. Neem (*Azadirachta indica*): Prehistory to contemporary medicinal uses to humankind. Asian Pac. J. Trop. Biomed. 3: 505–514.

Kutsuna, S., Y. Kato, M.L. Moi, A. Kotaki, M. Ota, K. Shinohara et al. 2015. Autochthonous dengue fever, Tokyo, Japan, 2014. Emerg. Infect. Dis. 21: 517–520.

La Forge, F. and W. Barthel. 1945a. Constituents of pyrethrum flowers. XVII. The isolation of five pyrethrolone semicarbazones. J. Org. Chem. 10: 106–113.

La Forge, F. and W. Barthel. 1945b. Constituents of pyrethrum flowers. XVIII. The structure and isomerism of pyrethrolone and cinerolone. J. Org. Chem. 10: 114–120.

La Forge, F. and W. Barthel. 1945c. Constituents of pyrethrum flowers. XIX. The structure of cinerolone. J. Org. Chem. 10: 222–227.

Lacroix, R., W.R. Mukabana, L.C. Gouagna and J.C. Koella. 2005. Malaria infection increases attractiveness of humans to mosquitoes. PLoS Biol. 3: e298.

Lam, T.T., W. Liu, T.A. Bowden, N. Cui, L. Zhang, K. Liu et al. 2013. Evolutionary and molecular analysis of the emergent severe fever with thrombocytopenia syndrome virus. Epidemics 5: 1–10.

Lazzari, C.R. 1992. Circadian organization of locomotion activity in the haematophagous bug *Triatoma infestans*. J. Insect Physiol. 38: 895–903.

Liu, Q., B. He, S.-Y. Huang, F. Wei and X.-Q. Zhu. 2014. Severe fever with thrombocytopenia syndrome, an emerging tick-borne zoonosis. Lancet Infect. Dis. 14: 763–772.

Lorenzo, M.G. and C.R. Lazzari. 1998. Activity pattern in relation to refuge exploitation and feeding in *Triatoma infestans* (Hemiptera: Reduviidae). Acta Trop. 70: 163–170.

Lusby, W.R., D.E. Sonenshine, C.E. Yunker, R.A. Norval and M.J. Burridge. 1991. Comparison of known and suspected pheromonal constituents in males of the African ticks, *Amblyomma hebraeum* Koch and *Amblyomma variegatum* (Fabricius). Exp. Appl. Acarol. 13: 143–152.

MacPherson, J. 1939. The eucalyptus in the daily life and medical practice of the Australian aborigines. Aust. J. Anthropol. 2: 175–180.

Maia, M.F. and S.J. Moore. 2011. Plant-based insect repellents: a review of their efficacy, development and testing. Malar. J. 10: S11.

Mands, V., D.L. Kline and A. Blackwell. 2004. *Culicoides* midge trap enhancement with animal odour baits in Scotland. Med. Vet. Entomol. 18: 336–342.

Manrique, G., A.C.R. Vitta, R.A. Ferreira, C.L. Zani, C.R. Unelius, C.R. Lazzari et al. 2006. Chemical communication in chagas disease vectors. Source, identity, and potential function of volatiles released by the Metasternal and Brindley's glands of *Triatoma infestans* adults. J. Chem. Ecol. 32: 2035–2052.

Marayati, B.F., C. Schal, L. Ponnusamy, C.S. Apperson, T.E. Rowland and G. Wasserberg. 2015. Attraction and oviposition preferences of *Phlebotomus papatasi* (Diptera: Psychodidae), vector of Old-World cutaneous leishmaniasis, to larval rearing media. Parasit. Vectors 8: 663.

Mascari, T.M., H.A. Hanafi, R.E. Jackson, S. Ouahabi, B. Ameur, C. Faraj et al. 2013. Ecological and control techniques for sand flies (Diptera: Psychodidae) associated with rodent reservoirs of leishmaniasis. PLoS Negl. Trop. Dis. 7: e2434.

McMahon, C., P.M. Guerin and Z. Syed. 2001. 1-Octen-3-ol isolated from bont ticks attracts *Amblyomma variegatum*. J. Chem. Ecol. 27: 471–486.

Menger, D.J., J.J.A. Van Loon and W. Takken. 2014. Assessing the efficacy of candidate mosquito repellents against the background of an attractive source that mimics a human host. Med Vet Entomol. 28: 407–413.

Mihok, S., D.A. Carlson and P.N. Ndegwa. 2007. Tsetse and other biting fly responses to Nzi traps baited with octenol, phenols and acetone. Med. Vet. Entomol. 21: 70–84.

Miller, L.H., D.I. Baruch, K. Marsh and O.K. Doumbo. 2002. The pathogenic basis of malaria. Nature 415: 673–679.

Miura, K., M. Kawada, T. Kakimoto, T. Watnabe, T. Hirose, T. Koyama et al. 2015. First confirmed autochthonous case of dengue fever in Japan in nearly 70 years, 2014. IASR 36: 35–37.

Moore, S.J., A. Lenglet and N. Hill. 2002. Field evaluation of three plant-based insect repellents against malaria vectors in Vaca Diez Province, the Bolivian Amazon. J. Am. Mosq. Control Assoc. 18: 107–110.

Morgan, E.D. 2009. Azadirachtin, a scientific gold mine. Bioorg. Med. Chem. 17: 4096–4105.

Mumcuoglu, K.Y., R. Galun and R. Ikan. 1986. The aggregation response of human body louse *Pediculus humanus* (Insecta: Anoplura) to its excretory products. Insect Sci. Appl. 7: 629–632.

Mumcuoglu, K.Y., R. Galun, U. Bach, J. Miller and S. Magdassi. 1996. Repellency of essential oils and their components to the human body louse, *Pediculus humanus humanus*. Entomol. Exp. Appl. 78: 309–314.

Musso, D., E.J. Nilles and V.-M. Cao-Lormeau. 2014. Rapid spread of emerging Zika virus in the Pacific area. Clin. Microbiol. Infect. 20: O595–O596.

Ndegwa, P.N. and S. Mihok. 1999. Development of odour-baited traps for *Glossina swynnertoni* (Diptera: Glossinidae). Bull. Entomol. Res. 89: 255–261.

Norval, R.A.I., T. Peter, C.E. Yunker, D.E. Sonenshine and M.J. Burridge. 1991a. Responses of the ticks *Amblyomma hebraeum* and *A. variegatum* to known or potential components of the aggregation-attachment pheromone. I. Long-range attraction. Exp. Appl. Acarol. 13: 11–18.

Norval, R.A.I., T. Peter, C.E. Yunker, D.E. Sonenshine and M.J. Burridge. 1991b. Responses of the ticks *Amblyomma hebraeum* and *A. variegatum* to known or potential components of the aggregation-attachment pheromone. II. Attachment stimulation. Exp. Appl. Acarol. 13: 19–26.

Novak, R.J. and E.J. Gerberg. 2005. Natural-based repellent products: efficacy for military and general public uses. J. Am. Mosq. Control Assoc. 21: 7–11.

Otálora-Luna, F., J.L. Perret and P.M. Guerin. 2004. Appetence behaviours of the triatomine bug *Rhodnius prolixus* on a servosphere in response to the host metabolites carbon dioxide and ammonia. J. Comp. Physiol. A 190: 847–854.

Peterson, C. and J. Coats. 2001. Insect repellents—past, present and future. Pestic. Outlook 12: 154–158.

Pickett, J.A., M.A. Birkett, S.Y. Dewhirst, J.G. Logan, M.O. Omolo, B. Torto et al. 2010. Chemical ecology of animal and human pathogen vectors in a changing global climate. J. Chem. Ecol. 36: 113–121.

Pontes, G.B., B. Bohman, C.R. Unelius and M.G. Lorenzo. 2008. Metasternal gland volatiles and sexual communication in the triatomine bug, *Rhodnius prolixus*. J. Chem. Ecol. 34: 450–457.

Poulin, R., B.R. Krasnov, G.I. Shenbrot, D. Mouillot and I.S. Khokhlova. 2006. Evolution of host specificity in fleas: is it directional and irreversible? Int. J. Parasitol. 36: 185–191.

Price, T.L., D.E. Sonenshine, R.A. Norval and M.J. Burridge. 1994. Pheromonal composition of two species of African *Amblyomma* ticks: similarities, differences and possible species specific components. Exp. Appl. Acarol. 18: 37–50.

Rayaisse, J.B., I. Tirados, D. Kaba, S.Y. Dewhirst, J.G. Logan, A. Diarrassouba et al. 2012. Prospects for the development of odour baits to control the tsetse flies *Glossina tachinoides* and *G. palpalis s.l.* PLoS Negl. Trop. Dis. 4: e632.

Remold, H. 1962. Ueber die biologische Bedeutung der Duftdrüsen bei den Landwanzen (Geocorisae). Z. Vergl. Physiol. 45: 636–694.

Schöni, R., E. Hess, W. Blum and K. Ramstein. 1984. The aggregation-attachment pheromone of the tropical bont tick *Amblyomma variegatum* Fabricius (Acari, Ixodidae): isolation, identification and action of its components. J. Insect Physiol. 30: 613–618.

Seigler, D.S. and R.L. Lampman. 2000. Isobutyric acid from the Brindley's glands of *Triatoma lecticularia*. J. Am. Mosq. Control Assoc. 16: 36–37.

Seki, N., Y. Iwashita, R. Moto, N. Kamiya, M. Kurita, N. Tahara et al. 2015. An autochthonous outbreak of dengue type 1 in Tokyo, Japan 2014. Jpn. J. Public Health 62: 238–250.

Sharma, U. and S. Singh. 2008. Insect vectors of *Leishmania*: distribution, physiology and their control. J. Vector Borne Dis. 45: 255–272.

Sharma, V.P., M.A. Ansari and R.K. Razdan. 1993. Mosquito repellent action of neem (*Azadirachta indica*) oil. J. Am. Mosq. Control Assoc. 9: 359–360.

Simarro, P.P., G. Cecchi, J.R. Franco, M. Paone, A. Diarra, J.A. Ruiz-Postigo et al. 2012. Estimating and mapping the population at risk of sleeping sickness. PLoS Negl. Trop. Dis. 6: e1859.

Sinden, R.E., Y. Alavi and J.D. Raine. 2004. Mosquito–malaria interactions: a reappraisal of the concepts of susceptibility and refractoriness. Insect Biochem. Mol. Biol. 34: 625–629.

Smallegange, R.C., G.-J. van Gemert, M. van de Vegte-Bolmer, S. Gezan, W. Takken, R.W. Sauerwein et al. 2013. Malaria infected mosquitoes express enhanced attraction to human odor. PLoS ONE 8: e63602.

Sonenshine, D.E. 2004. Pheromones and other semiochemicals of ticks and their use in tick control. Parasitology 129 Suppl.: S405–S425.

Sonenshine, D.E. 2006. Tick pheromones and their use in tick control. Annu. Rev. Entomol. 51: 557–580.

Späth, J. 1995. Olfactory attractants for West African tsetse flies, *Glossina* spp. (Diptera: Glossinidae). Trop. Med. Parasitol. 46: 253–257.

Staudinger, H. and L. Ruzicka. 1924a. Insektentötende Stoffe I. Über Isolierung und Konstitution des wirksamen Teiles des dalmatinischen Insektenpulvers. Helv. Chim. Acta 7: 177–201.

Staudinger, H. and L. Ruzicka. 1924b. Insektentötende Stoffe II. Zur Konstitution der Chrysanthemum-monocarbonsäure und -dicarbonsäure. Helv. Chim. Acta 7: 201–211.

Staudinger, H. and L. Ruzicka. 1924c. Insektentötende Stoffe III. Konstitution des Pyrethrolons. Helv. Chim. Acta 7: 212–235.

Stone, R. 2010. Rival teams identify a virus behind deaths in central China. Science 330: 20–21.

Sutton, B.D. and D.A. Carlson. 1997. Cuticular hydrocarbons of *Glossina*, III: Subgenera *Glossina* and *Nemorhina*. J. Chem. Ecol. 23: 1291–1320.

Syed, Z. and W.S. Leal. 2008. Mosquitoes smell and avoid the insect repellent DEET. Proc. Natl. Acad. Sci. USA 105: 13598–13603.

Tabata, J., M. Teshiba, N. Shimizu and H. Sugie. 2015. Mealybug mating disruption by a sex pheromone derived from lavender essential oil. J. Essent. Oil Res. 27: 232–237.

Takken, W. and B.G. Knols. 1999. Odor-mediated behavior of Afrotropical malaria mosquitoes. Annu. Rev. Entomol. 44: 131–157.

Taneja, J. and P.M. Guerin. 1995. Oriented responses of the triatomine bugs *Rhodnius prolixus* and *Triatoma infestans* to vertebrate odours on a servosphere. J. Comp. Physiol. A 176: 455–464.

Taneja, J. and P.M. Guerin. 1997. Ammonia attracts the haematophagous bug *Triatoma infestans*: Behavioural and neurophysiological data on nymphs. J. Comp. Physiol. A 181: 21–34.

Trigg, J.K. 1996. Evaluation of a eucalyptus-based repellent against Anopheles spp. in Tanzania. J. Am. Mosq. Control Assoc. 12: 243–246.

Tsitsanou, K.E., T. Thireou, C.E. Drakou, K. Koussis, M.V. Keramioti, D.D. Leonidas et al. 2012. *Anopheles gambiae* odorant binding protein crystal complex with the synthetic repellent DEET: implications for structure-based design of novel mosquito repellents. Cell. Mol. Life Sci. 69: 283–297.

van den Bossche, P. 1997. The control of *Glossina morsitans morsitans* (Diptera: Glossinidae) in a settled area in Petauke District (Eastern Province, Zambia) using odour-baited targets. Onderstepoort J. Vet. Res. 64: 251–257.

Verhulst, N.O., H. Beijleveld, B.G.J. Knols, W. Takken, G. Schraa, H.J. Bouwmeester et al. 2009. Cultured skin microbiota attracts malaria mosquitoes. Malaria J. 8: 302.

Verhulst, N.O., W. Takken, M. Dicke, G. Schraa and R.C. Smallegange. 2010. Chemical ecology of interactions between human skin microbiota and mosquitoes. FEMS Microbiol. Ecol. 74: 1–9.

Vezenegho, S.B., A. Adde, P. Gaborit, R. Carinci, J. Issaly, V.P. de Santi et al. 2014. Mosquito magnet® liberty plus trap baited with octenol confirmed best candidate for *Anopheles surveillance* and proved promising in predicting risk of malaria transmission in French Guiana. Malaria J. 13: 384.

Warmund, M. 2008. Pyrethrum daisies: 2000 years old and still growing! Missouri Environment and Garden. (https://ipm.missouri.edu/MEG/2008/6/Pyrethrum-Daisies-2000-Years-Old-and-Still-Growing/)

Whitehorn, J. and J. Farrar. 2010. Dengue. Br. Med. Bull. 95: 161-173.

Yu, X.-J., M.-F. Liang, S.-Y. Zhang, Y. Liu, J.-D. Li, Y.-L. Sun et al. 2011. Fever with thrombocytopenia associated with a novel bunyavirus in China. N. Engl. J. Med. 364: 1523–1532.

Index

A

Acalymma vittatum, 59, 60, 73-75, 79
Aggregation pheromone, 114, 120, 142
Alkaloid, 30
Allelochemicals, 29-31, 46
Ambrosia beetle, 142, 143
Ammonia, 276, 282, 284
Androconia, 172
Argentine ant *Linepithema humile*, 160, 162
Ascaroside, 96-99
Assassin bug, 272, 273, 282
Associative learning, 100
Attract and kill, 162, 163, 166, 167
Attract and reward, 234, 251

B

Bacterial wilt disease, 72, 75, 76
Behavioral plasticity, 100, 101, 103-105
Benzoxazinoids, 28, 33, 44
Biodiversity, 171
Biological control, 92, 101, 105, 107, 108
Biomass, 141
Biting midge, 273, 277, 278
Buchnera, 122
Buckwheat, 251, 252
Burnet moth, 175, 189

C

Camouflage, 199, 200, 211, 212, 217
Caste, 160
Chemical defense, 114
Chemical similarity, 3, 4
Chemosensory system, 7, 19
Cis-jasmone (CJ), 249
Clearwing moth, 170, 172, 173, 176, 177, 180, 183-186, 188, 189
Coccoid, 217, 218, 221, 223, 224
Co-infection, 72, 74, 77
Colony, 159-161
Communication, 197, 198, 210, 212, 213, 218, 225
Conservation, 170, 171, 187
Constitutively, 102
Contact sex pheromones, 114
Courtship, 171, 172, 174, 176
Cucumber beetle, 59, 60, 73, 75, 76, 78, 79
Cucumbers, 55, 61
Cucumis, 55, 56, 60, 61, 65, 66, 68-70, 73, 79
Cucurbita, 55, 56, 60-65, 69-73, 75, 77, 79
Cucurbitacins, 56-62, 66-68, 79
Cues, 94, 95, 100, 102-105, 107, 108
Cuticle, 137
Cuticular hydrocarbons, 160
Cutin, 137
Cytoplasmic incompatibility, 116, 117

D

DEET, 274, 275, 281
Dengue virus, 124, 269, 270
Diabrotica, 59-62, 64, 73, 74
DIMBOA, 33-35, 44
Diurnal moth, 170, 176, 186, 189

Drosophila C virus, 124

E

Elicitors, 28, 29, 38, 39
Emulsion, 201, 202
Endosymbionts, 98, 114, 115, 122
Entomopathologenic, 94
Environmental, 132, 133, 135, 137
Environmental protection, 171
Erwinia tracheiphila, 69, 72-75
Ethylene (ET), 248
Experience, 242, 255, 256

F

Farnesene, 162, 164
Fatty acid–amino acid conjugates, 29, 38
Feeding-deterrent, 8, 10
Feeding stimulants, 8, 11-13, 18
Feminization, 118, 119
Fermentation, 132, 133, 135-137, 139, 141
Flea, 271, 273, 280
Foraging, 234, 236-238, 241, 249, 250, 253, 254
Fungus, 131, 143

G

GABA, 37
β-Glucosidase, 29, 31, 33-38
Glucosinolates, 30-32, 42, 43
Glutamine synthetase, 40
Gourd, 55, 61, 74, 77
Gypsy moth, 201, 206, 208, 209, 224

H

Habitat, 234, 235, 240, 251, 257
Hamiltonella defensa, 122
HAVs, 243, 244, 246, 256
Herbivore, 29, 30, 35, 37, 38, 40, 43-45
Herbivore-induced plant volatiles, 237, 247
Hexadecenal, 161-166
(Z)-3-Hexenyl acetate, 248, 251
HIPVs, 237, 238, 242, 247-252, 256
Hologenomic concept, 114, 125
Host-associated volatiles, 243
Host selection, 3-7, 9, 10, 15, 18, 19
Host shift, 3-5, 9, 10, 15-19
Host usage, 4, 8, 15, 17, 19

I

Infective juveniles, 99, 103, 104
Intra-specific, 197
Intra-specifically, 101
Intraspecific communication, 161
Invasive species, 160, 161, 166
Iridoids, 28, 35, 36
Isothiocyanates, 28, 31-33, 42

J

Jasmonic acid (JA), 248

L

Lactic acid, 274-276, 284
Lavandulol, 218-220, 222
Leaf beetles, 57, 63, 80
Leishmaniasis, 278
Lepidoptera, 170, 171
Lice, 271, 273, 281
Life histories, 186

M

Malaria, 268, 271, 275-277, 284
Male-killing, 118, 119
Mating communication, 171, 174
Mating disruption, 159, 163, 197-202, 206, 208-225
Mealybug, 197, 217-224
MeSA, 248, 251, 252
Methyl jasmonate, 248
Methyl salicylate, 241, 248
Mimic, 173, 188
Mosquito, 269-277, 284
Multitrophic interactions, 101

N

Niche construction, 135
Nymphalidae, 3, 15

O

1-Octen-3-ol, 274, 275, 277, 280, 282, 284
OIPVs, 238, 239
Oviposition, 3-10, 12-19
Oviposition deterrent, 7, 14
Oviposition-induced plant volatiles, 237
Oviposition stimulant, 3, 7, 10, 12-15, 17, 18

P

Pantoea agglomerans, 120
Papilionidae, 3, 9, 10
Parasitic, 91-96, 98-102, 105-108
Parthenogenesis induction, 118
Pheromone, 96-99, 106, 107
Phyllosphere, 131, 137, 138
Pieridae, 3, 5, 9
Pieridae, Papilionidae, 3
Pierinae, 5, 6
Plant chemical defense, 28-30
Plant defensive chemicals, 3, 4
Pollination, 135, 136
Polyphagy, 29
Polyunsaturated hydrocarbons, 172
Predators, 37
Primary fatty alcohols, 172
Primary metabolite, 3, 18
Pumpkins, 55
Pyrethroid, 273, 281

R

Red imported fire ant *Solenopsis invicta*, 160
Repellent, 273, 274, 276, 281, 284
Resistance, 209-212, 224, 225

S

Saccharomyces cerevisiae, 132-135
Salicylic acid (SA), 248
Secondary metabolites, 3, 5, 6, 138
Sensory fatigue, 199, 200, 202
Sensory imbalance, 199, 200, 212
Sex pheromone, 170-190, 197, 198, 200, 208, 209, 211, 212, 214, 216, 218, 221, 223, 236, 237, 240, 241, 243-247
Sex ratio, 118, 119
Sexual behavior, 118, 119
Sexual dimorphisms, 170, 171
Signals, 94, 99, 100, 102-104, 106-108
Social insect, 160
Species diversification, 3, 4, 17
Spiroplasma poulsonii, 116
SPLAT®, 201, 202, 209
Squash, 55, 61, 63, 69, 72, 73, 77, 80
Synthetic pheromone, 199-202, 206, 210, 215, 222, 224

T

Tick, 270-273, 281, 283, 284
Trail pheromones, 159-167
Trypanosomiasis, 268, 271, 279, 280, 282

V

Vector, 268-270, 272, 273, 275-278, 280, 281, 284

W

Wireworm, 213, 214, 224
Wolbachia pipientis, 115, 117

Y

Yeast, 131-147
Yeast-like symbiont, 132, 145

Z

Zika virus, 268, 270
Zoonosis, 278-280, 282
Zucchini yellow mosaic virus (ZYMV), 71, 75